Scientific Essays in Honor of
H Pierre Noyes
on the Occasion of His 90th Birthday

K&E Series on Knots and Everything — Vol. 54

Scientific Essays in Honor of
H Pierre Noyes
on the Occasion of His 90th Birthday

Edited by

John C Amson
University of St Andrews, UK

Louis H Kauffman
University of Illinois at Chicago, USA

World Scientific

NEW JERSEY · LONDON · SINGAPORE · BEIJING · SHANGHAI · HONG KONG · TAIPEI · CHENNAI

Published by

World Scientific Publishing Co. Pte. Ltd.

5 Toh Tuck Link, Singapore 596224

USA office: 27 Warren Street, Suite 401-402, Hackensack, NJ 07601

UK office: 57 Shelton Street, Covent Garden, London WC2H 9HE

Library of Congress Cataloging-in-Publication Data
Scientific essays in honor of H. Pierre Noyes on the occasion of his 90th birthday / edited by John C. Amson
(University of St. Andrews, UK) & Louis H. Kauffman (University of Illinois at Chicago, USA).
pages cm -- (Series on knots and everything ; vol. 54)
Includes bibliographical references.
ISBN 978-981-4579-36-0 (hardcover : alk. paper)
1. Mathematical physics. 2. Physics. I. Noyes, H. Pierre, honouree. II. Amson, John C., editor of
compilation. III. Kauffman, Louis H., 1945– editor of compilation.
QC20.5.S35 2014
530.15--dc23

2013041168

British Library Cataloguing-in-Publication Data
A catalogue record for this book is available from the British Library.

Printed in Singapore by B & Jo Enterprise Pte Ltd

PREFACE

Those who have known Pierre Noyes, whether as a scholar, colleague, supervisor, teacher or friend, will have witnessed both his deep humanity and his keen excitement in the face of a world of physics that offers so much challenge to our understanding. Some of us have experienced it at close hand, others at a distance, and have all benefited each in our own way and come away the richer. No-one who has known Pierre can be unaware of that immense restless inquisitiveness, that urge to speculate, to explore, to appraise. Being close to Pierre in his most alert moments is like trying to catch a thirsty drink from a waterfall with a teaspoon or to contain the hot plasma of ball lightning in a towel. At worst you get soaked or scorched, at best you become inspired. Few are the scientists that have that knack of sharing and communicating their sense of purpose.

Of those with whom he roomed at Harvard in the early days are two who shared his questioning alertness, Thomas Kuhn and Phillip Anderson. The three of them, following their separate paths, each helped bring about a significant change in the way we are to think about the physical world we inhabit and explore: Kuhn with his paradigm shifts, Anderson with his insights into superconductivity, Pierre Noyes with his recognition of discreteness as an essential feature of observation and deduction.

Pierre has spent more than half a century of his professional life at Stanford Linear Accelerator Center, first as head of the theoretical physics division, then as Professor, and later as Emeritus Professor, much of the time concerned with the problems within and related to continuous quantum electrodynamics.

Much more could be said here about the many different facets of his professional career, but these are best summarised in the following section 'About Pierre'. What is not mentioned there is how much of Pierre's achievements and the benefits that have flowed to others have also been made possible by the sixty-two years he has shared with his 'fair, kind, and true' Mary. That bond we are all happy to recognise and to be thankful for.

How Pierre came to his appreciation of the essence of discreteness in physics may be worth telling here. It arose through the accidental discovery of what became known as the 'Combinatorial Hierarchy'.

Unlikely as it may sound the context in which the Combinatorial Hierarchy originated was not theoretical physics but machine translation, a subfield of computational linguistics. The place was the CLRU (Cambridge Language Research Unit[1]), the time was the early 1960s, and the catalyst was the coming together of three distinct ideas : a concept of order[2], Brouwerian fans and spreads[3], self-organisation and the notion of levels[4]. The principal agents of this mingling of ideas were Ted Bastin, Margaret Masterman, Frederick Parker-Rhodes, Clive Kilmister, John Amson. Each of those three distinct ideas had their own relevant roles in their own spheres. Each was perceived to be of help in struggling with the many exciting and innovative on-going investigations into the practical goals of automatic machine translation and artificial intelligence. Each was seen to be ably assisted by some of the earliest commercially and academically available computing machines and novel, dedicated software.

At some stage in this Machine Translation milieu it was noticed by Ted Bastin[5] the theoretical physicist, and Frederick Parker-Rhodes[6] the polymath, that the organisational structure of their very tentative binary-based system into successive levels was throwing up the iterated four numbers $3 = 2^2-1$, $7 = 2^3-1$, $127 = 2^7-1$ and $2^{127}-1$, the number of bit-strings in each of the first four levels. It was also noted that the particular algebraic construction employed ensured that the process could not continue beyond those four Levels. It was further noticed that their accumulated sums 3, 10, 137 and $(2^{127}-1) + 137 \approx 1.7 \times 10^{38}$ were surprisingly similar to the inverse dimensionless ratios of the super-strong, strong, electromagnetic, and gravitational coupling constants.

From that point on, 'theoretical physics' replaced 'artificial intelligence' as the prime motivator for further studies of what was then renamed the Combinatorial Hierarchy.

What was not appreciated at that time was the fact that an infinite hierarchy of sets with identical first four level-cardinalities can be constructed on purely set theoretical principles without reference to the algebraic rules deliberately preferred by the CLRU group at the time, and that related hierarchies[7] had already been introduced by Gödel in 1945, and by von Neumann in the 1950s. The latters' system (a 'cumulative hierarchy') had already been exploited to provide a logical framework for (ZF) Set Theory which together with assumption of the Axiom of Choice provided[8] a satisfactory foundation (ZFC) for 'doing mathematics'.

The Combinatorial Hierarchy differed from the cumulative hierarchies in two crucial aspects. The first was deliberately <u>introduced</u> by Parker-Rhodes : the empty-set (or the 'zero') was always excluded after the first level. This simple step produced the radically smaller numbers (<u>every</u> one now a prime number, the last three double-Mersenne primes) that related so directly to those four dimensionless physical constants, and also made possible connections with hierarchies of projective geometries[9]. The second was <u>discovered</u> by Parker-Rhodes : the growth of the hierarchy stopped at the fourth level (exponential growth trumps geometric growth). It seemed to suggest a new way for 'doing physics'.

The explorations by the CLRU group might well have jogged quietly along its own path had it not been for the fortunate accident that Ted Bastin, during visits to Stanford, in 1972 and 1973, had been invited by Patrick Suppes to talk to the Stanford Philosophy Department about the Combinatorial Hierarchy, and that in the audience a perhaps previously skeptical Pierre became persuaded to give discreteness a hearing. Pierre joined forces with the CLRU group and helped present its work at von Weizsäcker's Tutzing Conferences. With his renowned patience and characteristically energetic persuasion Pierre coerced the four 'B.A.N.K.' authors to publish their findings in their joint 1979 paper[10] *On the Physical Interpretation and the Mathematical Structure of the Combinatorial Hierarchy*, in the International Journal of Theoretical Physics. With this behind them, Pierre's next step was to suggest that, together with Parker-Rhodes, the appropriate thing to do was to promote the expansion of their ideas by forming a wider, international focus through the founding of an association unconstrained by the all too frequent dismissiveness of conventional establishment-based habits of thought.

Pierre had single-mindedly created ANPA, the Alternative Natural Philosophy Association.

Thirty four years later, many of his surviving colleagues have joined together to wish Pierre well on his 90th Birthday by contributing to this Festchrift.

In this volume twenty three authors contribute twenty-one papers.

1. Amson re-constructs the Generalized Kronecker Tensors in the functional analytic form of Unital Operators of every degree, explores their additional properties, and suggests how they can shed new light on existing and novel aspects of nonlinear physics.

2. Bowden makes first steps in reformulating the Combinatorial Hierarchy in the framework of Objects and Morphisms in an arbitrary Category, with special emphasis on the Level Change Operator.

3. Chew discusses quantum cosmological issues to which the special Mersenne primes may prove relevant; the largest of his quintet underpins a 'macroscopic' spacetime scale (the scale of life and consciousness) larger than particle scale by another factor of order 10^{19}.

4. Croll introduces his novel 'BiEntropy' algorithm that measures order and disorder in bitstrings, and provides examples that may not only apply to future bitstring physics, but also to number theory in that he shows that the sequence of primes is not periodic.

5. Deakin provides an introduction to his theory of constraints linking quantum and classical mechanics and shows how these constraints lead to new insights into

general relativity.

6. Doughty provides his own approach to the Combinatorial Hierarchy using the number theory of Double Fields, and develops his new concept of a catalogue of cosmic geometries.

7. Etter studies the structure of the Boolean lattice released from its particular top and bottom so that it relates to the geometry of the n-cube and investigates the relationship of this structure with link theory and quantum mechanics.

8. Gidney discusses the structure of consciousness in terms of a concept of relative existence.

9. Horner discusses the structure of the organization and productivity of the Alternative Natural Philosophy Association (ANPA) itself over its 33–year existence and how this relates to the Combinatorial Hierarchy.

10. Karmanov and Carbonell discuss two and three-body interactions, including relativistic effects. They note that the three-body system exists only for a limited range of two-body binding energy values, and that for stronger two-body interaction, the relativistic three-body system still collapses. They are led to the notion of critical stability.

11. Kauffman discusses the analysis of discrete physics using derivatives represented by commutators. This includes a dynamic interpretation of the square root of minus one as a clock, a reformulation of work with Noyes on the Feynman-Dyson derivation of electromagnetism and a generalization of Deakin's constraint algebra.

12. Kilmister's report retrospectively assesses in critical detail some of the history and aims of the ANPA research project.

13. Lindesay reflects on three and a half decades of his work with Noyes and its relationship with quantum effects in general relativity and the late time relationships with dark matter, as well as on Dyson's question as to whether intelligent life can be perpetual.

14. McGoveran discusses his Ordering Operator Calculus (OOC) and its applications to discrete physics, especially the notion that tensor equations can be interpreted as a special case of ordering operator equations.

15. Manthey and Matzke discuss the dynamical properties of geometric algebras and their analogies with particle physics, and conclude that driving the entropic expansion of the cosmos we find the Combinatorial Hierarchy.

16. Ord discusses new viewpoints on quantum mechanics and spacetime that emerge from his study of the discrete Dirac equation and the Feynman Checkerboard, allowing him to show that bit-strings provide a simple mechanism for the encoding of relativistic periodic processes, which he illustrates with simple model clocks.

17. Roscoe shows how fractal structure is intimately tied to cosmology, leading necessarily to his conclusion that a definitive signature of a Leibnizian universe is the perception that large scale structure in our experienced Universe is quasi-fractal of dimension 2.

18. Rowlands shows how naturally evolved algebras produce the Double Quaternions and lead naturally to the Dirac equation and to solutions to the Dirac equation from nilpotent elements in the algebra.

19. Shoup shows how the mathematics of special relativity emerges naturally in a discrete context, one in which the passage of time is derived from the motion itself, so that our clock does not tick when there is no motion.

20. Stein summarizes his notion of an oscillating cosmology in a set of thirty-eight concise, and often challenging principles.

21. Young describes the several decades of his work that led to his innovative approach to a new systems biology emerging from biology's surprising complexity, one in which many of its predictions have been verified.

22. In the unusual and unanticipated role as an 'epiloguer' to his own Festschrift Noyes gives voice to his present point of view and predictions arising from his new synthesis of fundamental processes in physics.

In concluding this preface, we wish to say how much this whole project of producing a book of essays for Pierre on his ninetieth birthday has been made possible by the commitment and enthusiasm of all of our contributors. We wish to acknowledge the earlier help of ANPA's current president Keith Bowden in approaching potential contributors. The idea of putting together this book of essays arose, unknown to Pierre, on his previous birthday. That it has turned so quickly from idea to fact in such a short space of time is entirely due to the the help we have received from our essayists at such short notice, and from the willingness and professionalism of our publishers and their editorial team to become involved with this time-constrained project.

Happy Ninetieth Birthday, Pierre, from all of us !

John C. Amson
St Andrews, Scotland, UK

Louis H. Kauffman
Chicago, Illinois, USA

September 2013

Notes to Preface

(1) The CLRU (1954–1981) was founded by Margaret Braithwaite Masterman (herself a distant relative of Lord Cavendish, 7th Duke of Devonshire, who funded the foundation of the Cavendish Laboratory in Cambridge in 1874 with James Clerk Maxwell as its first professor). Masterman went on to help found in 1965 the Lucy Cavendish College, Cambridge, of which she was its first President. [See *e.g.* Yorick Wilks, *Margaret Masterman*, in : *Early Years in Machine Translation*, edited by W. John Hutchins, Benjamin (2000) pp.279–297].

(2) E. W. Bastin, C. W. Kilmister, 'The Concept of Order : I The space-time structure; II Measurements; III General relativity as a technique for extrapolating over great distances'; Math. Proc. Camb. Philo. Soc. Vol.50(02) April, 278–286 (1954); Vol.51(03) July, 454–468 (1955); Vol.53(02) April, 462–472 (1957)

(3) (1) L. E. J. Brouwer. 'Points and Spaces', Canadian J. of Mathematics, vol.6, pp.26–34 (1954);
(2) L. E. J. Brouwer. *Brouwer's Cambridge Lectures on Intuitionism*, *Ed.* D. van Dalen, CUP (1981), re-issued (2011)
(3) Yorick Wilks, *Language, Cohesion and Form : Margaret Masterman* (Eleven of her papers as a posthumous tribute), C.U.P., Cambridge (2005). See: Paper No.2, 'Fans and Heads', fn.1.

(4) (1) Ted Bastin, Mary Hesse, Margaret Masterman, 'Self-Organization and the Notion of Level : A Summary of Discussion held in The Tower Mill, Overy Staithe, Norfolk, and in Cambridge, September – October 1960'. A Working Paper of the Information Structures Unit, 1960. Cambridge Language Research Unit, 20 Millington Road, Cambridge, UK. Recreated with permission, and edited by John Amson, ANPA Proceedings No.32, pp.3–23. (2012)
(2) E. W. Bastin, A. Woodside. 'A Theory of the Origin of Mass within a Control Model of the Elementary Particles', Information Structures Unit, February 20, 1962, Cambridge Language Research Unit. Recreated with permission, and edited, with a Commentary, by John Amson (unpublished, available as PDF) (2012)

(5) Ted Bastin, 'Mathematics of a Hierarchy of Brouwerian Operations', Information Structures Unit, September 01, 1964, Cambridge Language Research Unit. Recreated with permission, and edited by John Amson, ANPA Proceedings No.19, pp.5–22 (1998)

(6) A. F. Parker-Rhodes, J. C. Amson, 'Hierarchies of Descriptive Levels in Physical Theory', (original Parker-Rhodes unpublished paper dated 1964, posthumuous version edited and annotated by Amson) Intern. J. of General Systems, Vol.27, Nos.1,2,3, 57–80 (1998)

(7) Wikipedia has helpful information relating to:
(a) hereditarily finite sets, (b) von Neumann universes.

(8) (1) Keith J. Devlin. *The Axiom of Constructibility : a Guide for the Mathematician*, Lecture Notes in Mathsematics, No.617 Springer. (1977)
(2) Keith J. Devlin. *The Joy of Sets : Fundamentals of Contemporary Set Theory*, Second Edition, Undergraduate Texts in Mathematics, Springer-Verlag, Heidelberg (1993)
(3) Thomas Jech. *Set Theory : 3rd Millennium Edition, revised and expanded*, Springer-Verlag, Berlin (2003)

(9) John Amson. 'Discrimination Systems and Projective Geometries',
in *Discrete and Combinatorial Physics*, Proceedings of ANPA No.9, 158–189 (1988)

(10) Ted Bastin, H. Pierre Noyes, John Amson, Clive W. Kilmister, 'On the Physical Interpretation and the Mathematical Structure of the Combinatorial Hierarchy', (PITCH) Intern. J. of Theor. Phys., vol 18, No.7, pp.445–488 (1979)

ABOUT PIERRE

H. Pierre Noyes was born in 1923 in Paris, France, to the American chemist William Albert Noyes, Sr. and Katherine Macy, daughter of Jesse Macy. He grew up in a scientific environment. His father made pioneering determinations of atomic weights, chaired the Chemistry Department at the University of Illinois at Urbana-Champaign from 1907 to 1926, founded and edited several important chemical journals, and received the American Chemical Society's highest award, the Priestley Medal, in 1935. His older half-brothers, W. Albert Noyes, Jr. and Richard Macy Noyes, both became chemists.

Pierre received his baccalaureate degree in physics (*magna cum laude*) in 1943 from Harvard University. One of his room-mates during this time was Thomas Kuhn, author of *The Structure of Scientific Revolutions*; another was Phillip Anderson, author of *More and Different*, who later received the Nobel Prize in Physics for theoretical research into the electronic structure of magnetic and disordered systems. Before moving on to doctoral studies, Pierre spent a year at the Antenna Group at the MIT Radiation Laboratory and served in the US Navy for two years as an Aviation Electronics Technician Mate. Pierre earned his Ph.D. in theoretical physics from the University of California at Berkeley in 1950 doing research under the direction of Robert Serber with Geoffrey Chew as his advisor. Pierre's first doctoral problem was pion-pion scattering, followed by a second problem: meson production from proton-deuteron decay. His work under Chew was among the early applications of S-matrix theory. After earning his Ph.D., Pierre spent a postdoctoral year on a Fulbright scholarship at the University of Birmingham, UK, under the direction of Rudolf Peierls.

Pierre's career included several academic and research positions. He first worked as a post-doctoral fellow and then as assistant professor of Physics at the University of Rochester (1952–5). During that time, he compiled and edited the Proceedings of the 2nd, 3rd, 4th, and 5th Rochester Conferences on High Energy Physics. During the summers of those years, he worked at Project Matterhorn at Princeton, researching thermonuclear weapons (1952) and at the Brookhaven National Laboratory (1953). In the summer of 1954 he worked on calculating the binding energy of the triton using a particular non-relativistic, quantum mechanical potential model fitted to the low energy nucleon-nucleon parameters at the Lawrence Berkeley National Laboratory in Berkeley, California.

In 1955 Pierre joined the Theoretical Division of what was to become the

Lawrence Livermore National Laboratory. From 1956 to 1962 he served there as group leader of the General Research Group, under co-founder and director Edward Teller. During that time, he also served as co-chair of the design-review pre-mortem committee for the devices tested on Christmas Island in 1962 during Operation Dominic I, including the UGM-27 Polaris submarine-launched ballistic missile and Minuteman II intercontinental ballistic missile (ICBM) warhead prototypes.

During a sabbatical from his work at Lawrence Livermore in 1957 and 1958 Pierre was Leverhulme Trust Lecturer in the Experimental Physics Department of the University of Liverpool. He also worked as a consultant to General Atomics under Freeman Dyson and Ted Taylor for Project Orion (a nuclear explosion propelled space ship) from 1958 to 1961 at the invitation of Professor Dyson.

In 1961 Pierre served as AVCO visiting professor at Cornell University. Starting in 1962 he worked at SLAC as head of theoretical physics until he was replaced by Sidney Drell (who combined that responsibility with being Deputy Director of SLAC). He progressed from associate professor from 1962 through 1967 to professor (at SLAC, 1967–2002) and was awarded emeritus status in that rank on May 1, 2000. He collaborated with Richard Shoup at the Boundary Institute.

Pierre served as the Associate Editor of the Annual Reviews of Nuclear Science from 1962 until 1977. In 1979 he received an Alexander von Humboldt U.S. Senior Scientist Award, primarily to continue his theoretical work on the quantum mechanical three body problem for strongly interacting particles. In that same year he joined with John Amson, Ted Bastin, Clive W. Kilmister and A. Fredrick Parker-Rhodes, with whom he been collaborating on approaches to discrete physics *via* the 'Combinatorial Hierarchy' since 1973, to found the Alternative Natural Philosophy Association (ANPA), and was president of that organization until 1987.

In his early career Pierre primarily focused on nuclear forces from an elementary particle point of view. In *Inward Bound*, Abraham Pais had commented, correctly, that no one's work along this line lead to any fundamental new insights into elementary particle theory. Pierre's research into this area confirmed that the phenomenological analysis of nucleon-nucleon and pion-nucleon scattering, supplemented by an S-matrix based dispersion theory, shows that quantum field theory is roughly correct for two-particle scattering, and in some cases can be connected to the non-relativistic models used in nuclear physics. The research did not, however, lead to any unique, quantitative model of the strong interactions.

After leaving Lawrence Livermore Laboratory he started working on the quantum mechanical three body problem developed by Ludvig Faddeev, Alt, Grassberger, and Sandhas by reformulating it in the relativistic domain. In 1969 he concluded that any nonrelativistic quantum mechanical three body problem using strictly finite range forces between the pairs necessarily implies a non-local interaction in any three body system, which would extend to indefinitely large distances. A specific example of this is the fact that three identical particles with scattering lengths between the pairs that tend toward infinity will support an indefinitely large number of three body bound states with their radii increasing as the square of that number, as was shown independently by Vitaly Efimov in a specific model.

Driven by this success, an interest in John Stewart Bell's work, and Thomas Phipps' construction of a covering theory for both classical and quantum mechanics, Pierre was inspired to return his attention to the foundations of quantum mechanics. Around this time (1972-3) he heard a report from Ted Bastin on his combinatorial hierarchy work and met with Bastin and his collaborators (as noted above). The research conducted during this interaction resulted in the development of many papers on finite and discrete physics and cosmology, an area that became known as 'bit-string physics'. This work became Pierre's focus for much of the rest of the 20th century and into the present one. It continues to this day. His contributions to the new field include the following:

o   Pierre showed, thanks to a 1952 paper by Freeman Dyson, that the integer value of $\hbar c e^2 = 137$ given by the first three levels of the combinatorial hierarchy could be given physical interpretation as the maximum number of electron-positron pairs which could be distributed within a radius of $\hbar / 2me$, using renormalized quantum electrodynamics. Further, the rest energy of this system, $(137 \times (2\,m\,e\,c^2)) \approx m\,\pi$, could then suggest that the breakdown of quantum electrodynamics found by Dyson might be due to the strong interactions mediated by pions. The same argument extended to the fourth level suggested that the closure of the scheme at the fourth level, characterized by 2^{127}, could be understood as the formation of a black hole with the Planck mass by that number of baryons of protonic mass concentrated within a radius of $\hbar\,m\,p\,c$. Pierre, however, remained profoundly skeptical of these results until a decade later when David McGoveran showed that the scheme not only allowed one to derive the Sommerfeld-Dirac formula for the fine structure spectrum of hydrogen and then to correct the 137 approximation by correctly calculating the next four significant figures in the inverse fine-structure constant in agreement with experiment, but also to correct the value for Newton's gravitational constant and to compute several other elementary particle coupling constants and mass ratios.

o   In a joint paper with Louis Kauffman in 1996, published in the Proceedings of the Royal Society, London (A, Vol.452, 81-95), Pierre and Louis showed that the Feynman-Dyson derivation of electromagnetism from the formalism of quantum mechanics can be given a clear interpretation in terms of discrete physics. They also showed (Physics Letters A, No. 218 (1996), pp.139-146) how the Feynman Checkerboard model for the Dirac equation in 1+1 dimensions could be perfectly formulated in discrete physics with the correct continuum limit.

o   Work with Michael Manthey led to a cosmological model which long ago predicted that there was not enough matter to close the universe and that the ratio of dark matter to baryonic matter is 12.7. A consistent scheme developed by Ed Jones (Lawrence Livermore National Laboratories) also predicted a positive cosmological constant of the magnitude that was only observed in 2011.

○ Pierre compiled a selection of his papers into the volume
Bit–String Physics : A Finite and Discrete Approach to Natural Philosophy
(World Scientific, 2001) to serve as an introduction to this new field.

○ Some of Pierre's letters to Gregory Breit (1989–1981) are in the collection of the Yale University Library.

Pierre's honors include:

> Fulbright Scholarship (University of Birmingham, UK) (1950-1)
> Leverhulme Lecturer in the Experimental Physics Department of
> the University of Liverpool, UK (1957-8)
> AVCO visiting professor at Cornell University (1961)
> Alexander von Humboldt Senior Scientist Award (1979)
> First Annual Alternative Natural Philosopher Award (1989)
> 50-year Service Award, SLAC National Accelerator Laboratory (2013)

Pierre recieves the First ANPA Award, Palo Alto, 1989

The 1st ANPA Meeting, 1980, held at King's College, Cambridge

back row, l-r

Ian Ross   Ted Bastin* Irving Stein   Antony Deakin$^\circ$ Faruq Abdulla$^\times$

front row, l-r

Pierre Noyes* Clive Kilmister* John Amson* Frederick Parker-Rhodes*

Peter Marcer   Ugo Rota

* The Five Founder Members $^\circ$ Treasurer $^\times$ Secretary

LIST of CONTRIBUTORS

Geoffrey F. Chew	Lawrence Berkeley National Laboratory, California, USA
Irving Stein	San Francisco, California, USA
Clive W. Kilmister	(posthumuous) Sussex, UK
John C. Amson	St Andrews University, Scotland
Thomas Etter	(posthumuous) California, USA
Antony M. Deakin	Rowland's Castle, Hampshire, England
Michael Horner	Coppet, Vaud, Switzerland
Herb Doughty	Berkeley, California, USA
Louis Gidney	Strontian, Argyll, Scotland
David McGoveran	Alternative Technologies, Florida, USA
Michael Manthey	Colorado, USA
Louis H. Kauffman	University of Illinois at Chicago, USA
Richard Shoup	Boundary Institute, San Jose, California, USA
Jaume Carbonell	Institut de Physique Nucléaire, Paris, France
Vladimir A. Karmanov	Lebedev Physical Institute, Moscow, Russia
Douglas Matzke	Texas, USA
Garnet N. Ord	Ryerson University, Toronto, Canada
David F. Roscoe	Sheffield University, England
Peter Rowlands	Liverpool University, England
James Lindesay	Howard University, Washington, DC, USA
Fredric S. Young	Vicus Therapeutics, California, USA
Keith G. Bowden	Birkbeck College, London University, UK
Grenville J. Croll	Bury St Edmunds, Suffolk, England

Arranged in order of age seniority.

Brief biographies are appended to some Contributors' essays.

CONTENTS

Scientific Essays in Honor of

H Pierre Noyes

on the Occasion of His 90th Birthday

Unital Homogeneous Polynomial Operators on Hilbert Space

with applications to nonlinear theoretical physics

John C. Amson

School of Mathematics and Statistics,
University of St Andrews, St Andrews, Scotland,
E-mail: john.amson@uwclub.net

For Pierre Noyes on his 90-th birthday, in esteem and friendship.

The familiar notion of the linear unit operator on a Hilbert space is extended to its homogeneous polynomial operator analogues of arbitrary degree : 'unital' homogeneous polynomial operators, whose multimatrix representations are identified with generalized Kronecker symbols (tensors). Some properties are lost, many other novel structural properties are established.

The set of unital operators of all degrees on a Hilbert space constitutes a unique object, a 'Unital System'. It is unit-normed, closed and commutative under functional composition, and closed under functional polar composition. It is a multiplicatively and additively graded set, graded by the positive integer subset of the integer ring. It is the disjoint union of its homogeneous component 'grade' subsets each comprised of an equivalence class (under unitary similarity) of unital operators of same degree $\alpha \geq 1$, the first grade subset being the only one itself closed under composition.

Some operator equations of Fredholm and Characteristic type involving unital homogeneous polynomial operators are illustrated, their solubility described, and connections with an emerging theory of the spectrum of tensors and homogeneous polynomial operators are noted.

Possible applications to the theory of nonlinear phenomena in the presence of very high field strengths are indicated.

1. Introduction

The ubiquitous rôle of the unit operator on a Hilbert space in linear operator theory is so familiar that its indispensable importance is hardly ever recognized *per se*. It is only when one is obliged to venture into nonlinear operator theory that one is made aware of the need for an analogous nonlinear substitute. In many nonlinear situations theoretical physicists have long made use of the well-known Kronecker delta symbol and its many-dimensional forms as the Generalized Kronecker Contravariant Tensor $\delta^{k_1,\ldots,k_\rho}$ of order (*alias* rank) $\rho \geq 3$. This tensor has obviously taken over some of the properties of the linear unit operator, but not always consciously so.

In effect, this paper can be seen as a preliminary, functional analytic study and exploration of some of the properties of that Generalized Kronecker Contravariant Tensor in its new formal guise as a unital homogeneous polynomial operator in its own right.

It is now well established that the ordinary linear theories of quantum electrodynamics and electromagnetism cease to hold in the presence of very high and ultra high field strengths. It is enough for the purposes of this paper to mention just one area, that of nonlinear optics. Here the goal is to describe the behaviour of light in a media where the dielectric polarisation P reacts nonlinearly to the electric field E of the light (see *e.g.* [6]). Such nonlinearity is typically observed for example at the very high light intensity present in pulsed lasers, where the electric fields corresponds to interatomic electric fields of the order of 10^8 V/m. In quantum electrodynamics the 'Shwinger limit' indicates when the electromagnetic field is expected to become nonlinear. For the vacuum this is typically reported to occur at $\mathsf{E}_S = m_e^2\, c^3 / q_e\, \hbar \approx 1.3 \times 10^{18}$ V/m. In nonlinear optics, the superposition principle no longer holds; in functional analytic terms this means that operators are no longer additive : $A(x+y)$ and $A(x) + A(y)$ are no longer necessarily equal.

One is made aware of the nonlinearity by the appearance of 'second harmonic generation' and 'third harmonic generation'. The optical response can described by expressing the nonlinear Polarisation $\tilde{\mathsf{P}}(t)$ as a power series in the field strength $\tilde{\mathsf{E}}(t)$:

$$\tilde{\mathsf{P}}(t) \;=\; \epsilon_0 \left[\chi^1 \tilde{\mathsf{E}}(t) + \chi^2 \tilde{\mathsf{E}}^{(2)}(t) + \chi^3 \tilde{\mathsf{E}}^{(3)}(t) + \dots \right]$$

$$= \; \tilde{\mathsf{P}}^{(1)}(t) + \tilde{\mathsf{P}}^{(2)}(t) + \tilde{\mathsf{P}}^{(2)}(t) + \dots$$

Here, ϵ_0 is the permittivity of free space, and χ^1 is the linear susceptibility of free space, and χ^2, χ^3 are the 2nd- and 3rd-order susceptibilities of free space. In vector fields, the proportionality factors χ^1, χ^2, χ^3 are 2nd-, 3rd-, 4th-rank tensors. Note that the polarization at time t depends only on the instantaneous value of the electric field strength. The assumption that the medium responds instantaneously also implies through the Kramers–Kronig relations that the medium must be lossless and dispersionless [6]. In this case the tensors $\chi_{(r)}$ are real; otherwise they become complex.

Without entering into the more physical aspects of nonlinear optics it is sufficient to notice that the susceptibilities as written in a form such as, *e.g.* in the case of $\mathsf{P}^{(2)}$,

$$P_i^{(2)}(t) \;=\; \epsilon_0 \sum_{jk} \chi_{ijk}^{(2)}\, E_j(t)\, E_k(t)$$

reveals that the polarisation $\mathsf{P}^{(2)}$ is a homogeneous operator of degree 2, for which the above coordinate-wise expression in its multimatrix representation (apart from some minor notational differences) (see *e.g.* [2], and the details in the proof of Fact (4.5) below, *etc.*). Provided, of course that the tensor $\chi_{ijk}^{(2)}$ (which itself is the representation of a multilinear operator (see [12]) obeys a certain symmetry. The required symmetry is what I have called [2] 'pen-symmetry' *viz.* for each index

i, $\chi^{(2)}_{ijk}$ is symmetric in j, k. In the case of $\chi^{(2)}_{ijk}$ this is in fact true. In the case of $\chi^{(3)}_{ijk\ell}$ a stronger form of symmetry obtains, namely 'full symmetry' *viz.* symmetric w.r.t. all four indices i, j, k, ℓ, — which also happens to be true, and which in turn implies pen-symmetry of the related 4-dimensional multimatrix. See also related information in ref [13].

It is familiar (see *e.g.* [2, 10]) that to each each homogeneous polynomial operator hp of degree $\alpha > 1$ on a Hilbert space H, there corresponds a unique symmetric multilinear operator sm of degree α and a unique linear operator ℓ such that the following diagram is fully commutative.

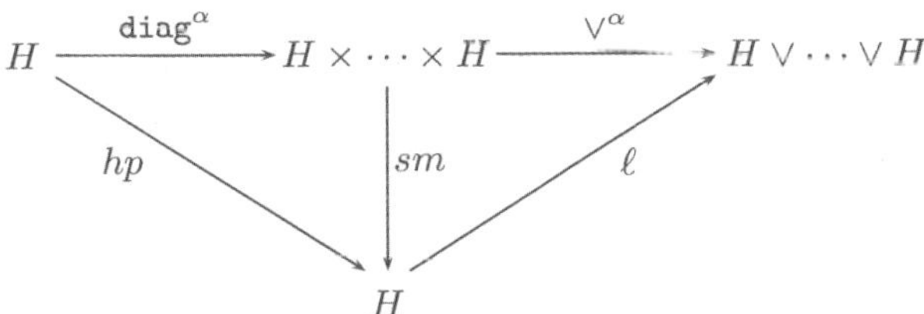

Here, $H \times \cdots \times H$ is the Cartesian product of α copies of H; $\vee^\alpha$ is the symmetric tensor product operator, and $H \vee \cdots \vee H$ is the symmetric tensor product of α copies of H. The latter is also often described as the *Bosonic Fock Space* component of dimension α. It follows that any such nonlinear homogenous polynomial operator as a Polarisation $\mathsf{P}^{(\alpha)}(t)$ of degree α is also uniquely identified with a unique linear operator from its Hilbert space domain into the Bosonic Fock Space component of dimension α. This seems to be a new, functional analytic way of looking at a known theoretical physics situation which deserves further investigation.

Since each polarisation $\mathsf{P}^{(\alpha)}(t)$ is a homogenous polynomial operator on an appropriate Hilbert space of electric fields $\mathsf{E}^{(\beta)}(t)$, we can make use at once of our novel unital homogeneous polynomial operators $\mathbf{I}_\alpha$ of degree α, to construct the 'Homogeneous Polynomial Characteristic Equation', *e.g.* in the case of $\alpha = 2$ (the 2nd-order, or quadratic case)

$$\mathsf{P}^{(2)}(t) = \lambda \mathbf{I}_2(t).$$

The solution of this homogenous polynomial equation of degree 2 leads to the novel concept of the *spectrum* of $\mathsf{P}^{(\alpha)}(t)$ — the true analogue of the linear spectrum — as illustrated in the last section of this paper.

One final caution needs to be made. In this introduction we speak of "polarization" of a material system. The same word "polarization" is also used historically and unavoidably in the definition of a homogenous polynomial operator (see § 2.1). It is not apparent at this stage of the exploration whether there is anything more than an accidental coincidence.

2. Unital homogeneous polynomial operators on a Hilbert space

2.1. *Context*

The context of this paper is a real or complex Hilbert space $\mathcal{H}$ of finite or infinite dimension $\dim \mathcal{H} = d$, familiar to theoretical physicists. It is equipped with an orthonormal basis $(\mathbf{b}^i)^{i \in I}$ where I is an index set of cardinality d. Thus each vector $\mathbf{x} \in \mathcal{H}$ has a unique (Fourier) representation $\mathbf{x} = \sum_{i \in I} \xi_i \mathbf{b}^i$, the scalars ξ_i being its unique 'basis-coordinates'. For notational convenience we assume $\mathcal{H}$ to be separable, and that $0 \leq d \leq \infty$.

For each integer $\alpha \geq 1$ we denote by $\mathcal{H}^\alpha = \mathcal{H} \times .^\alpha. \times \mathcal{H}$ the Hilbert space, Cartesian product space of α-many copies of a Hilbert space $\mathcal{H}$ equipped with the usual inner-product and norm. The subset

$$\mathrm{diag}(\mathcal{H}^\alpha) \overset{\mathrm{def}}{=} \{\, (\mathbf{x}_1, \ldots, \mathbf{x}_\alpha) \subset \mathcal{H}^\alpha \mid \mathbf{x}_1 = \cdots = \mathbf{x}_\alpha \,\},$$

is the 'diagonal axis' of $\mathcal{H}^\alpha$, and the mapping $\mathcal{H} \to \mathrm{diag}(\mathcal{H}^\alpha) : \mathbf{x} \mapsto (\mathbf{x}, .^\alpha., \mathbf{x})$ is a vector space isomorphism.

Homogeneous Polynomial operators, Symmetric Multilinear operators.
We remind ourselves that an operator $\mathrm{H} : \mathcal{H} \to \mathcal{H}$ is a homogeneous polynomial operator of integer degree $\alpha \geq 1$ if and only if there exists a multilinear operator $\mathrm{M} : \mathcal{H}^\alpha \to \mathcal{H}$ such that its restriction $\mathrm{M}^\nabla : \mathrm{diag}(\mathcal{H}^\alpha) \to \mathcal{H}$ and H have identical values, *i.e.* $\mathrm{M}^\nabla(\mathbf{x}) = \mathrm{M}(\mathbf{x}, \ldots, \mathbf{x}) = \mathrm{H}(\mathbf{x})$ for all $\mathbf{x} \in \mathcal{H}$. The restriction M^∇ of M is uniquely associated with H if and only if M is symmetric. In that case the multilinear operator M is called the *polar* of H and denoted by $\widehat{\mathrm{H}}$, and is uniquely determined by the 'Mazur-Orlitz' polarization identity ([11] §2 (22)), see also *e.g.* see also *e.g.* [5] Thm. A, or [7] Cor. 1.6 and notes on p.76), In effect, the relationship between a homogeneous polynomial operator H and its polar $\widehat{\mathrm{H}}$ is this :

$$\mathrm{H}(\lambda \mathbf{x} + \mu \mathbf{y}) \equiv \lambda^\alpha \mathrm{H}(\mathbf{x}) + \mu^\alpha \mathrm{H}(\mathbf{y})$$
$$+ \sum_{k=1}^{\alpha-1} \lambda^{\alpha-k} \mu^k \widehat{\mathrm{H}} \left(\mathbf{x}, {}^{\alpha-k}\!\ldots, \mathbf{x}, \mathbf{y}, .^k., \mathbf{y} \right), \tag{1}$$

for all vectors $\mathbf{x}, \mathbf{y} \in \mathcal{H}$, and for all scalars λ, μ in the scalar field for $\mathcal{H}$. By convention, the zero operator is a homogeneous polynomial operator of arbitrary degree.

It is familiar (see *e.g.* [1], [2]) that the degree of the functional composition of two homogeneous polynomial operators is the product (not the sum) of their degrees.

The completion, denoted by $\mathcal{P}(^\alpha \mathcal{H})$, of the supremum-normed vector space over the real or complex field $\mathbb{K}$ of all bounded homogeneous polynomial operators on $\mathcal{H}$ of degree $\alpha > 1$, and the completion, denoted by $\mathcal{L}^s(^\alpha \mathcal{H})$, of the supremum-normed vector space over the same field $\mathbb{K}$ of all bounded symmetric multilinear operators $\mathcal{H}^\alpha \to \mathcal{H}$ of degree $\alpha > 1$, are Banach spaces. They are isometric-isomorphic under the *Polarization Operator*

$$^\alpha\mathfrak{P} : \mathcal{P}(^\alpha \mathcal{H}) \to \mathcal{L}^s(^\alpha \mathcal{H}) : \mathrm{H} \mapsto \widehat{\mathrm{H}}; \tag{2}$$

its inverse $^\alpha\mathfrak{P}^{-1}$ is the *Axialization Operator*, denoted by $^\alpha\mathfrak{A}$.

A prototypical unital homogeneous polynomial operator.

Consider the degree-2 homogeneous polynomial operator H defined on the Hilbert space $\ell^2(\mathbf{N}, \mathbb{C})$ of square-summable complex sequences by the multimatrix $\mathbf{m} = (m_i^{j,k})$ (of degree 2 and dimension 3) whose scalar entries $m_i^{j,k}$ are given in terms of a generalized Kronecker Symbol $m_i^{j,k} = \delta^{i,j,k}$ which is 1 when all three indices are equal and 0 otherwise. Its action is given for $\mathbf{x} = (\xi_i)$, $\mathbf{x} = (\zeta_i)$ in $\ell^2(\mathbf{N}, \mathbb{C})$ by

$$H(\mathbf{x}) = \mathbf{x}, \quad \zeta_i = \sum_j \sum_k m_i^{j,k} \xi_j \xi_k \quad (\forall\, i \in \mathbf{N})$$

For each $i \in \mathbf{N}$ the vectors $\mathbf{b}^i \in \ell^2(\mathbf{N}, \mathbb{C})$ whose elements b_n^i are all 0 except where $n = i$ form the canonical orthonormal basis for $\ell^2(\mathbf{N}, \mathbb{C})$. It is obvious that $H(\mathbf{b}^i) = \mathbf{b}^i$ for every i, and that for the polar we have $\widehat{H}(\mathbf{b}^j, \mathbf{b}^k) = \mathbf{o}$ for every pair of unequal indices j and k. Also, for 'singleton' $\mathbf{x}$ (with all but one term $\xi_i = 0$) we have $H(\mathbf{x}) = \mathbf{x}$ whose corresponding term $\zeta_i = \xi_i^2$ with all other terms 0. This operator H is a degree-2 prototype of what we shall be calling a 'Unital' homogeneous polynomial operator.

2.2. *Unital homogeneous polynomial operators*

We introduce 'unital homogeneous polynomial operators' as a special sub-class of homogeneous polynomial operators.

Theorem 2.1. *Let H be a homogeneous polynomial operator of degree $\alpha > 1$ on a Hilbert space $\mathcal{H}$ of dimension $\dim \mathcal{H} = d$ $(1 \le d \le \infty)$, over the real or complex field $\mathbb{K}$, and let I be the ordered index set $\{i_1, i_2, \ldots\}$ of cardinality d.*

The following three conditions $[\,A\,]$, $[\,B\,]$, $[\,C\,]$, are equivalent :

$[\,A\,]$ *there exists an orthonormal basis $(\mathbf{b}^i)^{i \in I}$ for the Hilbert space $\mathcal{H}$ such that for all $i_1, \ldots, i_\alpha \in I^\alpha$,*

$$\widehat{H}(\mathbf{b}^{i_1, \ldots, i_\alpha}) = \begin{cases} \mathbf{b}^{i_1, \ldots, i_\alpha} & \text{if } \ i_1 = \ldots = i_\alpha \\ \mathbf{o} & \text{otherwise} \end{cases} \tag{3}$$

      whence $H(\mathbf{b}^i) = \mathbf{b}^i$ *for every* $i \in I$.

$[\,B\,]$ *there exists an orthonormal basis $(\mathbf{b}^i)^{i \in I}$ for the Hilbert space $\mathcal{H}$ such that, for every vector $\mathbf{x} \in \mathcal{H}$, whose (unique Fourier) representation is $\mathbf{x} = \sum_{i \in I} \xi_i \mathbf{b}^i$,*

$$H(\mathbf{x}) = \sum_{i \in I} (\xi_i)^\alpha \mathbf{b}^i \tag{4}$$

$[\,C\,]$ *there exists an orthonormal basis $(\mathbf{b}^i)^{i \in I}$ for the Hilbert space $\mathcal{H}$ such that, for all vectors $\mathbf{x}, \mathbf{y} \in \mathcal{H}$, whose (unique Fourier representations are $\mathbf{x} = \sum_{i \in I} \xi_i \mathbf{b}^i$, resp. $\mathbf{y} = \sum_{i \in I} \eta_i \mathbf{b}^i$, and for all scalars λ, μ,*

$$H(\lambda \mathbf{x} + \mu \mathbf{y}) = \lambda^\alpha H(\mathbf{x}) + \mu^\alpha H(\mathbf{y}); \tag{5}$$

Proof.

We first show that $[A] \iff [B]$.

$[A]$ implies $H(\mathbf{x}) = H(\sum_{i \in I} \xi_i \mathbf{b}^i) = \widehat{H}(\sum_{i_1 \in I} \xi_{i_1} \mathbf{b}^{i_1}, \ldots, \sum_{i_\alpha \in I} \xi_{i_\alpha} \mathbf{b}^{i_\alpha})$

$= \sum_{i_1 \in I} \xi_{i_1}, \ldots, \sum_{i_\alpha \in I} \xi_{i_\alpha} \widehat{H}(\mathbf{b}^{i_1}, \ldots, \mathbf{b}^{i_\alpha}) = \sum_{i \in I} (\xi_i)^\alpha H(\mathbf{b}^i)$

$= \sum_{i \in I} (\xi_i)^\alpha \mathbf{b}^i$, whence $[B]$.

Suppose $[B]$ holds. Simply for notational convenience we shall demonstrate the assertion in the case $\alpha = 2$; the general situations follows precisely the same argument. Let $(\mathbf{b}^i)^{i \in \{1,2\}}$ be an orthonormal basis for $\mathcal{H}$, and let $((\mathbf{b}^{(i_1)}, \mathbf{b}^{(i_2)}))^{(i_1, i_2) \in \{1,2\}^2}$ be the corresponding multi-basis for $\mathcal{H}^2$.

Then $\mathbf{x} = \xi_1 \mathbf{b}^{(1)} + \xi_2 \mathbf{b}^{(2)}$, and $H(\mathbf{x}) = \widehat{H}(\mathbf{x}, \mathbf{x})$

$\qquad = \widehat{H}(\xi_1 \mathbf{b}^{(1)} + \xi_2 \mathbf{b}^{(2)}, \xi_1 \mathbf{b}^{(1)} + \xi_2 \mathbf{b}^{(2)})$,

i.e. $H(\mathbf{x}) = \xi_1^2 H(\mathbf{b}^{(1)}) + \xi_2^2 H(\mathbf{b}^{(2)}) + 2\xi_1 \xi_2 \widehat{H}(\mathbf{b}^{i_1}, \mathbf{b}^{i_2})$, by (1).

But $H(\mathbf{x}) = \xi_1^2 \mathbf{b}^{(1)} + \xi_2^2 \mathbf{b}^{(2)}$, by hypothesis. That is to say,

$\mathbf{o} = \xi_1^2 (H(\mathbf{b}^{(1)}) - \mathbf{b}^{(1)}) + \xi_2^2 (H(\mathbf{b}^{(2)}) - \mathbf{b}^{(2)})$, for all scalars ξ_1 and ξ_2.

Whence, by the uniqueness of the Fourier representation with respect to the basis $(\mathbf{b}^i)^{i \in \{1,2\}}$, we have $\widehat{H}(\mathbf{b}^{i_1}, \mathbf{b}^{i_2}) \equiv \mathbf{o}$, $H(\mathbf{b}^{(1)}) = \mathbf{b}^{(1)}$, $H(\mathbf{b}^{(2)}) = \mathbf{b}^{(2)}$, and $[A]$ follows.

We now show that $[B] \iff [C]$.

$[B]$ implies $H(\lambda \mathbf{x} + \mu \mathbf{y}) = \sum_{i \in I} ((\lambda \xi_i)^\alpha + (\mu \eta_i)^\alpha) \mathbf{b}^i$

$= \lambda^\alpha \sum_{i \in I} (\xi_i)^\alpha \mathbf{b}^i + \mu^\alpha \sum_{i \in I} (\eta_i)^\alpha \mathbf{b}^i = \lambda^\alpha H(\mathbf{x}) + \mu^\alpha H(\mathbf{y},)$ whence $[C]$.

Suppose $[C]$ holds. By absolute convergence, each $\mathbf{x} = \sum_{i \in I} \xi_i \mathbf{b}^i$

$= \sum_{j \in J} \xi_j \mathbf{b}^j + \sum_{i \in I \setminus J} \xi_i \mathbf{b}^i$, for every finite subset of indices $J \subset I$.

In particular, $\mathbf{x} = \xi_1 \mathbf{b}^1 + \sum_{i>1} \xi_i \mathbf{b}^i$, whence by (5) $H(\xi_1 \mathbf{b}^1 + \sum_{i>1} \xi_i \mathbf{b}^i)$

$= \xi_1^\alpha H(\mathbf{b}^1) + H(\sum_{i>1} \xi_i \mathbf{b}^i) = \xi_1^\alpha \mathbf{b}^1 + H(\sum_{i>1} \xi_i \mathbf{b}^i)$ (by $[A]$).

That is to say : $H(\mathbf{x}) = \xi_1^\alpha \mathbf{b}^1 + H(\sum_{i>1} \xi_i \mathbf{b}^i)$.

Again, we have $\sum_{i>1} \xi_i \mathbf{b}^i = \xi_2 \mathbf{b}^2 + \sum_{i>2} \xi_i \mathbf{b}^i$, whence again by (5) and $[A]$,

$H(\xi_2 \mathbf{b}^1 + \sum_{i>2} \xi_i \mathbf{b}^i) = \xi_2^\alpha \mathbf{b}^2 + H(\sum_{i>2} \xi_i \mathbf{b}^i)$.

Hence it follows that : $H(\mathbf{x}) = \xi_1^\alpha \mathbf{b}^1 + \xi_2^\alpha \mathbf{b}^2 + H(\sum_{i>2} \xi_i \mathbf{b}^i)$.

By induction : $H(\mathbf{x}) = \xi_1^\alpha \mathbf{b}^1 + \xi_2^\alpha \mathbf{b}^2 + \xi_3^\alpha \mathbf{b}^3 + \ldots = \sum_{i \in I} (\xi_i)^\alpha \mathbf{b}^i$, the series being absolutely convergent since $\sum_{i \in I} (\xi_i) \mathbf{b}^i$ is; whence $[B]$. $\qquad \square$

Definition 2.1. A homogeneous polynomial operator H of degree $\alpha \geq 1$ on the Hilbert space $\mathcal{H}$, is said to be a *unital operator* if and only if it satisfies one (and hence all) of the three conditions $[A]$, $[B]$, $[C]$ in Theorem (2.1), and then it will be denoted by $\mathbf{I}_\alpha$.

Thus the following properties characterize a unital operator of degree α .

Fact 2.2. If $\mathbf{I}_\alpha$ be a unital operator of degree α, then there is an orthonormal basis $(\mathbf{b}^i)^{i \in I}$ for $\mathcal{H}$ such that

$$\mathbf{I}_\alpha(\mathbf{b}^i) \;=\; \mathbf{b}^i \qquad\qquad \text{all } i \in I\,, \tag{6}$$

$$\widehat{\mathbf{I}}_\alpha(\mathbf{b}^{i_1},\dots,\mathbf{b}^{i_\alpha}) \;=\; \mathbf{o} \qquad\qquad \text{all } i_1,\dots,i_\alpha \in I \text{ not all equal}\,, \tag{7}$$

$$\mathbf{I}_\alpha\Big(\sum_{i\in I} \xi_i\, \mathbf{b}^i\Big) \;=\; \sum_{i\in I}(\xi_i)^\alpha\, \mathbf{b}^i \qquad \text{all } \mathbf{x} = \sum_{i\in I} \xi_i\, \mathbf{b}^i \in \mathcal{H}\,. \tag{8}$$

and the polarization expansion (1) takes the simpler form :

$$\mathbf{I}_\alpha(\lambda\mathbf{x}+\mu\mathbf{y}) \;=\; \lambda^\alpha\, \mathbf{I}_\alpha(\mathbf{x}) \;+\; \mu^\alpha\, \mathbf{I}_\alpha(\mathbf{y})\,; \tag{9}$$

$$\text{all } \mathbf{x},\, \mathbf{y} \in \mathcal{H},\ \text{all scalars } \lambda,\, \mu.$$

Corollary 2.1. *For all x, y in $\mathcal{H}$,*

$$\mathbf{I}_\alpha(\mathbf{x}+\mathbf{y}) \;=\; \mathbf{I}_\alpha(\mathbf{x})+\mathbf{I}_\alpha(\mathbf{y})\,, \tag{10}$$

and if x, y, are linearly dependent vectors not both o in $\mathcal{H}$ then (for any α) $x = o$ or $y = o$ and (for odd α) $x + y = o$.

Proof. (10) follows immediately from (9). Suppose $\mathbf{x}$, $\mathbf{y}$, are linearly dependent vectors not both $\mathbf{o}$. W.l.o.g. we may assume $\mathbf{y} = \rho\mathbf{x}$ and $\mathbf{x} \neq \mathbf{o}$ (whence by (8) $\mathbf{I}_\alpha(\mathbf{x}) \neq \mathbf{o}$). From (9) we have $\mathbf{I}_\alpha(\mathbf{x}+\rho\mathbf{x}) = \mathbf{I}_\alpha(\mathbf{x})+\rho^\alpha\,\mathbf{I}_\alpha(\mathbf{x}) = (1+\rho^\alpha)\,\mathbf{I}_\alpha(\mathbf{x})$, whilst by homogeneity $\mathbf{I}_\alpha(\mathbf{x}+\rho\mathbf{x}) = (1+\rho)^\alpha\,\mathbf{I}_\alpha(\mathbf{x})$; hence $(1+\rho^\alpha) = (1+\rho)^\alpha$. But the latter equation's only real solutions are either $\rho = 0$ (α even) in which case $\mathbf{y} = \mathbf{o}$, and $\rho = 0$ or -1 (α odd) in which case $\mathbf{y} = \mathbf{o}$ or $\mathbf{x}+\mathbf{y} = \mathbf{o}$. $\qquad\square$

(Note also that the 'unital additive' condition (10) is not to be confused with the 'orthogonally additive' condition (see *e.g.* [4]) for a homogeneous polynomial operator H for which H($\mathbf{x}+\mathbf{y}$) = H($\mathbf{x}$)+H($\mathbf{y}$) for all 'orthogonal' vectors $\mathbf{x}$, $\mathbf{y}$, $|\mathbf{x}| \wedge |\mathbf{y}| = 0$, in a Banach lattice X.)

Fact 2.3.   Structural properties of $\mathbf{I}_\alpha$.
For each degree $\alpha > 1$:
[1] *the kernel of a unital operator $\mathbf{I}_\alpha$ is $\{\mathbf{o}\}$;*
[2] $\mathbf{I}_\alpha$ *is not injective.*
[3] *If (i) $\mathcal{H}$ has infinite dimension,*
or if (ii) $\mathcal{H}$ has finite dimension, is real, and α is even,
then $\mathbf{I}_\alpha$ is not surjective; otherwise $\mathbf{I}_\alpha$ is surjective.

Proof.
[1] By (8), for all nonzero scalars λ, we have $\mathbf{o} = \mathbf{I}_\alpha(\lambda\mathbf{x}) = \lambda^\alpha\mathbf{x}$; hence $\mathbf{x} = \mathbf{o}$.

[2] Non-Injectivity : Let $\mathbf{x}$ and $\mathbf{y}$ be two linearly independent vectors in $\mathcal{H}$. Suppose $\mathbf{I}_\alpha$ were injective, and that for two nonzero scalars λ, μ we have $\mathbf{I}_\alpha(\lambda\mathbf{x}) = \mathbf{I}_\alpha(\mu\mathbf{y})$. Then, by (8), $\lambda^\alpha\mathbf{x} = \mu^\alpha\mathbf{y}$, which implies $\mathbf{x}$ and $\mathbf{y}$ are linearly dependent, contrary to assumption.

[3] Case (i) : $\mathcal{H}$ has infinite dimension. Let $I = N = \{1, 2, \ldots\}$.
Take any $\mathbf{y} = \sum_{n \in N} \eta_n \, \mathbf{b}^n \in \mathcal{H}$ (hence $\sum_{n \in N} |\eta_n|^2$ convergent). Suppose there
exists $\mathbf{x} = \sum_{n \in N} \xi_n \, \mathbf{b}^n \in \mathcal{H}$ (hence $\sum_{n \in N} |\xi_n|^2$ convergent) and such that $\mathbf{y} = \mathbf{I}_\alpha(\mathbf{x})$. Then (by (8)) $\sum_{n \in N} \eta_n \, \mathbf{b}^n = \sum_{n \in N} (\xi_n)^\alpha \, \mathbf{b}^n$;   whence for each $n \in N$,
$\xi_n = \eta_n^{\,1/\alpha}$. Hence $\sum_{n \in N} |\xi_n|^2 = \sum_{n \in N} |\eta_n|^{2/\alpha}$. In particular, if each $\eta_n = 1/n$
and $\alpha \geq 2$ then $|\xi_n|^2 = 1/n^{2/\alpha} \geq 1/n$, hence $\sum_{n \in N} |\xi_n|^2$ is divergent, contrary to
hypothesis.

Case (ii) : $\mathcal{H}$ has finite dimension d. Let $\mathcal{H}$ be real, and α even.
Take any $\mathbf{y} = \sum_{n=1}^{d} \eta_n \, \mathbf{b}^n \in \mathcal{H}$. Suppose there exists $\mathbf{x} = \sum_{n=1}^{d} \xi_n \, \mathbf{b}^n \in \mathcal{H}$ and
such that $\mathbf{y} = \mathbf{I}_\alpha(\mathbf{x})$. Then (by (8)) $\sum_{n=1}^{d} \eta_n \, \mathbf{b}^n = \sum_{n=1}^{d} (\xi_n)^\alpha \, \mathbf{b}^n$;   whence
for each n, $\xi_n = \eta_n^{\,1/\alpha}$; but this is impossible if any $\eta_n < 0$. However, if finite-
dimensional $\mathcal{H}$ be complex then, for each degree $\alpha > 1$, every equation $\xi_n = \eta_n^{\,1/\alpha}$
is soluble in $\mathbb{C}$ and it follows that every $\mathbf{I}_\alpha$ is then surjective. $\qquad\square$

Fact 2.4.   Equivalence of unital operators.
*Every homogeneous polynomial operator of degree $\alpha > 1$ on $\mathcal{H}$ which is unitarily
similar to unital operator of degree $\alpha > 1$ on $\mathcal{H}$ is a unital operator on $\mathcal{H}$.*

*Whence, for each degree $\alpha > 1$, the unital operators of degree $\alpha > 1$ form an
equivalence class with respect to unitary similarity.*

Proof. Suppose a homogeneous polynomial operator H is unitarily similar to a
unital operator U of degree $\alpha > 1$. *i.e.*there exists a unitary linear operator L on
$\mathcal{H}$ such that $H = L \circ U \circ L^{-1}$.   Let $(\mathbf{b}^i)^{i \in I}$ and $(\mathbf{c}^i)^{i \in I}$ be orthonormal bases
for $\mathcal{H}$ such that (for each $i \in I$) $L(\mathbf{c}^i) = \mathbf{b}^i$, and (for each $\mathbf{x} \in \mathcal{H}$) we have
$\mathbf{x} = \sum_{i \in I} \eta_i \, \mathbf{c}^i$. Then, for each such $\mathbf{x}$,

$$
\begin{aligned}
H(\mathbf{x}) &= H\left(\sum_{i \in I} \eta_i \, \mathbf{c}^i \right) = \sum_{i \in I} (\eta_i)^\alpha \, H(\mathbf{c}^i) \\
&= \sum_{i \in I} (\eta_i)^\alpha \, (L \circ U \circ L^{-1})(\mathbf{c}^i) = \sum_{i \in I} (\eta_i)^\alpha \, (L \circ U)(\mathbf{b}^i) \\
&= \sum_{i \in I} (\eta_i)^\alpha \, L(U(\mathbf{b}^i)) = \sum_{i \in I} (\eta_i)^\alpha \, L(\mathbf{b}^i) = \sum_{i \in I} (\eta_i)^\alpha \, \mathbf{c}^i .
\end{aligned}
$$

Whence, by Theorem $(2.1)[\,B\,]$ above, H is also a unital operator, of degree $\alpha > 1$
on $\mathcal{H}$. $\qquad\square$

Fact 2.5.   Boundedness of $\mathbf{I}_\alpha$.
*Each unital operator $\mathbf{I}_\alpha$ of degree $\alpha \geq 1$ is bounded, with operator norm $\|\mathbf{I}_\alpha\| = 1$,
and hence continuous.*

Proof.   By definition of the operator norm of a homogeneous polynomial operator
we have $\|\mathbf{I}_\alpha\| = \sup \|\mathbf{I}_\alpha(\mathbf{x})\|$, the supremum being over all vectors $\mathbf{x}$ in the closed
unit ball $\mathbf{B}$ in $\mathcal{H}$. But $\mathbf{x} \in \mathbf{B}$ if and only if there exists a unit vector $\mathbf{u} \in \mathbf{B}$ and
a scalar $\lambda \leq 1$ such that $\mathbf{x} = \lambda \mathbf{u}$. Whence $\|\mathbf{I}_\alpha(\mathbf{x})\| = \|\mathbf{I}_\alpha(\lambda \mathbf{u})\| = |\lambda|^\alpha \|\mathbf{u}\|$, and

the supremum over all such $\mathbf{x}$ is $\|\mathbf{I}_\alpha\| = 1$. Thus $\mathbf{I}_\alpha$ is a bounded homogeneous polynomial operator and hence is continuous. $\qquad\square$

Fact 2.6.   Inferior boundedness of $\mathbf{I}_\alpha$.
Each unital operator $\mathbf{I}_\alpha$ of degree $\alpha \geq 1$ is bounded below, with inferior bound

$$\lfloor\mathbf{I}_\alpha\rfloor = \inf_{\|\mathbf{u}\|=1} \|\mathbf{I}_\alpha(\mathbf{u})\| = \begin{cases} 1/(\dim\mathcal{H})^{(\alpha-1)/2}, & (\dim\mathcal{H} < \infty) \\ 0, & (\dim\mathcal{H} = \infty) \end{cases}$$

For each fixed dimension $\dim\mathcal{H}$ the sequence $(\lfloor\mathbf{I}_\alpha\rfloor)_{\alpha=1,2,\dots}$ is a monotonic function of the degrees α, rapidly decreasing to 0.

Proof.   With respect to an orthonormal basis $(\mathbf{b}^i)^{i\in I}$ for $\mathcal{H}$, for each unit vector $\mathbf{u}$ we have $1 = \|\mathbf{u}\|^2 = \sum_{i\in I} |\xi_i|^2$, and by (4) (and, in effect, Parseval's Theorem) $\|\mathbf{I}_\alpha(\mathbf{u})\|^2 = \sum_{k\in I} |\xi_k|^{2\alpha}$.   The infimum of the sum $\sum_i |\xi_i|^{2\alpha}$ is achieved when $|\xi_1|^2 = |\xi_2|^2 = \dots = 1/\dim\mathcal{H}$, in case $\dim\mathcal{H} < \infty$, and 0 otherwise. In the finite dimensional case it then has the value $\dim\mathcal{H} \times 1/(\dim\mathcal{H})^\alpha = 1/(\dim\mathcal{H})^{\alpha-1}$; the assertions follow on taking square-roots. $\qquad\square$

Remark 2.1.   *Unital image of Unit Ball and Unit Cube.*
From Fact 2.5 and 2.6, in each Hilbert space of fixed finite dimension the image of the Unit Ball under a unital operator is a "smoothly dimpled ball" having smooth contact with the Unit Ball only at the basis vectors; as the degrees grow it has an increasingly "spikey (sea-urchin)" shape collapsing towards the origin between the fixed tips of the basis vectors.   However, the image of the Unit Cube is the Unit Cube itself; vertices map to vertices, edges to edges, (hyper-)faces to (hyper-)faces; but points in the Unit Cube and their images are nowhere in one-to-one correspondence except at the origin and each of its vertices.

Lemma 2.1.   *Using the notations from Fact (2.4), for each unital homogeneous polynomial operator $\mathbf{I}_\alpha$ of degree $\alpha > 1$ on $\mathcal{H}$, and for each $i \in I$,*

$$(\mathbf{b}^i \mid \mathbf{I}_\alpha(\mathbf{b}^i)) = (\mathbf{c}^i \mid \mathbf{I}_\alpha(\mathbf{c}^i))$$

Proof. $(\mathbf{b}^i \mid \mathbf{I}_\alpha(\mathbf{b}^i)) = (\mathbf{b}^i \mid \mathbf{b}^i) = 1 = (\mathbf{c}^i \mid \mathbf{c}^i) = (\mathbf{c}^i \mid \mathbf{I}_\alpha(\mathbf{c}^i))$ $\qquad\square$

This lemma allows us to introduce the *trace* $\operatorname{tr}\mathbf{I}_\alpha$ and *Hilbert-Schmidt norm* $\|\|\mathbf{I}_\alpha\|\|$ of a unital homogeneous polynomial operator $\mathbf{I}_\alpha$ on a finite dimensional Hilbert space $\mathcal{H}$, independently of the choice of orthonormal basis for $\mathcal{H}$:

Fact 2.7.   *Let $\mathbf{I}_\alpha$ be a unital operator of degree $\alpha > 1$ on a Hilbert space $\mathcal{H}$ of finite dimension $d > 1$. Then for every orthonormal basis $(\mathbf{b}^i)^{i \in I}$ for $\mathcal{H}$ we have :*

$$\operatorname{tr} \mathbf{I}_\alpha \overset{\text{def}}{=} \sum_{i \in I} (\mathbf{b}^i \mid \mathbf{I}_\alpha(\mathbf{b}^i)) = d, \tag{11}$$

$$\||\mathbf{I}_\alpha\|| \overset{\text{def}}{=} \left(\sum_{i \in I} \|\mathbf{I}_\alpha(\mathbf{b}^i)\|^2 \right)^{1/2} = \sqrt{d}. \tag{12}$$

$$\|\mathbf{I}_\alpha\| < \||\mathbf{I}_\alpha\|| < \operatorname{tr} \mathbf{I}_\alpha \tag{13}$$

Hence every unital operator $\mathbf{I}_\alpha$ on finite-dimensional $\mathcal{H}$ is of trace-class (i.e. nuclear), and compact.

Proof. By definitions, Lemma (2.1), and Fact (2.5). $\square$

3. Unital polynomial operators on a Hilbert space

Definition 3.1.   A polynomial operator of degree $\beta \geq 1$ on the Hilbert space $\mathcal{H}$, denoted by $\mathbf{J}_\beta$, is said to be the *polynomial unital operator* of degree β if and only if it is the sum

$$\mathbf{J}_\beta = \sum_{\alpha=1}^{\beta} \mathbf{I}_\alpha \tag{14}$$

of homogeneous unital operators $\mathbf{I}_\alpha$ $(\alpha = 1, \ldots, \beta)$ with respect to an orthonormal basis $(\mathbf{b}^i)^{i \in I}$ for $\mathcal{H}$. A polynomial unital operator is said to be *weighted* if each summand $\mathbf{I}_\alpha$ in (14) is replaced by $w_\alpha \mathbf{I}_\alpha$:

$$\mathbf{J}_\beta = \sum_{\alpha=1}^{\beta} w_\alpha \mathbf{I}_\alpha \tag{15}$$

where the 'weights' w_α belong to a sequence $\mathbf{w}$ of non-negative real numbers w_α, not all 0, which converges to 0.

Fact 3.1.   Properties of $\mathbf{J}_\beta$
  Let $\mathcal{H}$ be a Hilbert space of dimension d $(1 \leq d \leq \infty)$.
[a]   *For each degree $\beta > 1$, each $i \in I$, and each vector $\mathbf{x} = \sum_{i \in I} \xi_i \mathbf{b}^i$ in $\mathcal{H}$:*

$$\mathbf{J}_\beta(\mathbf{b}^i) = \beta \mathbf{b}^i, \quad \text{and} \quad \mathbf{J}_\beta(\mathbf{x}) = \sum_{i \in I} \Xi_i^{(\beta)} \mathbf{b}^i, \tag{16}$$

  where each $\Xi_i^{(\beta)}$ is the geometric sum $\xi_i^1 + \xi_i^2 + \ldots + \xi_i^\beta$.
[b]   *For each degree $\beta > 1$,*

$$\beta \geq \rfloor\lfloor \mathbf{J}_\beta \rfloor\lfloor \begin{cases} \geq \sum_{\alpha=1}^{\beta} 1/(\dim \mathcal{H})^{(\alpha-1)/2}, & (\dim \mathcal{H} < \infty) \\ = 0, & (\dim \mathcal{H} = \infty) \end{cases} \tag{17}$$

[c]   *For each degree* $\beta > 1$*, for each finite dimension* d,

$$\mathrm{tr}\,\mathbf{J}_\beta = \beta\,d, \quad \text{and} \quad \|\|\mathbf{J}_\beta\|\| = \beta\,\sqrt{d}. \tag{18}$$

[d]   *The sequence* $\left(\mathbf{J}_\beta\right)_{\beta=1,2,\dots}$ *is weakly-convergent to zero everywhere.*

[e]   *If* $\dim\mathcal{H}$ *is finite then* $\left(\mathbf{J}_\beta\right)_{\beta=1,2,\dots}$ *is strongly-convergent everywhere, otherwise it is strongly-convergent only in the interior of the Unit Cube.*

Proof.   [a] and [b] are immediate from the definitions, (8), absolute convergence, and exchanging the order of summation as required, and Facts 2.5 and 2.6.

[c] follows by Parseval's theorem and Fact 2.7.

[d] : let $\mathbf{x} = \sum_{i\in I} \xi_i\,\mathbf{b}^i$ and $\mathbf{y} = \sum_{j\in I} \eta_j\,\mathbf{b}^j$; then by (8)

$$\left(\mathbf{J}_\beta(\mathbf{x}) \mid \mathbf{y}\right) = \left(\sum_{i\in I} \frac{1-\xi_i^{\beta+1}}{1-\xi_i}\,\mathbf{b}^i \;\middle|\; \sum_{j\in I} \eta_j\,\mathbf{b}^j\right) = \sum_{i\in I} \frac{1-\xi_i^{\beta+1}}{1-\xi_i}\cdot\eta_i$$

and letting β tend to ∞ we have $\left(\mathbf{J}_\infty(\mathbf{x}) \mid \mathbf{y}\right) = \sum_{i\in I} \frac{\eta_i}{1-\xi_i} = 0$, i.e.$\mathbf{J}_\beta$ is wk-cgt to 0 everywhere.

[e] : By (8) and Parseval's theorem, $\|\mathbf{J}_\beta(\mathbf{x})\|^2 = \sum_{i\in I} |1-\xi_i^{\beta+1}|^2 / |1-\xi_i|^2$. Suppose $\dim\mathcal{H} = d < \infty$ and suppose $\mathbf{x} = \sum_{i=1,\dots,d} \xi_i\,\mathbf{b}^i$ with each $|\xi_i| < 1$; then in the limit as $\beta \to \infty$, $\|\mathbf{J}_\infty(\mathbf{x})\|^2 = \sum_{i=1,\dots,d} \frac{1}{|1-\xi_i|^2}$, and $\mathbf{J}_\infty(\mathbf{x}) = \sum_{i=1,\dots,d} \frac{1}{1-\xi_i}\,\mathbf{b}^{(i)}$, whence st-cgt in the interior of the Unit Cube.   If $\dim\mathcal{H} = \infty$ and $\mathbf{x} \neq \mathbf{o}$, then, as $\beta \to \infty$, $\|\mathbf{J}_\infty(\mathbf{x})\|^2$ exists only if the sum $\sum_{i\in I} \frac{1}{|1-\xi_i|^2}$ is convergent, i.e.only if $\lim_{i\to\infty} |\xi_i| = \infty$; but this is absurd. $\square$

Fact 3.2.   Structural properties of $\mathbf{J}_\beta$.

For each degree $\beta > 1$:

[1]   *the kernel of a polynomial unital operator* $\mathbf{J}_\beta$ *is* $\{\mathbf{o}\}$;

[2] $\mathbf{J}_\beta$ *is not injective;*

[3]   *If* $\mathcal{H}$ *has finite dimension and is complex then* $\mathbf{J}_\beta$ *is surjective, but not otherwise.*

Proof.

[1]   By (16), $\mathbf{o} = \mathbf{J}_\beta(\mathbf{x})$ implies $\mathbf{x} = \mathbf{o}$.

[2]   Non-Injectivity : Let $\beta > 1$; and let $\mathbf{x} = \sum_{i\in I} \xi_i\,\mathbf{b}^i$ and $\mathbf{y} = \sum_{i\in I} \eta_i\,\mathbf{b}^i$ be two distinct vectors in $\mathcal{H}$. If $\mathbf{J}_\beta(\mathbf{x}) = \mathbf{J}_\beta(\mathbf{y})$, then, by (16), we have $\sum_{i\in I} (\xi_i^1 + \xi_i^2 + \dots + \xi_i^\beta)\,\mathbf{b}^i = \sum_{i\in I} (\eta_i^1 + \eta_i^2 + \dots + \eta_i^\beta)\,\mathbf{b}^i$; whence (for each i) the two polynomials $\xi_i^1 + \xi_i^2 + \dots + \xi_i^\beta$ and $\eta_i^1 + \eta_i^2 + \dots + \eta_i^\beta$, both with the same unit coefficients, are equal and hence have the same roots. If (for each i) ζ_i be one of these common roots then $\mathbf{x} = \sum_{i\in I} \zeta_i\,\mathbf{b}^i$ and $\mathbf{y} = \sum_{i\in I} \zeta_i\,\mathbf{b}^i$ are identical, contrary to assumption.

[3]   Case (i) : $\mathcal{H}$ has infinite dimension.   Let $I = N = \{1,2,\dots\}$. Take any $\mathbf{y} = \sum_{n\in N} \eta_n\,\mathbf{b}^n \in \mathcal{H}$ (hence $\sum_{n\in N} |\eta_n|^2$ convergent).   Suppose there

exists $\mathbf{x} = \sum_{n \in N} \xi_n \mathbf{b}^n \in \mathcal{H}$ (hence $\sum_{n \in N} |\xi_n|^2$ convergent) and such that $\mathbf{y} = \mathbf{J}_\beta(\mathbf{x})$. Then (by (16)) $\sum_{n \in N} \eta_n \mathbf{b}^n = \sum_{n \in N} (\xi_n^1 + \xi_n^2 + \ldots + \xi_n^\beta) \mathbf{b}^n$; whence for each $n \in N$, $\eta_n = \xi_n^1 + \xi_n^2 + \ldots + \xi_n^\beta$. In particular, for each n, take each $\eta_n = 1/n$, and consider the polynomial equation :

$$\xi_n^\beta + \ldots + \xi_n^2 + \xi_n^1 = \frac{1}{n}. \tag{19}$$

This, for each $\beta \geq 2$, has a unique real solution $\xi_n = \sigma_n$ such that $1/(n\beta) < \sigma_n < 1/n$. But such solutions σ_n can be too large to allow the infinite sum $\sum_n |\sigma_n|^2$ to converge as required by hypothesis, as is shown by the following counter-example (*via*MAPLE).

Take $\beta = 2$, and consider $\xi_n^2 + \xi_n^1 = \frac{1}{n}$ $(n = 1, 2, \ldots)$. Each positive solution is $\sigma_n = \frac{1}{2}(-1 + (1 + 4/n)^{1/2})$; hence $|\sigma_n|^2 = \frac{n+2}{2n} + \left(\frac{n+4}{4n}\right)^{1/2}$. Let $f(t) = \frac{t+2}{2t} + \left(\frac{t+4}{4t}\right)^{1/2}$, for each real $t \geq 1$, a monotonically increasing function of t, with each $f(n) = |\sigma_n|^2$. Then

$$\int_1^n f(t)\, dt$$
$$= \sqrt{\frac{n+4}{n}} n \left(\sqrt{n^2 + 4n} + 2 \ln\left(n + 2 + \sqrt{n^2 + 4n}\right) \right) \frac{1}{\sqrt{(n+4)n}} - n$$
$$- \left(\sqrt{5} - 1 + 2 \ln\left(3 + \sqrt{5}\right) \right)$$

which is not bounded as $n \to \infty$. Hence, by the integral test, $\sum_n |\sigma_n|^2$ is not convergent. Hence $\mathbf{J}_2$ is non-surjective.

[3] Case (ii) : $\mathcal{H}$ has finite dimension d. Let $I = \{1, 2, \ldots d\}$. Take any $\mathbf{y} = \sum_{n=1}^d \eta_n \mathbf{b}^n \in \mathcal{H}$. Let $\mathbf{x} = \sum_{n=1}^d \xi_n \mathbf{b}^n \in \mathcal{H}$ and consider, for each n, the polynomial equation (*cf.*(19) :

$$\xi_n^\beta + \ldots + \xi_n^2 + \xi_n^1 = \eta_n. \tag{20}$$

If $\mathcal{H}$ be complex then this equation always has a solution ξ_n, one for which we have $\mathbf{y} = \mathbf{J}_\beta(\mathbf{x})$, so that each $\mathbf{J}_\beta$ is surjective. However, if $\mathcal{H}$ be real, then equation (20) may fail to have a real solution, and in this case each $\mathbf{J}_\beta$ is non-surjective. $\qquad\square$

4. Composition of homogenous unital operators

Fact 4.1. Compositional products of homogeneous unital operators. *Functional composition of unital homogeneous polynomial operators is commutative and associative; indeed, for all degrees $\alpha, \beta, \gamma \geq 1$,*

$$\mathbf{I}_\alpha \circ \mathbf{I}_\beta = \mathbf{I}_\beta \circ \mathbf{I}_\alpha = \mathbf{I}_{\alpha \times \beta}, \tag{21}$$

$$\mathbf{I}_\alpha \circ \left(\mathbf{I}_\beta \circ \mathbf{I}_\gamma \right) = \left(\mathbf{I}_\beta \circ \mathbf{I}_\alpha \right) \circ \mathbf{I}_\gamma = \mathbf{I}_{\alpha \times \beta \times \gamma}, \tag{22}$$

$$\|\mathbf{I}_{\alpha \times \beta}\| = \|\mathbf{I}_\alpha\| \times \|\mathbf{I}_\beta\| = 1, \tag{23}$$

$$0 < \|\mathbf{I}_{\alpha \times \beta}\| < \|\mathbf{I}_\alpha\| \times \|\mathbf{I}_\beta\| \quad (\text{if } \dim \mathcal{H} < \infty); \tag{24}$$

in particular, for all powers $n = 1, 2, \ldots$

$$\mathbf{I}_\alpha{}^n = \mathbf{I}_\alpha \circ \ldots_n \cdots \circ \mathbf{I}_\alpha = \mathbf{I}_{\alpha^n} \tag{25}$$

Proof.   That the degree of the composition of two homogeneous polynomial operators is the product of their degrees, is familiar.

For each $i \in I$, $(\mathbf{I}_\alpha \circ \mathbf{I}_\beta)(\mathbf{b}^i) = \mathbf{I}_\alpha(\mathbf{I}_\beta(\mathbf{b}^i)) = \mathbf{b}^i = (\mathbf{I}_\beta \circ \mathbf{I}_\alpha)(\mathbf{b}^i)$. The proof of commutativity is completed by similarly expanding the polars $\widehat{\mathbf{I}_\alpha \circ \mathbf{I}_\beta}$ and $\widehat{\mathbf{I}_\beta \circ \mathbf{I}_\alpha}$ in terms of the polars of $\widehat{\mathbf{I}_\alpha}$, and $\widehat{\mathbf{I}_\beta}$. Associativity follows at once from the associativity of composition of functions.

Although for two general bounded homogeneous polynomial operators P and Q we only have $\|P \circ Q\| \leq \|P\| \, \|Q\|^{deg(P)}$ for operator-norms, equality trivially holds for operator-norms of unit value. For the inferior bounds $\lfloor\!|.|\!\rfloor$, use Fact (2.6). $\qquad\square$

Remark 4.1.   By contrast with the homogeneous case, the functional composition of unital polynomial operators, though still associative, is not commutative in general. By (16), for each index $i \in I$, for each degrees $\beta \geq 2$ and $\gamma \geq 2$,

$(\mathbf{J}_\beta \circ \mathbf{J}_\gamma)(\mathbf{b}^i) = \mathbf{J}_\beta(\mathbf{J}_\gamma(\mathbf{b}^i)) = \mathbf{J}_\beta(\gamma \mathbf{b}^i) = \sum_{\alpha=1}^{\beta} \mathbf{I}_\alpha(\gamma \mathbf{b}^i)$

$= \left(\sum_{\alpha=1}^{\beta} \gamma^\alpha\right) \mathbf{b}^i$, whilst $(\mathbf{J}_\gamma \circ \mathbf{J}_\beta)(\mathbf{b}^i) = \mathbf{J}_\gamma(\mathbf{J}_\beta(\mathbf{b}^i)) = \mathbf{J}_\gamma(\beta \mathbf{b}^i) =$

$\sum_{\kappa=1}^{\gamma} \mathbf{I}_\kappa(\beta \mathbf{b}^i) = \left(\sum_{\kappa=1}^{\gamma} \beta^\kappa\right) \mathbf{b}^i$, which are identical if and only if

$\sum_{\alpha=1}^{\beta} \gamma^\alpha = \sum_{\kappa=1}^{\gamma} \beta^\kappa$. Whence $\mathbf{J}_\beta \circ \mathbf{J}_\gamma = \mathbf{J}_\gamma \circ \mathbf{J}_\beta$ if and only if $\beta = \gamma$. If that be the case then $\mathbf{J}_\beta{}^2(\mathbf{b}^i) \overset{\text{def}}{=} (\mathbf{J}_\beta \circ \mathbf{J}_\beta)(\mathbf{b}^i) = \sum_{\alpha=1}^{\beta} \beta^\alpha \mathbf{b}^i = \mathbf{J}_\delta(\mathbf{b}^i)$, where $\delta = \sum_{\alpha=1}^{\beta} \beta^\alpha$; thus $\mathbf{J}_\beta{}^2 = \mathbf{J}_{\beta(\beta^\beta-1)/(\beta-1)}$; *etc.*

Fact 4.2.   More compositional product properties of $\mathbf{I}_\alpha$.

Let H_β, H'_β, H''_β be homogeneous polynomial operators of degree $\beta \geq 1$; let O be the zero homogeneous polynomial operator of arbitrary degree.   Then :

$[1] \quad H_\beta \circ \mathbf{I}_\alpha \neq \mathbf{I}_\alpha \circ H_\beta \quad$ *in general;* $\qquad$ *[general commutativity fails]*

$[2] \quad H_\beta \circ \mathbf{I}_\alpha = O \implies H_\beta = O \qquad\qquad$ *[left 'divisor of zero']*

$[3] \quad \mathbf{I}_\alpha \circ H_\beta = O \implies H_\beta = O \quad$ *if H_β and $\mathbf{I}_\alpha$ commute;*

$\qquad\qquad\qquad\qquad\qquad\qquad\qquad\qquad$ *[conditional right 'divisor of zero']*

$[4] \quad H_\beta \circ \mathbf{I}_\alpha = \mathbf{I}_{\alpha\beta} \implies H_\beta = \mathbf{I}_\beta \qquad\qquad$ *[left 'uniqueness']*

$[5] \quad \mathbf{I}_\alpha \circ H_\beta = \mathbf{I}_{\alpha\beta} \implies H_\beta = \mathbf{I}_\beta \quad$ *if H_β and $\mathbf{I}_\alpha$ commute;*

$\qquad\qquad\qquad\qquad\qquad\qquad\qquad\qquad$ *[conditional right 'uniqueness']*

$[6] \quad H'_\beta \circ \mathbf{I}_\alpha = H''_\beta \circ \mathbf{I}_\alpha \implies H'_\beta = H''_\beta \qquad$ *[left 'cancellation']*

$[7] \quad \mathbf{I}_\alpha \circ H'_\beta = \mathbf{I}_\alpha \circ H''_\beta \implies H'_\beta = H''_\beta \quad$ *if H_β and $\mathbf{I}_\alpha$ commute;*

$\qquad\qquad\qquad\qquad\qquad\qquad\qquad\qquad$ *[conditional right 'cancellation']*

Proof.

$[1]$: For simple counter-examples, take $\mathcal{H} = \mathcal{E}^{[2]}$, Euclidean 2-space, and take H_β

to be a quadratic operator Q .

Then $\mathbf{x} = [x, y]^\mathsf{T}$, $Q(\mathbf{x}) = [ax^2 + 2bxy + cy^2, ex^2 + 2fxy + gy^2]^\mathsf{T}$; $\mathbf{I}_2(\mathbf{x}) = [x^2, y^2]^\mathsf{T}$.

Whence $(Q \circ \mathbf{I}_2)(\mathbf{x}) = [ax^4 + 2bx^2y^2 + cy^4, ex^4 + 2fx^2y^2 + gy^4]^\mathsf{T}$,

and $(\mathbf{I}_2 \circ Q)(\mathbf{x}) = [(ax^2 + 2bxy + cy^2)^2, (ex^2 + 2fxy + gy^2)^2]^\mathsf{T}$;

but here $(Q \circ \mathbf{I}_2)(\mathbf{x})$ and $(\mathbf{I}_2 \circ Q)(\mathbf{x})$ are identically equal only for particular values of the Q coefficients a, b, c, d, e, f.

$[\,2\,]$: By (4), $\mathbf{o} \equiv H_\beta\big(\mathbf{I}_\alpha(\mathbf{x})\big) = H_\beta\big(\mathbf{I}_\alpha\big(\sum_{i \in I} \xi_i\, \mathbf{b}^i\,\big)\big) = H_\beta\big(\sum_{i \in I} (\xi_i)^\alpha\, \mathbf{b}^i\big)$
$= \sum_{i \in I} (\xi_i)^{\alpha\beta}\, H_\beta(\,\mathbf{b}^i\,)$ for all scalars ξ_i;   but H_β is uniquely determined by the values it takes on an orthonormal basis; hence every $H_\beta(\,\mathbf{b}^i\,) = 0$ and $H_\beta = \mathbf{O}$.

$[\,3\,]$: follows from $[\,2\,]$ and $[\,1\,]$.

$[\,4\,]$: By (4), $\sum_{i \in I} (\xi_i)^{\alpha\beta}\, \mathbf{b}^i = \mathbf{I}_{\alpha\beta}(\mathbf{x}) = H_\beta\big(\mathbf{I}_\alpha(\mathbf{x})\big) = \ldots$
$\ldots = \sum_{i \in I} (\xi_i)^{\alpha\beta}\, H_\beta(\,\mathbf{b}^i\,)$ for all scalars ξ_i;   but H_β is uniquely determined by the values it takes on an orthonormal basis; hence every $H_\beta(\,\mathbf{b}^i\,) = \mathbf{b}^i$ for every i, hence $H_\beta = \mathbf{I}_\beta$.

$[\,5\,]$: follows from $[\,2\,]$ and $[\,1\,]$.

$[\,6\,]$: As in the proof of $[\,2\,]$, $H'_\beta\big(\mathbf{I}_\alpha(\mathbf{x})\big) = \sum_{i \in I} (\xi_i)^{\alpha\beta}\, H'_\beta(\,\mathbf{b}^i\,)$ and
$H''_\beta\big(\mathbf{I}_\alpha(\mathbf{x})\big) = \sum_{i \in I} (\xi_i)^{\alpha\beta}\, H''_\beta(\,\mathbf{b}^i\,)$. Hence identical equality of the LeftHand members implies $\sum_{i \in I} (\xi_i)^{\alpha\beta}\, H'_\beta(\,\mathbf{b}^i\,) \;=\; \sum_{i \in I} (\xi_i)^{\alpha\beta}\, H''_\beta(\,\mathbf{b}^i\,)$ for all scalars ξ_i and every i; but both H'_β and H''_β are uniquely determined by the values they take on an orthonormal basis and hence are one and the same homogeneous polynomial operator.

$[\,7\,]$: follows from $[\,6\,]$ and $[\,1\,]$. $\square$

Fact 4.3.   Prime compositional product properties.
A unital operator $\mathbf{I}_\alpha$ of prime degree α is not the compositional product of any other unital operators of degree greater than 1.
Every unital operator $\mathbf{I}_\alpha$ of non-prime degree α is the (not necessarily unique) compositional product of finitely-many unital operators $\mathbf{I}_{\alpha_j}$ such that $\mathbf{I}_\alpha = \mathbf{I}_{\alpha_1} \circ \mathbf{I}_{\alpha_2} \circ \cdots \circ \mathbf{I}_{\alpha_n}$ where $\alpha = \prod_{j=1}^{n} \alpha_j$ and the degrees α_j are the prime-factors of α.

Proof. Immediate. $\square$

Polarized functional compositional products.

To pass from the functional compositional products to their polarized forms, we shall use (see [2]) the *split-diamond* (split-$\lozenge$) notation $A\langle B_1, \ldots, B_\alpha\rangle$ for the 'polar composition' of homogeneous polynomial operators $A, B_1, \ldots, B_\alpha$ of degrees $\alpha, \beta_1, \ldots, \beta_\alpha$. That is to say, the bounded homogeneous polynomial operator C such that $C(\mathbf{x}) = \widehat{A}(B_1(\mathbf{x}), \ldots, B_\alpha(\mathbf{x}))$ $(\mathbf{x} \in \mathcal{H})$ and
$\|C\| \leq \|A\|\,\|B_1\| \cdots \|B_\alpha\|$;   and such that whenever $B_1 = \cdots = B_\alpha = B$ of degree β, we have

$$\mathrm{A}\big\langle \mathrm{B}\ldots,\mathrm{B}\big\rangle \equiv \mathrm{A}\Diamond\mathrm{B} \overset{\mathrm{def}}{=} \mathrm{A}\circ\mathrm{B}\ ; \quad \text{and}\quad \|\mathrm{A}\Diamond\mathrm{B}\| \le \|\mathrm{A}\|\,\|\mathrm{B}\|^{\deg(\mathrm{A})}\ .$$

Fact 4.4.   Polar compositional products of unital operators.

For all degrees $\alpha,\ \beta_1,\ \ldots,\ \beta_\alpha \ge 1,$ *and for any scalars* $a,\ b_1,\ldots,b_\alpha,$ *we have*

$$\mathbf{I}_\alpha\big\langle \mathbf{I}_{\beta_1}\cdots\mathbf{I}_{\beta_\alpha}\big\rangle = \mathbf{I}_{\beta_1+\cdots+\beta_\alpha}\ , \tag{26}$$

$$\big\|\mathbf{I}_\alpha\big\langle \mathbf{I}_{\beta_1}\cdots\mathbf{I}_{\beta_\alpha}\big\rangle\big\| = \|\mathbf{I}_\alpha\|\,\|\mathbf{I}_{\beta_1}\|\cdots\|\mathbf{I}_{\beta_\alpha}\| = \|\mathbf{I}_{\beta_1+\cdots+\beta_\alpha}\|\ , \tag{27}$$

$$a\,\mathbf{I}_\alpha\big\langle b_1\,\mathbf{I}_{\beta_1}\cdots b_\alpha\,\mathbf{I}_{\beta_\alpha}\big\rangle = a\,b_1\cdots b_\alpha\,\mathbf{I}_{\beta_1+\cdots+\beta_\alpha}\ . \tag{28}$$

Proof.   First, we calculate from the definitions :

$$\mathbf{I}_\alpha\big\langle \mathbf{I}_{\beta_1}\cdots\mathbf{I}_{\beta_\alpha}\big\rangle(\mathbf{x}) = \widehat{\mathbf{I}_\alpha}\big(\mathbf{I}_{\beta_1}(\mathbf{x}),\ldots,\mathbf{I}_{\beta_\alpha}(\mathbf{x})\big)$$

$$= \widehat{\mathbf{I}_\alpha}\big(\mathbf{I}_{\beta_1}(\textstyle\sum_{i_1}\xi_{i_1}\mathbf{b}^{i_1}),\ldots,\mathbf{I}_{\beta_\alpha}(\textstyle\sum_{i_\alpha}\xi_{i_\alpha}\mathbf{b}^{i_\alpha})\big) =$$

$$\textstyle\sum_{i_1\ldots i_\alpha}\xi_{i_1}{}^{\beta_1}\ldots\xi_{i_\alpha}{}^{\beta_\alpha}\,\widehat{\mathbf{I}_\alpha}\big(\mathbf{I}_{\beta_1}(\mathbf{b}^{i_1}),\ldots,\mathbf{I}_{\beta_\alpha}(\mathbf{b}^{i_\alpha})\big) =$$

$$\textstyle\sum_{i_1\ldots i_\alpha}\xi_{i_1}{}^{\beta_1}\ldots\xi_{i_\alpha}{}^{\beta_\alpha}\,\widehat{\mathbf{I}_\alpha}\big(\mathbf{b}^{i_1}\ldots,\mathbf{b}^{i_\alpha}\big) = \textstyle\sum_k\xi_k{}^{\beta_1+\cdots+\beta_\alpha}\,\mathbf{b}^k$$

$$= \mathbf{I}_{\beta_1+\cdots+\beta_\alpha}(\mathbf{x}).$$

The second assertion follows since every unital operator has unit norm; the third is immediate from multilinearity. $\square$

Remark 4.2.   Quite unlike the compositional product of arbitrary homogeneous polynomial operators or unital operators of arbitrary degrees, the compositional product of unital operators all of the same degree is both commutative and associative.

By *compositional commutativity* we mean that for a list of $(\alpha+1)$–many unital operators $(\ \mathrm{U}^{(i)}\ )^{i=1,\ldots,(\alpha+1)}$ all of the same degree $\alpha > 2$, we have, for every permutation π :

$$\mathrm{U}^{(1)}\big\langle \mathrm{U}^{(2)},\ldots,\mathrm{U}^{(\alpha+1)}\big\rangle = \mathrm{U}^{(\pi(1))}\big\langle \mathrm{U}^{(\pi(2))},\ldots,\mathrm{U}^{(\pi(\alpha+1))}\big\rangle \tag{29}$$

By *compositional associativity* we mean, for instance, that for a list of seven unital operators $(\ \mathrm{U}^{(i)}\ )^{i=1,\ldots,7}$ all of the same degree $\alpha > 2$, we have :

$$\Big(\ \mathrm{U}^{(1)}\big\langle \mathrm{U}^{(2)},\ \mathrm{U}^{(3)}\big\rangle\ \Big)\big\langle \mathrm{U}^{(4)},\ \mathrm{U}^{(5)},\ \mathrm{U}^{(6)},\ \mathrm{U}^{(7)}\big\rangle$$

$$= \mathrm{U}^{(1)}\big\langle\ \mathrm{U}^{(2)}\big\langle \mathrm{U}^{(4)},\ \mathrm{U}^{(5)}\big\rangle,\ \mathrm{U}^{(3)}\big\langle \mathrm{U}^{(6)},\ \mathrm{U}^{(7)}\big\rangle\ \big\rangle. \tag{30}$$

Note that the first term $\big(\ \mathrm{U}^{(1)}\big\langle \mathrm{U}^{(2)},\ \mathrm{U}^{(3)}\big\rangle\ \big)$ is a unital operator of degree 2×2 and so requires 4 operands $\mathrm{U}^{(4)},\ \mathrm{U}^{(5)},\ \mathrm{U}^{(6)},\ \mathrm{U}^{(7)}$, whilst both $\mathrm{U}^{(2)},\ \mathrm{U}^{(3)}$ require only 2 each.   Note also, that since polar compositional products are 'symmetric' in that, for example $\mathrm{U}^{(1)}\big\langle \mathrm{U}^{(2)},\ \mathrm{U}^{(3)}\big\rangle \equiv \mathrm{U}^{(1)}\big\langle \mathrm{U}^{(3)},\ \mathrm{U}^{(2)}\big\rangle$, there are many other possible rearrangements of the $\mathrm{U}^{(i)}$ on the Right Hand Side of (30), every one of which must also equal the Left Hand Side, in order to satisfy associativity.

It is a matter of straightforward but tedious "symbol manipulation" to verify the assertion of commutativity and associativity in the special case of unital operators of same degree. They can clearly fail when their degrees are unequal since the number of operators inside the polar brackets $\langle\ \rangle$ is equal to the degree of the operator currently outside to the left of the brackets. Those properties are even more regrettably absent in the case of general homogeneous polynomial operators as simple counter-examples in 2-dimensional real Hilbert space can demonstrate, using their multimatrix representations.

Fact 4.5.   Multimatrix Representations.
Unital operators $\mathbf{I}_\alpha$ on a separable Hilbert space $\mathcal{H}$ of dimension $\dim \mathcal{H} = d$ have a representation as a pen-symmetric multimatrix of type (d^α, d) with respect to an orthornomal basis for $\mathcal{H}$, a representation which corresponds to a generalized Kronecker Symbol of degree α.

Proof. It is familiar (see *e.g.* [2]) that every bounded homogeneous polynomial operator H of degree $\alpha \geq 1$ on a separable Hilbert space $\mathcal{H}$ has a representation, with respect to an orthonormal basis $(\mathbf{b}^i)$ for $\mathcal{H}$, as a (real or complex) square-summable multimatrix $\left(m_i^{j_1,\dots,j_\alpha} \right)$ where $i \in I$, $(j_1, ..., j_\alpha) \in I^\alpha$, with I an index set of cardinality equal to the dimension d of $\mathcal{H}$, and is then said to be of *type* (d^α, d). Its elements are given for each coordinate-index $i \in I$ and each upper α-list $(j_1, \dots, j_\alpha) \in I^\alpha$ as the Fourier coefficients w.r.t. the inner-product $(\,.\,|\,.\,)$ by :

$$m_i^{j_1,\dots,j_\alpha} = \left(\widehat{\mathrm{H}}(\,\mathbf{b}^{(j_1)}, \dots, \mathbf{b}^{(j_\alpha)}\,) \ \middle| \ \mathbf{b}^{(i)} \right) \tag{31}$$

This representation is unique whenever the multimatrix is 'pen-symmetric' (*i.e.* for each lower index i it is symmetric in the upper indices $j_1, ..., j_\alpha$).

When we replace H by $\mathbf{I}_\alpha$ and use the defining conditions from Def.2.1, the defining relationship (31) becomes

$$m_i^{j_1,\dots,j_\alpha} = \begin{cases} 1 & \text{if} \quad i = j_1 = \dots = j_\alpha, \\ 0 & \text{otherwise,} \end{cases} \tag{32}$$

corresponding to a generalized Kronecker Symbol $\delta^{i,j_1,\dots,j_\alpha}$ (a covariant tensor of order $\rho = \alpha{+}1$) with only 1s down the main 'diagonal' and 0s elsewhere. $\square$

We note that multimatrices are very closely related to tensors (see *e.g.* [12]). For instance our pen-symmetric multimatrices become equivalent to 'super-symmetric tensors' (see *e.g.* [8]) when further constrained to be symmetric in all its lower and upper indices — as indeed are our unital operators. The functional compositions of unital operators can readily be computed using multimatrix multiplications (*cf.* tensor multiplications).

5. The unital system on a Hilbert space

Remark 5.1. *The Unital System.*

The set $\mathcal{U}$ of all unital operators of every degree on a Hilbert space $\mathcal{H}$ constitutes a unique object, a *Unital System*, that is to say, a unit-normed set, closed and commutative under (functional) composition, and closed (but neither commutative nor associative) under (functional) polar composition. It is a multiplicatively and additively graded set, graded by the positive integer subset of the integer ring. (It is a particularly special example of a 'poly-groupoid', a sub-system of a 'poly-algebra', a graded system characterised by the polar composition of homogeneous polynomial operators, as defined in [2].) It is the disjoint union

$$\mathcal{U} \;=\; \bigsqcup_{\alpha \geq 1} \mathcal{U}_\alpha \,. \tag{33}$$

of its homogeneous component 'grade' subsets $\mathcal{U}_\alpha$, each comprised of the unitarily similar equivalence class of unital operators of same degree $\alpha \geq 1$. We note that the first grade subset $\mathcal{U}_1$ in $\mathcal{U}$ is the only one itself closed under composition.

Remark 5.2. *Unital Subsystems.*

The set $\mathcal{U}$ of all unital operators of every degree on $\mathcal{H}$, being closed under both composition and polar composition, admits subsystems generated by one or more unital operators. We introduce the first three of them.

[1] : *Unital Power-Tower.*

   The unital *power-tower* of a unital operator $\mathbf{I}_\alpha$ of degree $\alpha > 1$ is the subset of $\mathcal{U}$, denoted by $[[\,\mathbf{I}_\alpha\,]]$, and defined iteratively by

$$\mathbf{I} = \mathbf{I}_1 \in [[\,\mathbf{I}_\alpha\,]]; \quad \mathbf{I}_\alpha \in [[\,\mathbf{I}_\alpha\,]];$$
$$\mathrm{H} \in [[\,\mathbf{I}_\alpha\,]] \;\Longrightarrow\; \mathbf{I}_\alpha \circ \mathrm{H} \in [[\,\mathbf{I}_\alpha\,]]. \tag{34}$$

The power-tower $[[\,\mathbf{I}_\alpha\,]]$ is indexed by the 'power' integer subset $(\alpha^n)_{n=0,1,2,\ldots}$.

[2] : *Unital Polar-Tower.*

   The unital *polar-tower* of a single unital operator $\mathbf{I}_\alpha$ of degree $\alpha > 1$ is the subset of $\mathcal{U}$, denoted by $\langle\langle\,\mathbf{I}_\alpha\,\rangle\rangle$, and defined iteratively by

$$\mathbf{I} = \mathbf{I}_1 \in \langle\langle\,\mathbf{I}_\alpha\,\rangle\rangle; \quad \mathbf{I}_\alpha \in \langle\langle\,\mathbf{I}_\alpha\,\rangle\rangle;$$
$$\mathrm{H}_{i_1},\ldots,\mathrm{H}_{i_\alpha}, \in \langle\langle\,\mathbf{I}_\alpha\,\rangle\rangle \;\Longrightarrow\; \mathbf{I}_\alpha\langle\,\mathrm{H}_{i_1}\cdots\mathrm{H}_{i_\alpha}\,\rangle \in \langle\langle\,\mathbf{I}_\alpha\,\rangle\rangle. \tag{35}$$

The polar-tower $\langle\langle\,\mathbf{I}_\alpha\,\rangle\rangle$ is indexed by the '1st–order' integer subset :
$$(p + q\alpha)_{p,q=0,1,2,\ldots}.$$

[3] : *Unital Polar-Tower of 2nd–order.*

   The unital polar-tower of a pair of unital operators $\mathbf{I}_\alpha$, $\mathbf{I}_\beta$ of degrees $\alpha, \beta > 1$, a 2nd–order *polar-tower*, is the subset of $\mathcal{U}$, denoted by $\langle\langle\,\mathbf{I}_\alpha, \mathbf{I}_\beta\,\rangle\rangle$, and defined iteratively by

$$\mathbf{I} = \mathbf{I}_1 \in \langle\langle\, \mathbf{I}_\alpha, \mathbf{I}_\beta \,\rangle\rangle; \quad \mathbf{I}_\alpha \in \langle\langle\, \mathbf{I}_\alpha, \mathbf{I}_\beta \,\rangle\rangle; \quad \mathbf{I}_\beta \in \langle\langle\, \mathbf{I}_\alpha, \mathbf{I}_\beta \,\rangle\rangle;$$

$$H_{i_1}, \ldots, H_{i_\alpha}, \in \langle\langle\, \mathbf{I}_\alpha, \mathbf{I}_\beta \,\rangle\rangle \implies$$

$$\mathbf{I}_\alpha \langle\, H_{i_1} \cdots H_{i_\alpha} \,\rangle \in \langle\langle\, \mathbf{I}_\alpha, \mathbf{I}_\beta \,\rangle\rangle \text{ and } \mathbf{I}_\beta \langle\, H_{i_1} \cdots H_{i_\alpha} \,\rangle \in \langle\langle\, \mathbf{I}_\alpha, \mathbf{I}_\beta \,\rangle\rangle. \tag{36}$$

This 2nd–order polar-tower $\langle\langle\, \mathbf{I}_\alpha, \mathbf{I}_\beta \,\rangle\rangle$ is indexed by the '2nd–order' integer subset :

$$(\, p + q\alpha + r\beta \,)_{p,q,r=0,1,2,\ldots}.$$

We note that polar-towers of nth–order, for all other integer orders $n > 2$, analogously exist, are increasingly elaborate subsets of the unital system $\mathcal{U}$, and play analogous rôles to these lower order ones.

6. Equations involving unital homogeneous polynomial operators

6.1. *Fredholm equations with unital homogeneous polynomial operators*

Equations such as Fredholm equations occupy an important place in the theory of linear operator equations in Hilbert space. They are often only soluble iteratively. Similar equations in which the operator in question is a homogeneous polynomial operator can sometimes also be solved iteratively. When the operator is a unital homogeneous polynomial operator they are often more amenable to solubility, due to their 'coordinate-based' descriptions. Here, in this present paper, there is room for but a few representative examples.

Example 6.1.   In our first example we consider the Fredholm equation of the Second kind $\mathbf{x} = \mathbf{y} + \lambda\,A(\mathbf{x})$, on a separable Hilbert space $\mathcal{H}$ where $\mathbf{y}$ is given, λ is a characteristic parameter and $\mathbf{x}$ is the unknown, which can also be written as :

$$\lambda\,A(\mathbf{x}) - \mathbf{I}(\mathbf{x}) = -\mathbf{y}. \tag{37}$$

In the familiar case where A is linear it can be solved iteratively, leading (*viaa* geometric series) to the classical Neumann-Liouville solution. In the case where A is a homogeneous polynomial operator of degree $\alpha > 1$ it is known ([1]) that it can also be solved iteratively, leading (*viaa* hypergeometric-Lagrange series) to a solution in the form of a norm-convergent series of bounded homogeneous polynomial operators.

Indeed, if A is *itself* a unital operator $\mathbf{I}_\alpha$ of degree $\alpha > 1$ then a sufficient condition for a solution to exist is simply that

$$\|\lambda\|\,\|\mathbf{I}_\alpha\|\,\|\mathbf{y}\| = \|\lambda\|\,\|\mathbf{y}\| < (\alpha - 1)^{(\alpha-1)}/\alpha^\alpha . \tag{38}$$

The solution $H(\mathbf{y})$ (a small solution that tends to zero with $\mathbf{y}$) is a norm-convergent bounded homogeneous polynomial operator power series $\sum_{k=0,1,2,\ldots} \lambda^k\,H^{(k)}$ where the $H^{(k)}$ are defined by the $(\alpha)^{th}$-order algorithm:-

$$H^{(0)} = \mathbf{I} = \mathbf{I}_1; \quad H^{(1)} = \mathbf{I}_\alpha; \quad H^{(k)} = \sum_{[k-1]} \mathbf{I}_\alpha \langle\, H^{(k_1)} \cdots H^{(k_1)} \,\rangle \tag{39}$$

and where the (Abel) summation convention is indicated by

$$[k-1] = \{\, k_i = 0,1,2\ldots \;\Big|\; \textstyle\sum_{i=1}^{\alpha} k_i = k-1 \,\}\,.$$

It is immediately evident that the bounded homogeneous polynomial operator terms $\mathbf{I}_\alpha \langle\, \mathrm{H}^{(k_1)} \cdots \mathrm{H}^{(k_1)} \,\rangle$ are precisely the sort of terms that form the polar-tower $\langle\!\langle\, \mathbf{I}_\alpha \,\rangle\!\rangle$ (see Remark 5.2).   For each small enough λ and $\mathbf{y} = (\,\eta_i\,)_{i=0,1,2,\ldots}$, the solution $\mathbf{x} = (\,\xi_i\,)_{i=0,1,2,\ldots}$ is :

$$\mathbf{x} \;=\; c_0 \mathbf{y} + \sum_{n=1}^{\infty} c_n\, \lambda^n\, \mathbf{I}_\alpha(\,\mathbf{y}\,) \tag{40}$$

where the c_n ($n=0,1,2,\ldots$) are the Catalan numbers (1, 1, 2, 5, 14, 42, $\ldots$) . Note that the operator power series $\sum_{n=1}^{\infty} c_n\, \lambda^n\, \mathbf{I}_\alpha$ appearing here is norm-convergent : it is an example of a *holonomial operator* (see [1]) which is holomorphic on the open ball of radius equal to its radius of absolute uniform convergence in $\mathcal{H}$:

$$\rho_u = \liminf{}_n(\, c_n\, |\lambda|^n\, \|\mathbf{I}_\alpha\| \,)^{1/n} = \liminf{}_n(\, c_n\, |\lambda|^n \,)^{1/n}$$

In effect the solution of the unital homogeneous polynomial operator equation (37) involves a convergent series whose partial sums are again weighted unital polynomial operators (see (15)) with weights $c_n\, \lambda^n$.

As an alternative approach in this particular situation, if we choose an ortho-normal basis $(\,\mathbf{b}^i\,)_{i\in I}$ for $\mathcal{H}$, then equation (37) becomes (by (4))

$$\sum_{i\in I} \xi_i\, \mathbf{b}^i \;=\; \sum_{i\in I} \eta_i\, \mathbf{b}^i + \lambda \sum_{i\in I} (\xi_i)^\alpha\, \mathbf{b}^i\,. \tag{41}$$

That is to say, for each $i \in I$, the coordinates ξ_i must satisfy the well-known Lagrangian nonlinear complex equation

$$\xi_i \;=\; \eta_i + \lambda\, (\xi_i)^\alpha \,; \tag{42}$$

and the solution (40) becomes

$$\xi_i \;=\; c_0\, \eta_i + \sum_{n=1}^{\infty} c_n\, \lambda^n\, (\eta_i)^n \quad \text{(for each } i \in I) \tag{43}$$

Example 6.2.   For our second example we consider a generalized (polynomial) Fredholm equation of the second kind $\mathbf{x} = \mathbf{y} + \lambda_1\, \mathbf{I}_{\alpha_1}(\,\mathbf{x}\,) + \lambda_2\, \mathbf{I}_{\alpha_2}(\,\mathbf{x}\,)$, on a separable Hilbert space $\mathcal{H}$, where $\mathbf{y}$ is given, λ_1, λ_2 are characteristic parameters and $\mathbf{x}$ is the unknown. This can also be re-written as :

$$\left(\, -\mathbf{I}_1 + \lambda_1\, \mathbf{I}_{\alpha_1} + \lambda_2\, \mathbf{I}_{\alpha_2} \,\right)(\,\mathbf{x}\,) \;=\; -\mathbf{y} \tag{44}$$

Here the operator $-\mathbf{I}_1 + \lambda_1\, \mathbf{I}_{\alpha_1} + \lambda_2\, \mathbf{I}_{\alpha_2}$ is a fairly simple example of a weighted unital polynomial operator (see (15)). Here again, it is known (*ibid.*) that a sufficient condition for a solution to exist is that

$$\|\mathbf{y}\| \;<\; \inf_{|z|=\gamma} |\mathrm{f}(\mathrm{z})| \tag{45}$$

where $f(z) = z - (\,\lambda_1\, \|\mathbf{I}_{\alpha_1}\|\, z^{\alpha_1} + \lambda_2\, \|\mathbf{I}_{\alpha_2}\|\, z^{\alpha_2}\,) = z - (\,\lambda_1\, z^{\alpha_1} + \lambda_2\, z^{\alpha_2}\,)$ $(z \in \mathbb{C})$, and γ is the least number $|z_0|$ such that $z_0 \neq 0$ and $f'(z_0) = 0$.

Again, it is known that the solution $H(\mathbf{y})$ (a small solution that tends to zero with $\mathbf{y}$) is another norm-convergent bounded homogeneous polynomial operator power series

$$\sum_{k=0,1,2,\ldots} \lambda^k \, H^{(k)} \tag{46}$$

generated by a more elaborate $(\alpha_1, \alpha_2)^{th}$-order algorithm version of (39), in which the polar composition terms belong to the 2nd-order polar-tower $\langle\langle \, \mathbf{I}_{\alpha_1}, \, \mathbf{I}_{\alpha_2} \, \rangle\rangle$ (see (36)). The solution, generalizing (40) and (43)), is a double power series in λ_1, λ_2, ξ_i whose coefficients are the 2nd-order Catalan numbers :

$$c_{n_1,n_2} \;=\; \frac{(n_1\alpha_1 + n_1\alpha_1)\,!}{[\,n_1(\alpha_1 - 1) + n_2(\alpha_2 - 1)\,]\,!\, n_1\,!\, n_2\,!}$$

the coefficients of the complex Lagrange series majorizing in norm the unital operator solution power series.

6.2. *Characteristic equations with unital homogeneous polynomial operators*

Example 6.3.   In our third example we consider an equation on the specific Hilbert space $\ell^2(\mathbf{N}, \mathbb{C})$

$$\mathbf{L}_\alpha\,(\,\mathbf{x}\,) \;=\; \lambda\,\mathbf{I}_\alpha\,(\,\mathbf{x}\,) \tag{47}$$

involving the homogeneous polynomial Left Shift operator $\mathbf{L}_\alpha$ of degree $\alpha > 1$ and the unital operator $\mathbf{I}_\alpha$, both with respect to the canonical orthonormal basis for $\ell^2(\mathbf{N}, \mathbb{C})$ Here we have

$$\mathbf{x} = (\, \xi_0, \xi_1, \xi_2, \xi_3, \ldots \,)$$

$$(\xi_1^\alpha, \xi_2^\alpha, \ldots) \stackrel{\text{def}}{=} \mathbf{L}_\alpha(\mathbf{x}) = \lambda\,\mathbf{I}_\alpha(\mathbf{x}) = \lambda\,(\,\xi_0^\alpha, \xi_1^\alpha, \xi_2^\alpha, \ldots\,)\,; \quad \text{(by (4))}$$

thus $\quad \xi_1^\alpha = \lambda\xi_0^\alpha$, and $\quad \xi_n^\alpha = \lambda\xi_0^\alpha$ for all n, $\quad$ *i.e.* $\xi_n = \lambda^{1/\alpha}\xi_0$ for all $n = 1, 2, \ldots$. Whence every solution $\mathbf{x}$ lies on the ray through the origin $\mathbf{o}$ and the vector $\mathbf{x}_0 = (\, 1, \lambda^{1/\alpha}, \lambda^{2/\alpha}, \ldots \,)$, which exists for every λ with $|\lambda| < 1$.

Example 6.4.   For our fourth example we consider a companion equation to our last one, again on the specific Hilbert space $\ell^2(\mathbf{N}, \mathbb{C})$,

$$\mathbf{R}_\alpha\,(\,\mathbf{x}\,) \;=\; \lambda\,\mathbf{I}_\alpha\,(\,\mathbf{x}\,) \tag{48}$$

involving the homogeneous polynomial Right Shift operator $\mathbf{R}_\alpha$ of degree $\alpha > 1$ and the unital operator $\mathbf{I}_\alpha$, both with respect to the canonical orthonormal basis for $\ell^2(\mathbf{N}, \mathbb{C})$. Here we have

$$\mathbf{x} = (\, \xi_0, \xi_1, \xi_2, \xi_3, \ldots \,)$$

$$(\,0, \xi_0^\alpha, \xi_1^\alpha, \xi_2^\alpha, \ldots\,) \stackrel{\text{def}}{=} \mathbf{R}_\alpha(\mathbf{x}) = \lambda\,\mathbf{I}_\alpha(\mathbf{x}) = \lambda\,(\,\xi_0^\alpha, \xi_1^\alpha, \xi_2^\alpha, \ldots\,)\,; \quad \text{(by (4))}$$

thus $\quad 0 = \lambda\,\xi_0^\alpha$, and $\quad \xi_n^\alpha = \lambda\,\xi_{n-1}^\alpha$ for all $n > 0$; whence either $\lambda = 0$ or $\xi_0 = 0$.

If $\lambda = 0$ then $\xi_n^\alpha = 0$ for every $n = 1, 2, \ldots$ and ξ_0 is arbitrary; if $\lambda \neq 0$ then $\xi_0 = 0$ and hence every $\xi_n = 0$. Whence in the case $\lambda = 0$, every solution $\mathbf{x}$ lies in the one-dimensional subspace $\{\,(\,\xi_0, 0, 0, 0, \ldots\,)\,\}$.

Example 6.5.   For our next to last example we consider an equation on the 2-dimensional Hilbert space $\mathcal{E}^{[2]}$ (real or complex Euclidean plane)

$$Q\,(\,\mathbf{x}\,) \;=\; \lambda\,\mathbf{I}_2\,(\,\mathbf{x}\,) \tag{49}$$

involving the degree 2 (quadratic) homogeneous polynomial operator Q and the degree 2 unital operator $\mathbf{I}_2$, both with respect to the canonical orthonormal basis for $\mathcal{E}^{[2]}$. Here we have, for $\mathbf{x} = [\,x_1,\, x_2\,]^\mathsf{T}$,

$$\mathbf{o} \;=\; (\,Q - \lambda\mathbf{I}_2\,)\,(\,\mathbf{x}\,) \;=\; \begin{bmatrix} (a - \lambda)\,x_1^2 \,+\, 2\,b\,x_1\,x_2 \,+\, & c\,x_2^2 \\ e\,x_1^2 \,+\, 2\,f\,x_1\,x_2 \,+\, & (g - \lambda)\,x_2^2 \end{bmatrix}.$$

By Sylvester's dialytic elimination, the necessary and sufficient condition for this equation to have non-zero solution $\mathbf{x}$ is that the Sylvester Resultant Determinant

$$\mathrm{RES} \;=\; \begin{vmatrix} (a - \lambda) & 2b & c & 0 \\ 0 & (a - \lambda) & 2b & c \\ e & 2f & (g - \lambda) & 0 \\ 0 & e & 2f & (g - \lambda) \end{vmatrix},$$

a quartic polynomial in λ, shall vanish. It can be shown (very easily, using *e.g.* MAPLE software) that this resultant determinant is a classical invariant, that is to say its values under any non-singular linear change-of-basis transformation with matrix $[\,[p, q],\, [r, s]\,]$ changes only by the modulus-factor which in this case is equal to the 4th–power of the transformation's determinant $ps - qr$. Hence the invariant roots of the determinant's quartic polynomial are the eigenvalues of the quadratic operator Q .

Illustration   To illustrate this situation, consider a numeric example on the real Euclidean plane in which we arbitrarily take $a = 1, \quad b = 1, \quad c = 0$, and $e = 1, \; f = 0, g = -2$, (these might well have been scaled values for some 3rd–order permittivity tensor in nonlinear optics). An example for which we have this $2 \times 2 \times 2$ pen-symmetric multimatrix display

$$Q \;\overset{\mathrm{disp}}{=}\; \left[\, \begin{bmatrix} 1 & 1 \\ 1 & 0 \end{bmatrix},\; \begin{bmatrix} 1 & 0 \\ 0 & -2 \end{bmatrix} \,\right]^\mathsf{T}$$

$$\text{and} \quad Q - \lambda\,,\mathbf{I}_2 \;\overset{\mathrm{disp}}{=}\; \left[\, \begin{bmatrix} (1 - \lambda) & 1 \\ 1 & 0 \end{bmatrix},\; \begin{bmatrix} 1 & 0 \\ 0 & (-2 - \lambda) \end{bmatrix} \,\right]^\mathsf{T}$$

Using MAPLE software we faitly quickly compute the following POINT SPECTRUM SCHEDULE for Q :

1. $\rfloor\!\lfloor Q \rfloor\!\lfloor = 0.236448$, $\| Q \| = \rceil\!\lceil Q \rceil\!\lceil = 2$.

2.   The Sylvester Resultant Determinant is

$$\text{RES} = \begin{vmatrix} (1-\lambda) & 2 & 0 & 0 \\ 0 & (1-\lambda) & 2 & 0 \\ 1 & 0 & (-2-\lambda) & 0 \\ 0 & 1 & 0 & (-2-\lambda) \end{vmatrix} = (\lambda-2)(\lambda+2)(\lambda+1)^2.$$

3.   The point spectrum has 3 distinct eigen-values :
$$\lambda_1 = -2, \ \lambda_2 = 2, \ \lambda_3 = -1 \ (\text{twice}).$$

4.   The bounds on point spectrum are :
$$0.236448 = \rfloor\!\lfloor Q \rfloor\!\lfloor \le |\lambda| \le 2^{1/2} \times \rceil\!\lceil Q \rceil\!\lceil = 2.8284271.$$

5.   The pairs of simultaneous homogeneous quadratic equations associated with each of the three eigen-values are :
$$\begin{aligned}
[\lambda_1] \quad &: \quad 3x^2 + 2xy = 0; \qquad x^2 = 0; \\
[\lambda_2] \quad &: \quad 2x^2 + 2xy = 0; \qquad x^2 - y^2 = 0; \\
[\lambda_3] \quad &: \quad -x^2 + 2xy = 0; \qquad x^2 - 4y^2 = 0.
\end{aligned}$$

6.   The three eigenvector-sets each are comprised of a punctured (origin-less) straight-line in $\mathcal{E}^{[2]}$; they are disjoint and (unlike in the linear theory) they are mutually non-orthogonal. They, and their geometric ('Apparent') dimensions, are :
$$\begin{aligned}
\mathrm{EigVec}_{\lambda_1}(Q) &= \ \{\,(x,y) \mid x=0, y=arb.\,\} \setminus (0,0); \qquad & App.dim = 1 \\
\mathrm{EigVec}_{\lambda_2}(Q) &= \ \{\,(x,y) \mid y=-x\,\} \setminus (0,0); \qquad & App.dim = 1 \\
\mathrm{EigVec}_{\lambda_3}(Q) &= \ \{\,(x,y) \mid y=\tfrac{1}{2}x\,\} \setminus (0,0). \qquad & App.dim = 1
\end{aligned}$$

7.   Their three associated Projective Varieties are these three distinct pairs of points (designated here by rotational angle θ) on the unit circle in $\mathcal{E}^{[2]}$ (the Projective Hilbert space $\mathbf{P}(\mathcal{E}^{[2]})$) :
$$\begin{aligned}
\mathbf{PV}(\mathrm{EigVec}_{\lambda_1}(Q)) \quad &: \quad \theta = \pi/2, \ 3\pi/2 \\
\mathbf{PV}(\mathrm{EigVec}_{\lambda_2}(Q)) \quad &: \quad \theta = 3\pi/4, \ 7\pi/4 \\
\mathbf{PV}(\mathrm{EigVec}_{\lambda_3}(Q)) \quad &: \quad \theta = \arctan(\tfrac{1}{2}), \ \arctan(\tfrac{1}{2}) + \pi
\end{aligned}$$

9.   Their three reduced Gröbner bases, and their Hilbert dimensions, are :
$$\begin{aligned}
\mathrm{rGB}_{\lambda_1}(Q) &= \ \left[\,xy, \ x^2\,\right]; \qquad & Hilb.dim = 1 \\
\mathrm{rGB}_{\lambda_2}(Q) &= \ \left[\,y^2 + xy, \ x^2 - y^2\,\right]; \qquad & Hilb.dim = 1 \\
\mathrm{rGB}_{\lambda_3}(Q) &= \ \left[\,xy - 2y^2, \ x^2 - 4y^2\,\right]. \qquad & Hilb.dim = 1
\end{aligned}$$

Example 6.6.   The final illustration here again reveals how quickly our computational difficulties can increase as the degree of our homogeneous polynomial operators rises.

Let us now consider a cubic homogeneous polynomial operator K on Euclidean 3-space $\mathcal{E}^{[3]}$. It is represented by a triple of real homogeneous ternary cubic forms :

$$x' = a_1 x^3 + b_1 y^3 + c_1 z^3$$
$$+ 3d_1 x^2 y + 3e_1 x^2 z + 3f_1 xy^2 + + 3g_1 y^2 z + 3h_1 xz^2 + 3j_1 yz^2$$
$$+ 6k_1 xyz$$
$$y' = a_2 x^3 + b_2 y^3 + c_2 z^3 + [\ldots etc.\ldots] + 6k_2 xyz$$
$$z' = a_3 x^3 + b_3 y^3 + c_3 z^3 + [\ldots etc.\ldots] + 6k_3 xyz \tag{50}$$

As before, we form

$$K(\mathbf{x}) = \lambda \mathbf{I}_3(\mathbf{x}), \tag{51}$$

the Homogeneous Characteristic Operator equation for K, of degree 3. We seek those scalars λ for which this has non-zero solution eigen-vectors $\mathbf{x}$.

The (pen-symmetric) multimatrix representation of this Cubic Operator $K - \lambda\,\mathbf{I}_3$ (equivalent to a 4th-order tensor) is :

$$\left[\begin{array}{c} \left\{ \begin{bmatrix} a_1-\lambda & d_1 & e_1 \\ d_1 & f_1 & k_1 \\ e_1 & k_1 & h_1 \end{bmatrix} \begin{bmatrix} d_1 & f_1 & k_1 \\ f_1 & b_1 & g_1 \\ k_1 & g_1 & j_1 \end{bmatrix} \begin{bmatrix} e_1 & k_1 & h_1 \\ k_1 & g_1 & j_1 \\ h_1 & j_1 & c_1 \end{bmatrix} \right\} \\[4ex] \left\{ \begin{bmatrix} a_2 & d_2 & e_2 \\ d_2 & f_2 & k_2 \\ e_2 & k_2 & h_2 \end{bmatrix} \begin{bmatrix} d_2 & f_2 & k_2 \\ f_2 & b_2-\lambda & g_2 \\ k_2 & g_2 & j_2 \end{bmatrix} \begin{bmatrix} e_2 & k_2 & h_2 \\ k_2 & g_2 & j_2 \\ h_2 & j_2 & c_2 \end{bmatrix} \right\} \\[4ex] \left\{ \begin{bmatrix} a_3 & d_3 & e_3 \\ d_3 & f_3 & k_3 \\ e_3 & k_3 & h_3 \end{bmatrix} \begin{bmatrix} d_3 & f_3 & k_3 \\ f_3 & b_3 & g_3 \\ k_3 & g_3 & j_3 \end{bmatrix} \begin{bmatrix} e_3 & k_3 & h_3 \\ k_3 & g_3 & j_3 \\ h_3 & j_3 & c_3-\lambda \end{bmatrix} \right\} \end{array}\right]$$

$$\tag{52}$$

[Writing it in this 'display' format perhaps makes it easier to visualize any explicit numerical instantiation. Note, for instance, the pen-symmetric dispositions of *e.g.* the 6 repetitions of the k_i coefficients in each sub-row-block.]

This is a $3\times3\times3\times3$ pen-symmetric multimatrix with 81 terms of which at most 30 can be distinct; and it has a $3\times3\times3\times3$ multideterminant. The full expansion of this multideterminant appears not yet to have been achieved using computer software [15]. A semi-anonymous comment included in the website [16] claims that its degree has been computed to be 810, in which case its number of terms may be incomprehensibly large, and its expansion physically unprintable.

However, a versatile MAPLE software script [14] for computing the Sylvester Resultant of three ternary cubics such as we have here at (50), can be utilized. The Resultant will be a factor of the larger hyperdeterminant, and its vanishing is a sufficient condition for our three cubics (50) to have a common non-zero root.

We can modify these three cubics by attaching the parameter $-\lambda$ to each of the leading terms x^3, y^3, z^3 (in effect, by subtracting $\lambda \mathbf{I}_3$ from K, (as seen in (52)). The computed Sylvester Resultant is a 15×15 determinant, which, equated to zero, delivers a polynomial of degree 27 in λ. Its (real) solutions are the eigen-values we are seeking for the cubic homogeneous polynomial operator K.

For example, arbitrarily choosing the coefficients of the three ternary cubics (50) to be :

	a	b	c	d	e	f	g	h	j	k
eqn.1	1	2	3	−1	0	2	0	2	0	0
eqn.2	−1	0	0	0	0	0	3	0	4	1
eqn.3	0	0	2	2	0	1	0	5	0	0

(these might well have been scaled values for some 4th–order permittivity tensor in nonlinear optics) the program delivered this 'Characteristic Polynomial' for K :

$$\lambda^{27} - 27\,\lambda^{26} + 360\,\lambda^{25} - 9135\,\lambda^{24} + 35217\,\lambda^{23} + 259776\,\lambda^{22}$$

$$+2643717\,\lambda^{21} + 270330390\,\lambda^{20} - 189842805\,\lambda^{19} - 4151398197\,\lambda^{18}$$

$$\dots \quad etc. \quad \dots$$

$$-2405297184053735016537\,\lambda^3 + 5557980464326475115066\,\lambda^2$$

$$-384361176611500537824\,\lambda + 738347509741784948 4554 \tag{53}$$

There happens to be 7 real roots in this instance :

$$1.571454383,\ -4.445341471,\ 6.534273619,\ ,7.201697960,$$
$$14.42621084,\ 14.66259075,\ 22.83094269 \quad \text{(our latent-values)},$$
$$\text{and 10 pairs of complex roots.}$$

The absolute values of all 27 roots lie in the interval $[\,1.091843331, \dots 22.83094269\,]$. The inferior and superior bounds of the cubic homogeneous polynomial operator K are computer-estimated to be $\rfloor\!\lfloor K \rfloor\!\lfloor = 0.323792$ and $\rceil\!\lceil K \rceil\!\lceil = 8.784240$.

It is known [3] that in finite-dimensional spaces, the eigen-values of a bounded homogeneous polynomial operator are bounded by its inferior and superior operator bounds (or a fixed multiple thereof). Hence we have these spectral bounds for K :

$$0.323792 \ \le \ |\lambda| \ \le \ (3^{2/2}) \times 8.784240 \ = \ 26.35720$$

which certainly bound the roots of the Characteristic Polynomial (53) for K.

A full POINT SPECTRUM SCHEDULE for this cubic operator K could also be produced (as in the previous example) but there is insufficient space to report it here.

Concluding Remarks

The last four Examples in the previous section are drawn from the author's spectral theory of homogeneous polynomial operators and other nonlinear operators [3] which is closely related to the rapidly evolving spectral theory of tensors (see *e.g.* [8], and the bibliography [9]).

Equation (47) is the 'characteristic homogeneous polynomial operator equation' for the α–degree left shift $\mathbf{L}_\alpha$ whose spectrum consists of all λ in the open unit disc in $\mathbb{C}$, and whose eigenvectors lie on the afore-mentioned ray, excluding the origin; on the other hand equation (48) is the 'characteristic homogeneous polynomial operator equation' for the α–degree right shift $\mathbf{R}_\alpha$ whose spectrum consists of the singleton set the origin in $\mathbb{C}$, and whose eigenvectors lie in the afore-mentioned one-dimensional subspace, excluding the origin.

Equations (49) and (51) are the respective 'characteristic homogeneous polynomial operator equations' for the quadratic operator Q and the cubic operator K whose detailed 'spectrum schedules' can readily be computed, using appropriate algebraic-geometrical software, as shown in the typical illustrations accompanying those examples. Spectrum schedules for other examples involving higher degree operators on higher-dimensional Hilbert spaces have also been shown to be computable. The now all-too-familiar, and extraordinary 'explosive' increase in the computational complexity unavoidable with increasingly higher degrees is apparently restricting explorations to homogenous polynomial operators of at most 5th degree at present.

What the present investigations are now revealing is the relative ease with which otherwise intractable problems relating to the novel concept of the spectral theory of, for example the optical response of the dielectric polarisation to a nonlinear electric field, can be converted to a functional analytic problem. One in which the novel idea of unital homogeneous polynomial operators play a vital part, together with the now obvious way in which the constitutive homogeneous polynomial operators can be directly associated through their unique multimatrix representations with, say, the permittivity tensors of higher order.

In the case of tensors that arise in other branches of theoretical physics where our required kinds of symmetry are replaced by other less tractable ones, it seems that certain extensions to the analogous situation involving unital multilinear linear and other constitutive multilinear operators (*e.g.* representing more general tensors of mixed contravariant and covariant type) may provide an equally suitable methodology. One in which extensions from homogenous to non-homogenous algebraic geometry may play an important rôle. The possibilities seem to justify much further study and many more hard-data based exemplars.

References

[1] J. C. AMSON, 'On polynomial operator equations in Banach space'. J. London Math. Soc. (1975) (**2**) (10), 171–174.

[2] J. C. AMSON, 'Multimatrix polyalgebra representations of the polar composition of polynomial operators'. Proc. London Math. Soc. (1972) (3) (25), 465–485.

[3] J. C. AMSON, *Spectral Theory of Homogenous Polynomial and Multilinear Operators in Hilbert Space.* (In preparation, 2013)

[4] Y. Benyamini, S. Lasalle, J.G. Llavona, 'Homogeneous Orthogonally Additive Polynomials on Banach Lattices', Bull. London Math. Soc.(2006) 38, 3, 459–469.

[5] J. Bochnak, J. Siciak, 'Polynomials and multilinear mappings in topological spaces'. Studia Math. (1971) 39, 59–76.

[6] Robert W. Boyd. *Nonlinear Optics*, Academic Press, Elsevier, 3rd Edition (2008)

[7] S. Dineen. *Complex Analysis on Infinite Dimensional Spaces*, (Springer Monographs in Science, Springer, 1999, reprinted in paper-back 2012)

[8] Liqun Qi. 'Eigenvalues and invariants of tensors'. J. Math. Anal. Appl. 325 (2007) 1363–1377.

[9] Liqun Qi. 'Bibliography of Eigenvalues of tensors'. January 9, 2012.
http://www.polyu.edu.hk/ama/staff/new/qilq/biblio.pdf

[10] Werner Greub. *Multilinear Algebra*, Universitext, Springer-Verlag, Heidelberg, (2nd. ed.) (1978)

[11] S. Mazur, W. Orlicz. 'Grundlegende Eigenschaften der Polynonomische Operationen'. Studia Math. (1934) 5, 50–68, 179–189.

[12] Nadar Jeevanjee. *An Introduction to Tensors and Group Theory for Physicists*, (Birkhäuser, Springer, 2011) — Especially : Ch.3. Tensors.

[13] J. F. Nye. *Physical Properties of Crystals : Their representation by Tensors and Matrices*, (Oxford Science Publications), O.U.P. (1st ed, 1957), reprinted (2011)

[14] B. Sturmfels. Private communication, (2011)
Maple software script : 'ThreeTernaryCubics'.

[15] J. Weyman. Private communication, (2008)

[16] website (2011) :
http://hyperdeterminant.wordpress.com/2008/09/29/cayleys-hyperdeterminant

....

Brief Biography

Born 1927, three years later than Pierre Noyes. Enjoyed many different careers, some simultaneously, since leaving school in Liverpool aged 15 during WW2.

Initially trained as Chartered Land Surveyor (1942-1948). Senior Town Planning Assistant, North Riding of Yorkshire County Council, England (1949-1955). Member, London Mathematical Society (1954 to date). Honours & Prizes in Special Mathematics via National Mature Student Scholarship at Reading University (1955-1958). Senior Research Officer & Team Leader (theoretical & experimental) in Physics of Electric Welding Arc, BWRA (Brit.Weld.Res.Assoc.) (1958-1962). Contributor to NASA (Apollo) contract studies of welding arc in low pressure & space environments, 1961. Co-organizer of 2nd British Commonwealth Welding Physics Conference, London, 1962. British Delegate then Expert, Physics of Welding Arc Committee, International Institute of Welding (1962-1975). Keynote Lecturer, I.I.W. Annual University Research Conference of Welding Research Council, New York 1961. Invited to collaborate with Prof. N.N.Rykalin, Chairman of USSR Welding Committee, Moscow, 1962. Awarded The Churchill College Research Studentship (1962-1965) for Cambridge PhD in Mathematics: Functional Analysis. Fellow of the Cambridge Philosophical Society (1965 to date). Member, Edinburgh

Mathematical Society (1965 to date).   Senior Lectureship in Mathematics, St Andrews University, specialising in Real, Complex & Functional Analysis, Banach Algebras (1965-1982).   Primary research interest : Spectral Theory of Homogeneous Polynomial Operators in Hilbert Space.   Awarded Social Sciences Research Council Grants: Advanced Mathematics in Urban Studies (1970-1971, 1977-1978). Introduced discontinuous and catastrophic Evolutionary Dynamics into Urban-&-Regional Systems Studies (1971-1978).   Sole proprietor, Navigational Software Firm (1982-1999), created the first commercial Tidal Prediction Software Package, Europe, 1989.   Royal Yachting Association Scotland Volunteer of the Year Award, 2004.   Fellow of the Royal Astronomical Society (1988 to date).

Collaborated with Ted Bastin and Frederick Parker-Rhodes over Brouwerian Hierarchies (1964-1966), then also with Clive Kilmister over Combinatorial Hierarchies & Discrimination Systems (1965-2010), and with Pierre Noyes over Bitstring Physics (1973 to date). Founder member, with P.N., T.B., F.P-R, C.K., of ANPA (Alternative Natural Philosophy Association), 1979.   First Parker-Rhodes Memorial Lecturer, 1989.

Honorary Member (School of Mathematics & Statistics), St Andrews University (1982 to date).   2013 – still not retired.

Towards a Generalised Combinatorial Hierarchy

Keith G. Bowden

Theoretical Physics Research Unit,
Birkbeck College, University of London,
Malet St, London WC1E 7HX
E-mail: k.bowden@physics.ac.uk

The Combinatorial Hierarchy is a four level bit string hierarchy discovered by Frederick Parker-Rhodes in 1963. Associated with the four levels is the number sequence 3, 10, 137, 1.7×10^{38}, the last two closely resembling two of the Structure Constants of Physics. Much has been made of this phenomenon; however this paper concentrates on furthering the understanding of the mathematical structure of the Hierarchy by generalising it to Objects and Morphisms in an arbitrary Category. The author has long argued that if the Combinatorial Hierarchy has any meaning then the important part of it is not the structure of the levels of the Hierarchy, but that of the Level Change Operator, that maps Objects from one level of the Hierarchy to the next. We conclude that if a connection to Quantum Nonlocality is to be made it is more likely to be through the structure of the Level Change Operator on a Topos than through traditional means.

Introduction

The author has long held that if the Combinatorial Hierarchy has any meaning then the important part of it is not the structure of the Hierarchy itself, but that of the Level Change Operator, that maps Objects from one Level of the Hierarchy to the next. John Amson started the process of formally generalising the Level Change Operator in terms of Group Theoretic operations [4]. The work was continued by Amson and Clive Kilmister, Amson and me (informally), and then Kilmister and me [9]. This paper aims at formally specifying the Level Change Operator in terms of Morphisms between a sequence of Objects in an arbitrary Category.

We show that such a Level Change Operator is of the form "Autos mod Monos", or, less mnemonically, that in an arbitrary Category the Level Change Operator would generate a new Object (the next Level) by taking the (generalised) induced closure of (Representative Elements of) the Equivalence Class of Automorphisms mod Subobjects of a given Object. We speculate that an arbitrary Topos admits such a Level Change Operator, but not, perhaps, an arbitrary Category.

Motivation

The Combinatorial Hierarchy has its origins in Ted Bastin's philosophical ideas about Physics coming from Hierarchies of Levels of Information (or Structure).

Frederick Parker-Rhodes extended these ideas to Hierarchies in which each Level was generated from the "parts" of the previous Level. A simple case would be the Hierarchy in which each Level is the set of subsets of the previous Level. Then using bit strings as the most primitive information carrying entities out of which to build the Levels he discovered the Combinatorial Hierarchy, a four level bit string Hierarchy, in 1963. Associated with the four levels is the number sequence 3, 10, 137, 1.7×10^{38}, the last two closely resembling two of the structure constants of Physics.

Much has been made of this phenomenon. Typically the idea might be that the four levels could be associated with the gluons (strong force), W and Z particles (weak force), the photon (electromagnetism) and the Higgs particle (gravity). By adding extra rules (or weakening existing ones), Clive Kilmister, before his untimely death in 2010 (the year before Ted died), produced a series of corrections to the number associated with the third level, until it corresponded, within experimental accuracy, to the measured value for the Fine Structure Constant. He and Ted also worked on an ontological explanation for the numerical phenomenon. Pierre Noyes at SLAC mapped the bit string structure to the Standard Model of Quarks and Leptons. Ultimately the Physics community, with some notable exceptions, exhibited a marked lack of interest in these results.

In this paper the author makes no claims about the Physics but seeks to better understand the mathematical structure of the Combinatorial Hierarchy itself. In particular, following John Amson's 1979 rewrite of the Hierarchy in terms of Group Theory, we seek to reformulate the Hierarchy in terms of Category Theory in an attempt to understand the essence of Frederick Parker-Rhodes' construction. Rewriting the Hierarchy in this way is a generalisation that omits the particular structure resulting from the use of bit strings or Groups, but retains the essence of Ted and Fred's ideas of constructing a Hierarchy of Levels in which the elements of each Level are the *parts* of the previous Level.

Ironically, reformulating the Hierarchy in terms of Categories loses all notion of the concept of elements as, in general, we are no longer dealing with Categories of structured Sets, so that all information about Objects is held as Morphisms (mappings) between the Objects in the Category. Indeed ultimately the Objects themselves are described in this way. We could say that Category Theory is *Set Theory from the outside*, rather than the inside. Nevertheless we maintain that this approach is entirely consistent with Ted's and Fred's approach of reducing things to their essence, in order to understand their meaning. It has the added advantage that there is a plethora of machinery — theory, structure, and theorems — in Category Theory that we can use to our advantage.

It is also consistent with our claim that the essence of the Hierarchy is in the Level Change Operator and not in the Levels themselves. In describing the Hierarchy using Category Theory we do not have to specify what the structure of the (Objects representing the) Levels is but only the way in which (the Morphisms of) the Level Change Operator map(s) between the Levels. We discover that the structure of the Level Change Operator can be specified in two parts. First we specify how the Phixers (Representative Automorphisms) of a Level are generated by its Subobjects, and then how the next level is built from the Phixers. The first half of this process we achieve in an arbitrary Category, for the first time, in this paper; but the second half proves harder for Objects in an arbitrary Category. We speculate that an arbitrary Topos (and hence an arbitrary non-Boolean Topos) admits such a Level Change Operator. This would be a strong and useful result as it may in the future allow a direct connection between the Hierarchy and Quantum Mechanics, particularly with regard to the work of Kochen and Specker. We now see the Hierarchy on an arbitrary nonBoolean Topos as being potentially Quantum Mechanical, but in any other Category as being essentially classical.

The Classical Hierarchy

The levels of the classical Combinatorial Hierarchy are formed from (contain) "Discriminately Closed Subsets" (DCS's) of bit strings of a fixed length (initially 2) under the discrimination (Exclusive OR, XOR) operator as addition (and AND as multiplication). The elements of each DCS, are *"Phixed"* ("fixed" and "unfixed") by an Automorphism A (premultiplication by a nonsingular bit matrix) so that $Ax = x$ ("fixing") for each element x in the DCS (where bit multiplication is AND) and $Ax \neq x$ ("unfixing") for every element outside the DCS. x is a *fixed point*. The elements x are then clearly eigenvectors of the matrix A which has all unit eigenvalues (being a *non-singular* bit matrix).

The elements of the next level are obtained by choosing a representative Automorphism (nonsingular bit matrix) fixing each DCS, and slicing each (2×2) matrix A into a 4-vector by concatenating its column vectors in order. The next level is the closure of these vectors under Exclusive OR (the "induced operator") and new DCS's are the Subgroups of this closure. And so on.

Notes

1. The slicing of the matrices into vectors is simply to enable matrix multiplication at the next level of the Hierarchy. It has no meaning in any of the generalisations.

2. John Amson has emphasised Frederick's condition that the new generator set should be linearly independent, but Clive questioned this with us. Certainly linear independence is the most "efficient" means of information preservation between the levels. In the linear case a linearly independent set becomes a minimal generator set of the new Group. Otherwise the condition is needed only to generate the cutoff point at Level Four. However the span of $\{A, B, A + B\}$ is of course the same as the span of $\{A, B\}$. Further, it seems to me that choosing Automorphisms that Phix distinct Subgroups is introducing a strong element of independence between the Automorphisms. This should be understood better.

3. John also emphasised that the new generator set must be chosen from elements of the previous level and not just arbitrary members of the Group in which it

lives. This is taken care of automatically in an iterative construction such as that described here.

Generalising to Groups

That the DCS's obey the structure rules for a Group can be easily seen by considering $A(x + y) = x + y = Ax + Ay$ and $A(0) = 0$. As each level is the closure under the induced operator $+$ of the Set of Representative Automorphisms inherited from the previous level, then each level is a Group and the DCS's are its Subgroups.

When starting from an arbitrary Group say Q, rather than Linear Algebra on bit strings, we have to be more careful because, although each Automorphism Phixes a Subgroup, not every Subgroup is necessarily Phixed by an Automorphism, and some are Phixed by more than one. Nevertheless, when we can, for each Subgroup we choose an Automorphism that Phixes it. This gives us a set of Representative Elements for the Equivalence Class of Automorphisms   mod Subgroups. The next level of the Hierarchy is given by the closure of this set under the Induced Operator $(A + B)(\_) = A(\_) + B(\_)$ inherited from the original Group. For Abelian Groups this closure consists of elements of the Endomorphism Group of the Level, for non Abelian Groups it is a just a subset of the elements of the Permutation Group.

Notes

4. It is not clear that choosing different Representative Elements for the same Equivalence Class won't lead to different Groups at the next level. Clive had no further comment on this.

5. Consider the Quaternion Group Q. In the usual notation there are many different Automorphisms that Phix the Subgroup $(\{1, -1\}, \times)$.

6. Clive had examples of Subgroups that are not fixed by <u>any</u> Automorphism.

7. Are PGS's and AGS's [9] the same? Consider subgroups generated from a subset of the generators. Then $(\{1, -1\}, \times)$ is a subgroup of Q which is generated by more than one Automorphism, but it is <u>not</u> generated by a subset of any minimal generator set of Q. However all subgroups of Q are fixed by some Automorphism. There are DCS's which are AGS's but not PGS's.

Generalising to Categories

A *Category* is a collection of *Objects* and *Arrows* (Morphisms) between the Objects subject to a number of rules. As such is it a special case of a Directed Graph and can be shown to be a generalisation of collections of structured Sets and Homomorphisms such as Groups (Monoids, *etc.*) and Group (Monoid, *etc.*) Homomorphisms *etc.* An Arrow e maps from an Object dom(e), the domain (or source) of e to an Object cod(e), the codomain (or target) of e. The rules are:

1. Every Object should have a unique identity self-loop (the id or 1 Morphism).

2. The composition of any two consecutive compatible morphisms f and g exists and is written $e = g\,f$ or

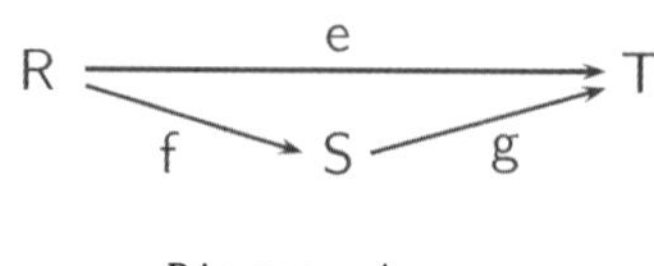

Diagram 1.

We say that the diagram commutes. Composition with a self-loop does not change any Morphism so $h1 = 1h = h$ for any Morphism h and the appropriate compatible self-loop 1. (By compatible Morphisms we mean those for which $\mathrm{cod}(f) = \mathrm{dom}(g)$ for a composition of the form gf).

3. Composition of Morphisms is associative, although that does not affect anything in this paper, so we might well be working with Lou Kauffman's "Conceptions" [10], which are Categories with the Associativity restriction removed.

The data of (information in) a Category is the set of relationships between the arrows as described in 2. above. For the Category of Finite Groups (for instance) it is equivalent (maybe homomorphic) to the data in the Cayley tables or the sets of generators and relations or the set of specifications of the Groups using matrix representations.

Our Level Change Operator in a Category is now constructed from the Equivalence Class of Automorphisms of a Level (or Object) mod "Phixing" the Subobjects of that Level (Object). This is done in two stages, first the concept of Phixer is specified as a commutative diagram in a Category, then we show how to construct the next Level out of the Phixers of the previous Level for particular classes of Category, and discuss a route to doing this in general.

Morphisms [Wikipedia]

A *Epimorphism* (also called an *Epic Morphism* or, colloquially, an *Epi*) is a Morphism $f : X \to Y$ that is right-cancellative in the sense that, for all Morphisms $g_1, g_2 : Y \to Z$,

$$g_1 \circ f = g_2 \circ f \quad \Longrightarrow \quad g_1 = g_2 .$$

Epimorphisms are analogues of surjective functions, but they are not exactly the same. The dual of an Epimorphism is a Monomorphism (*i.e.* an Epimorphism in a category C is a Monomorphism in the dual Category C^{op}).

A *Monomorphism* (also called a *Monic Morphism* or a *Mono*) is a left-cancellative Morphism, that is, an arrow $f : X \to Y$ such that, for all Morphisms $g_1, g_2 : Z \to X$,

$$f \circ g_1 = f \circ g_2 \quad \Longrightarrow \quad g_1 = g_2 .$$

Monomorphisms are a Categorical generalisation of injective functions (also called "one-to-one functions"); in some Categories the notions coincide, but Monomorphisms are more general.

An *Isomorphism* is a Morphism that is both Epi and Mono.

An *Automorphism* is an Isomorphism whose domain and codomain are identical (*i.e.* a self-loop).

Subobjects [Wikipedia]

Given two Monomorphisms f : R → T and g : S → T with codomain T we say that f ≤ g if f factors through g, that is if there exists some w : R → S such that

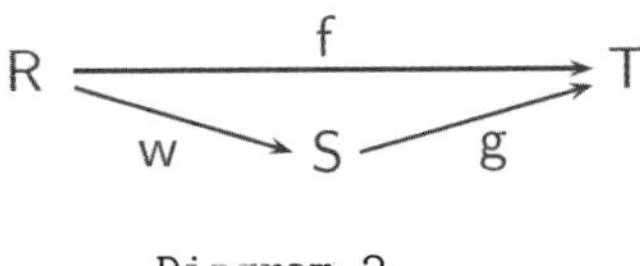

Diagram 2.

commutes. The binary relation ≡ defined by f ≡ g iff f ≤ g and g ≤ f, is an Equivalence Relation on the Monomorphisms with codomain T, and the corresponding Equivalence Classes of these Monomorphisms are the *Subobjects* of T. If two Monomorphisms represent the same Subobject of T, then their domains are Isomorphic. The collection of Monomorphisms with codomain T under the relation ≤ forms a <u>preorder</u>, but the definition of a Subobject ensures that the collection of Subobjects of T is a <u>partial order</u>. It can easily be seen that for Groups this partial order is 1 : 1 with the partial order on Subgroups.

Phixers

For Categories of structured Sets a *Phixer* is an Automorphism A, on an Object O, that fixes and "unfixes" (the domain of a representative element of) a Subobject S of O, that is, the elements of S are (respectively) all (and only) the fixed points of A in O. We often omit the brackets before "a Subobject". Remember also that these domains are isomorphic.

If A fixes S then, still for Categories of structured Sets, A restricted to S = $\{r \mid r \equiv s\}$, where s is a representative Monomorphism of S, is the identity map on S, that is $A|_S = \mathrm{id}_S$, therefore

$$A|_S \circ s = \mathrm{id}_S \circ s = s$$

so that the following diagram commutes

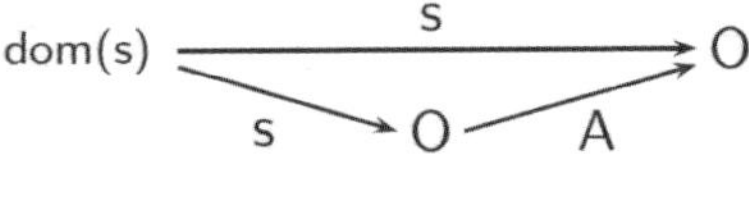

Diagram 3.

and conversely if the diagram commutes then A must be the identity map on S and is therefore a fixer of S, so that *for a general Category* we can define that A fixes $S = \{r \mid r \equiv s\}$ iff the diagram commutes.

Now if A also unfixes S in O, S must *also* be the largest Subobject "in" (*i.e.* of) O that is fixed by A so that it must also be true that there is no (nontrivial) decomposition of s into a Monomorphism w and a Monomorphism t such that A fixes the domain of t and that the following diagram commutes

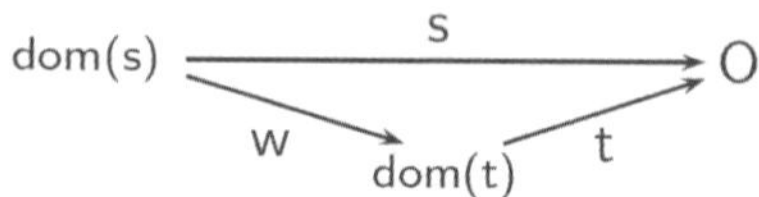

Diagram 4.

that is, that there is no $t \geq s$. Conversely if S is the largest Subobject in O that is fixed by A then A also unfixes S.

Conversely, again, we could say that A unfixes S in O iff for all Monomorphisms $t \to O$ such that A fixes dom(t) there is a w such that the following diagram commutes

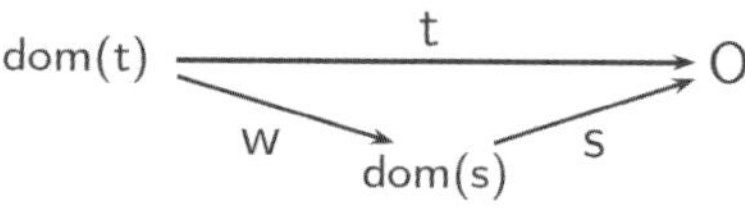

Diagram 5.

that is $t \leq s$.

So now we can define a *Phixer* A of a Subobject $S = \{r \mid r \equiv s\}$ in O as an Automorphism A of O such that (1) the following diagram commutes (A is a fixer of S)

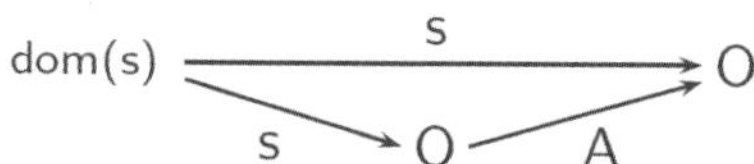

Diagram 6.

and (2) that there are *no* Monomorphisms t, w, such that the following diagram commutes (A is an unfixer of S)

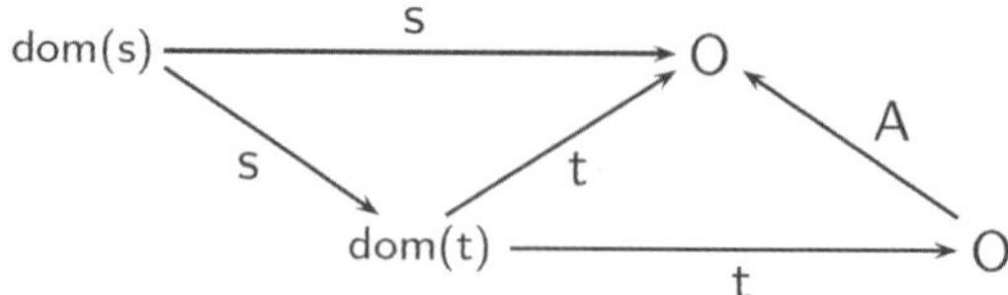

Diagram 7.

(or that the converse isn't true).

Alternatively we can combine the two diagrams and say that, still *for a general Category*, a *Phixer* A of a Subobject S = { r | r ≡ s } in O is an Automorphism A of O such that 1. there is no pair of Monomorphisms w and t such that the following diagram commutes

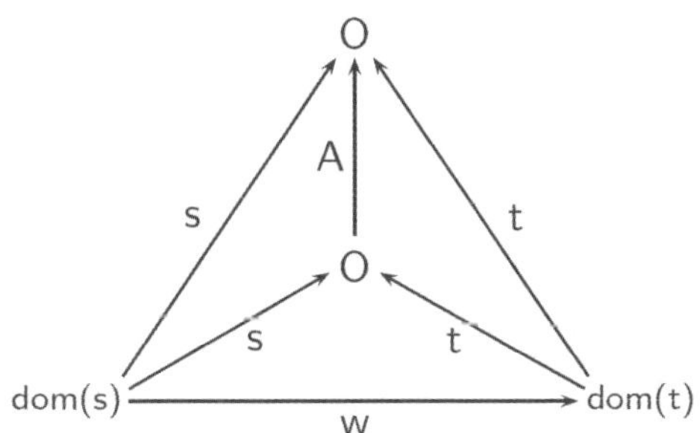

Diagram 8.

and that 2. the upper left triangle does commute.

It is however perhaps neater to say that a *Phixer* A of a Subobject S = { t | t ≡ s } in O is an Automorphism A of O such that for any Monomorphism t → O such that A fixes dom(t) there is a w such that the following diagram commutes

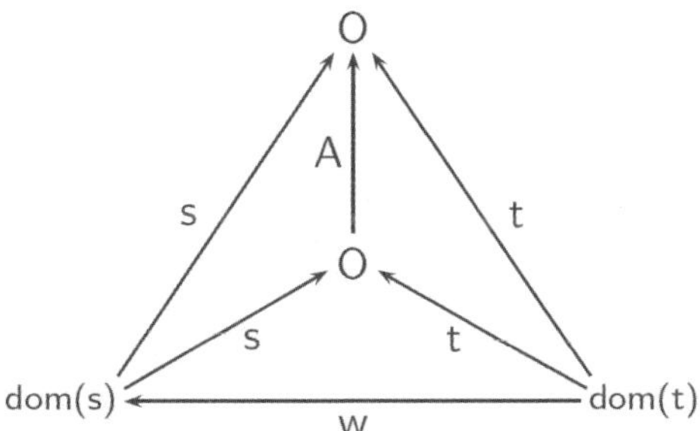

Diagram 9.

We say A Phixes S = { r | r ≡ s } iff for any t such that A fixes dom(t) there is a w such that the diagram commutes.

Now consider the partial order on all the Subobjects of O. Replacing each Subobject by the domain of a representative element induces a partial order on the domains which is unique up to isomorphism (as the domains of the elements of each Subobject are isomorphic). For Categories in which all the Isomorphisms are Automorphisms (self-loops) the partial order is unique. This partial order can be written as a set of inclusion mappings (Monomorphisms) between the Objects (*i.e.* the domains) of the Category (and eventually extended to all the Objects of the Category). Now consider mappings from this partial order to the Automorphisms A : O → O in which every Subobject is mapped to *each* of the two copies of the

Object O by a Monomorphism. The resulting complex diagram (which we will not attenpt to draw here!) contains all the information about this level of the Hierarchy and can, in the obvious way, be extended to contain all the information about all the levels of the Hierarchy.

Examples

We will look at some examples involving the partial order $C_2 \leq C_4 \leq Q$, with the Quaternion Group written in the usual way as $Q = (\{1, i, j, k, -i, -j, -k, -1\}, \times)$ and its 24 Automorphisms written as the images of the mappings from the generator set $\{i, j\}$ into Q. The rest of the image of each Automorphism can be obtained by closure under the structure relations for a Group. Thus the Automorphism $A = \{j, i\}$ means that $A : \{i, j\} \to \{j, i\}$ which expands to the Automorphism

$$A : (\{1, i, j, k, -i, -j, -k\}, \times) \to (\{1, j, i, -k, -j, -i, k, -1\}, \times)$$

under the structure rules.   Note that this example A Phixes the Subgroup $C_2 = (\{1, -1\}, \times)$.   Exercises: show that the Automorphisms $\{i, -j\}$, $\{-i, -j\}$ Phix the Subgroups $(\{1, i, -i, -1\}, \times)$ and $(\{1, k, -k, -1\}, \times)$ respectively. From now on we will omit the $\times$. Note that these are just two of the different ways that the Cyclic Group C_4 is Phixed by Q under different Automorphisms. A full table of the Automorphisms is given on page 40 of Proc. ANPA 28 [12].

Consider the first example above. $A = \{j, i\}$ Phixes $C_2 = \{1, -1\}$ in Q. In Categorical language we can say A Phixes the (domain of a representative element of) the Subobject C_2 in Q. Here we use C_2 to label both the Object and the corresponding Subobject. Thus we have the diagram

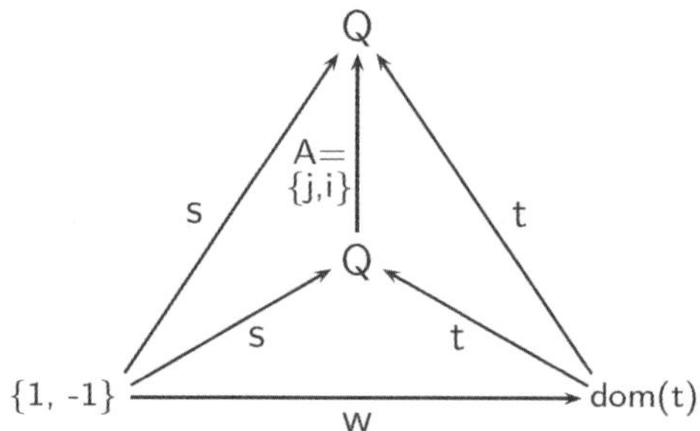

Diagram 10.

It can easily be seen that $A = \{j, i\}$ ("swap i and j") is not going to affect Monomorphisms from $\{1, -1\}$ to Q so that the top left triangle commutes. Furthermore there is no t that is bigger than s inside Q subject to $t = At$ so there is no pair of Monomorphisms t and w such that the whole diagram commutes. C_4 is not a viable candidate for dom(t) because it is not fixed by A.

Now consider a second example : $A = \{i, -j\}$ Phixes $C_4 = \{1, i, -i, -1\}$. It also fixes $C_2 = \{1, -1\}$, which is a Subobject of C_4, **but** does **not** unfix it (because i is fixed by A but is not in $\{1, -1\}$) so the diagram for $A = \{i, -j\}$ and C_2 (*sic*) is

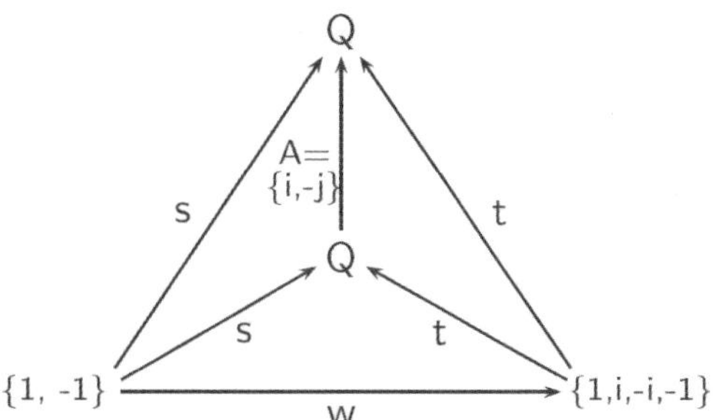

Diagram 11.

and there clearly is a decomposition of s into w and t such that the diagram commutes, so that A does **not** Phix C_2 (because it Phixes C_4). Remember that the Group $\{1, -1\}$ here is an abbreviation for $(\{1, -1\}, \times)$. The (inner) Automorphism $\{i, -j\}$ is an abbreviation for A $:$ $(\{1, i, j, k, -i, -j, -k, -1\}, \times)$ $\rightarrow$ $(\{1, i, -j, -k, -i, j, k, -1\}, \times) = i \_ i^{-1}$.

Combining the Automorphisms

So for Categories in which all Finite Products exist we have a generalisation of the induced operator $A \times A \rightarrow A$ and can use this to combine pairs of Representative Automorphisms in the usual way and use the closure of this operation to identify a new Object for the next Level of the Hierarchy. But, for an arbitrary Category, we have to decide what it means to close the set of Representative Elements into a new Object (Level) when we do not even necessarily have an Operator to induce! Clearly what we have to do is enrich this set in some analogous way to closing a Group under an induced operator, so that it becomes isomorphic to some Object already in the Category.

In order for this to work, the Representative Automorphisms (as the only route) have to pass information from the lower Object to the upper one. Now, Set Theory and Algebra are the study of Objects from their inside. Category theory and Higher Algebra are the study of Objects from their outside, in terms of their Morphisms only. All the information about an Object is held in its Morphisms. The part that is transmitted to the next level of the Combinatorial Hierarchy is the part of this that is held in its (Representative) Automorphisms, so this is not unreasonable.

Consider working in the Category of Directed Graphs. In order for the Morphisms of this Category to be combined into a new Object (Directed Graph) the Morphisms must presumably look a bit like Directed Graphs themselves, which they may. Where have we seen something like this before? Consider working on dynamic systems in the time domain. We have a set of signals which are functions of t. We have a set of differential equations which we may write down as functions of t, but may also write down as polynomials in s, the Laplace transform. We then realise that we can also write the signals down as polynomials in s. We have started with

Objects and Morphisms, then found a way to transform the Morphisms into a form
into which we can also transform the Objects.

Perhaps this is an approach to solving this problem in general but for now these
are speculations and the best we find that we can do is to work in a Category that
has all Finite Products to guarantee that an induced Operator (*i.e.*, the product)
can be inherited from one Level of the Hierarchy to the next. Thus we would have
a generalisation of the Level change Operator to, for instance, Cartesian Closed
Categories and (thus) Elementary Topoi, but not to arbitrary Categories as yet.

Conclusions

Firstly we mention an interesting alternative generalisation. The philosopher, Eric
Steinhart [13] has (roughly speaking) defined "combinatorial hierarchies" as mul-
tilevel, hierarchical representations of ontologies, with the levels ordered by complexity. It is not known if the name was taken from the Combinatorial Hierarchy.
There is an "initial" level, a level change operator ("successor rule") and an upper
("limit or final") level for each hierarchy. The levels are built out of the subsets
and/or (mereological) parts of the lower levels. Steinhart lists a large collection of
hierarchies discussed by recent philosophers including Quine and David Lewis (as
described in "Parts of Classes"). He lists the hierarchies in order from the simplest
("very few things exist"), to the richest ("most generous point of view") which
includes all elements from all other hierarchies.

At first sight Steinhart's scheme would include the Combinatorial Hierarchy,
however Steinhart's view of "parthood" seems too weak to include structures built
out of Subobjects (although it is an easy generalisation). However the reverse is
also true. The Combinatorial Hierarchy, generalised to arbitrary Categories, would
be general enough to describe all of the hierarchies in Steinhart's scheme. It should
also be noted that Dan Kurth's *Topos of Emergence* [14] describes a similar but
different scheme invoking Categorification in the Level Change. We look forward
to making connections with Dan's work. Unpublished work from two decades ago
by Mike Wright [15] also confirms the underlying Topos nature of the Levels of
Parker-Rhodes Hierarchy *via* his "triparitous" logic.

It is instructive to consider Hierarchies of Categorical Objects (perhaps Topoi)
with alternative Level Change Operators. We will describe in a future paper how
Parker-Rhodes Level Change Operator forces a nonBoolean Logic on the Subobjects
of the Hierarchy thus making the context of a Category of Topoi inevitable. This is
what led him to define his triparitous logic even though he apparently knew nothing
about Topoi.

We have generalised the Level Change Operator of the Combinatorial Hierarchy
in two parts. Firstly we looked at the generalisation of the Representative Automorphisms of a Level and found that this could be done with commutative diagrams
in an arbitrary Category. Then we looked at generalisation of the process of constructing a new Categorical Object for the next Level from these Representative

Automorphisms and concluded that, as yet at least, we only know how we might do this in Categories with all Finite Products. Clearly all Cartesian Closed Categories would then admit an FPR type Level Change Operator, and specifically one could be constructed in an arbitrary Elementary Topos (and hence an arbitrary Grothendieck Topos). This has implications for building connections between the Hierarchy and Quantum Mechanics, particularly *via* the Kochen-Specker Theorem.

This generalisation allows the full apparatus of Category Theory to be used on the Combinatorial Hierarchy. Ideas that have been discussed vaguely before — such as the idea that the Abelianisation operator (for instance from Clive and Arleta's NonCommutative Hierarchies to the classical one) should commute with the Level Change Operator — can now be made precise. There is a great deal more that could be said, but that is for a later paper.

Dedication

This work is dedicated to Pierre Noyes on the occasion of his 90th birthday.

Acknowledgements

I am indebted to John Amson, Pierre Noyes and to the late Clive Kilmister and Ted Bastin who all contributed so much to this work. Also to Basil Hiley, Arleta Ford, Lou Kauffman, and all the members of ANPA and TPRU with whom I have discussed this topic, and particularly for mathematical help and encouragement to Steve Vickers and Chris Isham and to Charlotte D'Sa for her careful proofreading of the final copy.

References

[1] A. F. Parker-Rhodes, J. C. Amson, 'Hierarchies of Descriptive Levels in Physical Theory', (original unpublished paper dated 1964)
Intern. J. of General Systems, Vol.27, Nos.1,2,3, 57–80 (1998)

[2] John Amson, Frederick Parker-Rhodes. 'Essentially Finite Chains', (original unpublished paper dated 1965)
International Journal of General Systems, Vol.27, No.1,2,3, 81–92 (1998)

[3] Ted Bastin, 'On the origins of the Scale Constants of Physics', Studia Philosophica Gandenzia, 4, Gent (1966)

[4] Ted Bastin, H. Pierre Noyes, John Amson, Clive W. Kilmister,
'On the Physical Interpretation and the Mathematical Structure of the Combinatorial Hierarchy', (PITCH)
Intern. J. of Theor. Phys., vol 18, No.7, pp.445–488 (1979)

[5] John Amson. 'Discrimination Systems and Projective Geometries',
in *Discrete and Combinatorial Physics*, Proceedings of ANPA9, 158–189 (1988)

[6] Clive Kilmister, 'Discrimination with Aspect',
in *Aspects I*, Proceedings of ANPA19 (1997)

[7] Arleta Ford, 'On the NonCommutative Combinatorial Hierarchy'
in *Aspects II*, Proceedings of ANPA20 (1998)

[8]   Clive W Kilmister, 'The Fine Structure Constant'
        in *Spin*, Proceedings of ANPA25 (2003)
[9]   Keith Bowden and Clive Kilmister, 'The Essential Frederick Construction',
        in *Conceptions*, Proceedings of ANPA27 (2005)
[10] Lou Kauffman, 'Concepts and Conceptions',
        in *Conceptions*, Proceedings of ANPA27 (2005)
[11] Clive Kilmister, 'New Foundations for the Combinatorial Hierarchy',
        in *Foundations*, Proceedings of ANPA28 (2006)
[12] Keith Bowden, 'The Internal Structure of the Level Change Operator',
        in *Foundations*, Proceedings of ANPA28 (2006)
[13] Eric Steinhart, *More Precisely: the Math you need to do Philosophy*,
        Broadview Press (2009)
[14] Dan Kurth, 'The Topos of Emergence',
        in *Boundaries*, Proceedings of ANPA24 (2002)
[15] Mike Wright, Private Communication (2013)

Brief Biography

My first "proper" job was designing and building technology for lunar sample analysis for NASA in the Department of Space Physics at the University of Sheffield in the Summer of 1972. Since then I have been in a wide variety of freelance, consulting and academic roles in Science, Engineering, Computing, Quantum Computation, Theoretical Physics, Music and Video (for Yamaha in Tokyo amongst others), Science Fiction and various genres of Journalism. I once interviewed Paul Williams, Timothy Leary's campaign manager in his stand against Ronald Reagan for the Governator's job, about their meeting with John and Yoko during the Montreal bed-in and his publication of the International Bill of Human Rights and subsequent meeting with Jimmy Carter. I plan to republish this remarkable story shortly. From 1979 to 1997 I was also Council Member and Secretary to the Science Fiction Foundation and from 1982 to 1992 I was simultaneously the British Agent for the Philip K Dick Society. From 1997 to the present day I have held a Research Fellowship in the Theoretical Physics Research Unit, Birkbeck College, University of London, working on de Broglie-Jessel Theory, amongst other things.

Quantum Cosmology and Special Mersenne Primes

Geoffrey F. Chew

Theoretical Physics Group, Physics Division
Lawrence Berkeley National Laboratory
Berkeley, California 94720, USA
E-mail: geoffreyfchew@gmail.com

Summary

We here strive to increase understanding of Milne quantum cosmology (MQU) by attention to the ordered set of 'special' Mersenne primes, 2, 3, 7, 127, $2^{127}-1$. Each special prime (after the first) is related to its predecessor by Mersenne's prescription. MQU is a T.O.E. founded on an 'extended-Lorentz' Lie-group symmetry that defines, for individual universe 'constituents', conserved momentum, angular momentum, 'spirality', electric charge and energy. Any particle, such as a photon, is a composite of several different MQU constituents.

MQU's 9-parameter symmetry group, with separate 'left' complex-number displacement and 'right' Lorentz subgroups, immediately invokes the primes 2 and 3. Because the particle-theoretic Standard Model (SM) flattens MQU's curved 3-space by contracting the Lorentz subgroup to the Euclidean group — suppressing redshift and replacing MQU symmetry by that of the gauge-augmented Poincaré group — MQU capabilities associating to the remaining special Mersenne primes are susceptible to 'SM obscuration'. We here speculate about capacity of two 'electromagnetic' special Mersenne primes, the pair 7 and 127, (that follow 2 and 3) to reduce SM arbitrariness while underpinning its perturbative recipe for comparison with data. The prime 7 may provide *raison d'être* for quarks as well as leptons and three generations of 'elementary fermions', while assigning 'Majorana' status to neutrinos.

MQU unification of electromagnetism with gravity allows the huge remaining member of Mersenne's special prime quintet to be called 'gravitational' — quantifying the meaning of 'inflation' by specifying MQU 'age at big bang'. Particle-energy scale then (in accord with SM) lies in the Gev neighborhood — below the Planck energy ($\sim 10^{-5}$ gm) by a factor of order 10^{19}. The huge special prime further underpins a 'macroscopic' spacetime scale —- the scale of life and consciousness — larger than particle scale by another factor of order 10^{19}.

Introduction

Prior to the age of 60, the author's theoretical-physics research recognized neither number theory nor Lie group theory. Analytic functions (of complex variables) were heavily invoked together with Dirac's Hilbert-space-based quantum theory as taught by my mentor, Enrico Fermi. In the course of shaping me as a physicist I was endowed by Fermi with a pragmatic, philosophy-scorning, view of quantum mechanics.

Fermi led me in my youth to join the 'shut up and calculate' ranks. But, during the final decades of the twentieth century my friend Pierre Noyes grabbed my attention by stressing physicist dependence, usually without appreciation, on mysterious universal dimensionless parameters. The values of these parameters seem essential to the high accuracy sometimes achieved by a Galilean enterprise founded on the inherently-approximate idea of 'reproducible measurement'.

Pierre introduced me to ANPA — the 'alternative natural-philosophy association' that had been inspired by the science-fiction-evocative notion that special Mersenne primes might underlie the success of physics. ANPA for a decade or so held meetings at Stanford and for three decades at Cambridge (England). So far as I was made aware, ANPA was gestated by the 5-member sequence of special Mersenne primes, 2, 3, 7, 127, $2^{127}-1$. Each sequence member after the first relates to its predecessor by the rule that defines a Mersenne prime : 2 raised to a prime number, minus 1.

In recent decades the author has pursued quantum cosmology where, around 1980, a foundational notion of 'inflation' had become appreciated, with no explanation of the huge dimensionless parameter quantifying that notion. 'Inflation' describes the ratio between Planck scale and that of particle physics. I have come to suspect the huge member of the foregoing 'Mersenne quintet' as specifying not only the 'extent' of inflation but the ratio between Planck's spacetime scale and the 'macro' scale of life and consciousness. (The scales of molecular, atomic and nuclear physics all locate between macro scale and particle scale.)

My candidate quantum cosmology, dubbed 'MQU' — shorthand for 'Milne quantum universe' — is based on Milne's classical cosmology, which in turn was based on the Lorentz group where the primes 2 and 3 are central.[1] Invoking the group U(1) to represent electric charge (following Kaluza-Klein, not SM), I have been led (by Gelfand-Naimark unitary Hilbert-space representations of the Lorentz group[2]) to postulate for the universe an 'extended Lorentz symmetry' associated to a 9-parameter Lie group[3].

This group's generators (the group algebra) represent below-listed conserved attributes of discrete universe 'constituents'. An individual MQU constituent the author has chosen to call '*quc*' (a pronounceable acronym).[3]

A single *quc* is never a 'particle' — being, *alone*, neither spatially localizable nor temporally stable even though characterizable as 'fermionic' and attributed by the symmetry-group's algebra with conserved discrete electric charge, energy and angular momentum as well as with conserved continuous momentum. Any 'objective reality' such as a 'particle', is a temporally-stabilizing *relationship* — a 'marriage' — between several collaborating *quc*s.

MQU 'relativity' means the universe remains unchanged when any symmetry-group element is applied to the *entire* fixed finite (constant) set of *quc*s. *Vanishing* are total-universe values of electric charge, energy, momentum, angular momentum and 'spirality'. We below, in addressing photons, leptons and quarks, superficially discuss the latter *quc* attribute, which allows any *quc* to be described as 'fermionic'.

Mach's principle is obeyed. *Negative-energy quc*s underpin the cosmological phenomenon called 'dark energy'. (4) The author believes they also underpin the 'hidden reality' that has required a probabilistic (Copenhagen') interpretation for any quantum theory which ignores negative energy. The S matrix, on which the author's Fermi-inspired career was based, is an example of such a theory.

ANPA attention concentrated on possible connection between the prime 127 and the fine-structure constant, whose mysteriously small value had become appreciated as essential to the accurate atomic spectroscopy (during the twentieth century's first quarter) that led to physical quantum theory. Why is the fine-structure constant 'small'?

In the present informal paper I shall refer to 127 as the 'large' member of the 'special' Mersenne-prime set. I share ANPA feelings about the large prime, although applying these feelings to quantum cosmology rather than to (less fundamental) particle physics.

I believe the large special Mersenne prime 127 to prescribe 'smallness' for the discrete electric charges carried by *quc*s (a smallness prerequisite both to atomic physics and to the Standard Model's perturbative rules for S-matrix computation). I believe that the special prime 7, the large special prime's predecessor in the Mersenne sequence, prescribes the number of different possibilities (all 'small') for the electric charge that, without ever changing as the universe evolves, an individual *quc* forever carries.

This paper will show how the simplest *quc* representation of 'elementary fermion' — by a 3-*quc* composite — falls into one of 3 'generations' that share a common set of spins, electric charges and baryon numbers while differing in the *distribution* of charges between the *quc* constituents of *different-generation* fermions. 'Elementary' 3-*quc* neutrinos appear to be of 'Majorana' (as opposed to 'Dirac') type.

We shall also here address quantum-cosmological 'inflation' — universe 'big-bang-beginning' at a spacetime scale hugely larger than that of Planck — a scale corresponding to particle energies in the Gev range. We associate inflation to the *huge* special Mersenne prime. Gravity-electromagnetism unification by an MQU Schrödinger equation is essential to our reasoning. The author supposes 'objective reality', such as elementary-particle masses, to have been created immediately after the big bang, within a multi-*quc* wave function, by Schrödinger-equation MQU dynamics. The *age* of the universe at big-bang, in Planck time units, provides MQU its meaning for 'inflation'.

The MQU wave function *started* at 'big bang', in absence of any objective reality such as particles but with an energy scale (in Gev neighborhood) set by huge-prime-specified starting-universe age and with a set of *quc*s each destined to 'live forever'. Despite the author's anticipation of *eventual* special-prime specification of the (forever-constant) total number of *quc*s, the value of this latter number currently remains unknown. (A sixth 'special prime' might be involved.)

Remark: The Dirac self-adjoint hamiltonian operator that evolves the MQU multi-*quc* wave function, although not yet shown to be 'total universe energy', nevertheless exhibits redshift (inverse proportionality to universe age) and comprises 3 components that may be called 'kinetic *quc* energy', 'electromagnetic *quc* potential energy' and 'gravitational *quc* potential energy'. Each *quc*'s potential energy is generated by the set of *all quc*s. There seems no room for further MQU-hamiltonian components. Atomic nuclei as well as 'elementary' bosons and leptons we presume generated, as stable *quc* composites, by gravity and electromagnetism working, in tandem, with kinetic energy.

Preliminaries

Any *quc* is distinguished from all others by its set of 3 integer labels — a 2-valued (± 1) spirality integer (manifesting Lorentz-group *double* covering by the MQU symmetry group), a 7-valued ($0, \pm 1, \pm 2, \pm 3$) electric-charge integer, and a ($2\mathsf{M}_{max} + 1$)-valued energy integer that takes the values ($0, \pm 1, \pm 2, \ldots, \mathsf{M}_{max}$). The value of M_{max}, although we suppose specified by special Mersenne primes, is not yet known. The total number of *quc*s comprising MQU is then $14(2\mathsf{M}_{max} + 1)$.

Conserved spirality — a *quc* attribute without physics precedent although confusable with certain physics usages of the term, 'chirality' — is completely specified by its 2-valued integer whereas electric charge and energy each have units. The electric charge unit we suppose proportional (following ANPA thinking) to the inverse square root of the large special Mersenne prime 127. The coefficient we believe to involve the irrational number π as well as the prime integers 2 and 3. (Hilbert-space unitary representation of the Lorentz group involves π.)

The MQU energy unit, with $\hbar = c = 1$, is $(2\tau)^{-1}$, where τ is Milne-universe age — an invariant under the 9-parameter symmetry group. Redshift *automatically* accompanies this energy unit. The factor 2 in the energy unit associates to Lorentz-group *double* covering by the MQU symmetry group.[3]

Milne's universe occupies the interior of a forward lightcone, with the forwardly-arrowed invariant age of any spacetime location its Minkowski distance from the lightcone vertex. The importance (and novelty) of Milne meaning for 'age' is difficult to over-emphasize. General relativity lacks any such feature.

The Newtonian meaning of a single time — applicable to 'everything anywhere within the universe' is sustained by Milne age — a symmetry-group invariant. Designating *big-bang* age by the symbol τ_o, the scale ratio we call 'inflation' (leading from Planck scale to particle scale) is $\tau_0/G^{1/2}$, where G is the gravitational constant.

The hamiltonian (a self-adjoint Dirac operator acting on a multi-*quc* Hilbert space) that determines universe evolution *after* τ_0, depends on age and on all five special Mersenne primes. The electromagnetic *quc* potential energy, for example, is inversely proportional to the large special Mersenne prime.

The corresponding gravitational potential energy is inversely proportional to the square of universe age and thereby at big bang to τ_0^{-2}. With $\hbar = c = 1$ the gravitational constant has the dimensions of time squared. We associate τ_0^2/G to the huge special Mersenne prime ($\sim 10^{38}$).

Elementary Leptons and Quarks as 3-Quc Composites

A 3-*quc* SM elementary-fermion (a lepton or a quark) comprises a 2-*quc* spirality-zero, electrically-neutral but energy-endowed 'core' plus a single 'valence' *quc* that carries the fermion's charge and spirality plus a portion of its energy. Stability of the composite is provided by a balancing of hamiltonian positive kinetic energy against negative (attractive') gravitational potential energy in the sense familiar from atomic physics. (The gravitational attraction is here between the 2-*quc* core and the valence *quc*.)

The elementary fermion's 2-*quc* core differs from the 2-*quc* photon in its zero spirality. The photon has spirality ± 2. But in both cases a positively-charged *quc* is paired with a partner whose charge is of equal magnitude while of opposite sign; fermion-core stabilization and photon stabilization plausibly both depend on electromagnetic attraction between a pair of oppositely-charged *quc*s.

The remarkable uniqueness and stability of the (valence-less) photon and its QED couplings to charged leptons and quarks we suppose result from a photon being a spirality ± 2 quantum superposition of ± 1, ± 2, ± 3 charged-*quc* pairs. Charged-lepton valence *quc*s would have charge integers ± 3 while quark valence-*quc*s would have charge integers ± 1, ± 2. Neutrino valence *quc*s have zero charge integer while sharing the ± 1 spirality of all 'elementary fermions'.

The latter have *in common* a set of 3 *different* possible zero-spirality zero-charge cores. Electromagnetic interaction between 'valence *quc*' and 'dipole core' — a 'perturbation' of the gravitational attraction between core and valence *quc* — might account for mass differences between different fermion 'generations'.

Although the author may live insufficiently long to know the result, the foregoing speculations are subject to verification by fermion-mass calculations that use the MQU Hamiltonian and Schrödinger equation. One guess is that *lower* fermion masses accompany larger absolute values for core-*quc* electric charges. Another is that Majorana's meaning for 'neutrino' will prove closer to that of **MQU** than Dirac's meaning.

Electromagnetism-Gravity Unification and Huge-Prime Inflation Specification

The MQU Hamiltonian's potential energy comprises electromagnetic and gravitational components proportional, respectively, to *quc* electric charge and to *quc* energy. Adhering to SM Lagrangian practice, the author has denoted the dimensionless coefficient of the MQU electromagnetic potential-energy component by the symbol g. For the gravitational component the corresponding dimensionless coefficient is unambiguously $G^{1/2}/\tau$, where G is the gravitational constant and τ the universe age. (SM ignores gravity.)

Because g^2 we have above associated to the reciprocal of the large special Mersenne prime, a desire to unify electromagnetism and gravity has led the author to equate G/τ_0^2 with the reciprocal of the huge Mersenne prime. The big-bang energy scale $\hbar/\tau_0$ then locates in the (Gev) neighborhood (consistently with the set of SM arbitrary energy parameters that have been chosen to match observed particle properties such as lepton and quark masses).

It remains to be seen whether dynamical calculations with the MQU hamiltonian of early-universe evolution (immediately following big bang) will explain the observed 'elementary-particle' masses. We nevertheless feel encouraged by capacity of the huge special Mersenne prime to explain the overall scale of 'particle physics'.

Note: A tentative MQU Hilbert-space ('superselection') constraint on subspaces with numbers of *quc*s smaller than the total number (such as the 3-*quc* subspace considered above) distinguishes electric charge from energy by forbidding superposition of wave functions with *differing* (total) electric charge.

The Macroscopic Spacetime Scale of Life and Consciousness

Particle physics, without appreciation thereof, depends in its practice on a huge ratio between particle spacetime scale (above related to Planck's time scale $G^{1/2}$ by the square root of the huge special Mersenne prime) and 'laboratory' spacetime scale. Although recognition of the particle to lab ratio appears in S matrix definition *via* an energy often denoted by the symbol ϵ, the S matrix is defined by a limit in which ϵ approaches zero. (Interpretation of all particle-physics measurements employs the S matrix.) The energy ϵ is 'tiny' because of its inverse proportionality to the 'size' of laboratory apparatus that is 'huge' on particle scale.

The meaning of 'laboratory' intertwines with that of 'life and consciousness'. We here represent the scale of such notions by the single adjective 'macroscopic'. Note that, although macro scale is far above particle scale — by a factor $\sim 10^{19}$ — it is by a comparable factor below the 'Hubble scale' set by present-universe age.

The author has felt frustrated by inability, so far, to uncover a plausible connection between macro scale and the huge special (gravitational) Mersenne prime, despite striking similarity between Hubble-macro, macro-particle and particle-Planck spacetime-scale ratios. Employment of the term 'life' in place of the term 'macro' associates the macro-scale mystery to that of life. The molecular-biology-achieved understanding of 'life' indicates the latter's possibility only within a very-special cosmologically-narrow interval of kinetic-energy density.

Conclusion

A survey has here been given of quantum-cosmological issues to which special Mersenne primes may prove relevant. Arbitrariness in Standard-Model particle-theoretic elementary-fermion representation promises to be reduced by future Schrödinger-equation calculations of a type familiar in atomic physics.

Acknowledgment

A lengthy helpful discussion with Herb Doughty preceded
this paper's preparation.

References

[1] E. A. Milne, *Relativity, Gravitation and World Structure,* Clarendon Press, Oxford, (1935)
[2] M. Naimark, *Linear Representations of the Lorentz Group,* MacMillan, New York, (1964)
[3] G. F. Chew, `arXiv 1107.0492` (2011) and `arXiv 1508.4366` (2013); 'Schrödinger Equation for Milne Quantum Universe', to be published, (2013)
[4] G. F. Chew, `arXiv 1107.0492` (2011)

BiEntropy – the Measurement and Algebras of Order and Disorder in Finite Binary Strings

Grenville J. Croll

Woolpit, Bury St Edmunds, Suffolk, UK
grenvillecroll@gmail.com

We introduce the BiEntropy algorithm which measures the order and disorder within a finite binary string of arbitrary length. The BiEntropy algorithm is a weighted average of the Shannon Entropies of the string and all but one of the binary derivatives of the string. The BiEntropy algorithm has been successfully tested empirically in a number of diverse domains. We summarise the history behind and the development of the algorithm and provide some simple algebras to demonstrate its future use in bit-string physics.

1. Introduction

Noyes [1997] is a leading contributor to an emerging body of theory which considers the universe to be a continuing sequence of operations on finite binary strings. The essence of the theory is a progression of Binary Vector Spaces (Levels) more commonly known or referred to as the Combinatorial Hierarchy composed of bit-strings of length 2, 4, 16 and 256. The number of members within each Level is as follows:

$$2^2 - 1 = 3$$
$$2^3 - 1 = 7$$
$$2^7 - 1 = 127$$
$$2^{127} - 1 \cong 10^{38}$$

Each member of each Level can be represented by a binary string. The sum of the sizes of the first three Levels, $1+3+7+127 = 137$, corresponds closely to the inverse of the fine structure constant. The size of the last Level corresponds closely to the order of magnitude of the inverse gravitational coupling constant. The Combinatorial Hierarchy terminates naturally at this last Level. There appear to be a number of correspondences between the Combinatorial Hierarchy and physical phenomena. In studying the physical relevance of a universe represented by finite binary strings, consideration could reasonably be given to some measure of the order and disorder of the bits comprising each binary string within each of the Levels. The purpose of this paper therefore is to briefly summarise the background, development and testing of the BiEntropy algorithm and demonstrate some simple algebraic properties of BiEntropy which may be of use in the physical sciences.

2. Historical Background

There is a long history to the development of tests, measures and algorithms to detect order, disorder, randomness or entropy within strings of various types. Some of the earliest work was that of [Shannon, 1948] who introduced the concept of Binary Entropy, which is based upon the log to the base 2 of the proportion of 1's observed within a binary string. Other very early tests of disorder observe simple statistical features such as sums, averages and runs characterised by the work of [Marsaglia, 1968] amongst many others. More recently, [Pincus, 1991, 1996, 1997] introduced the first measure of serial irregularity which was expressed through his Approximate Entropy (ApEn) algorithm. This algorithm has subsequently been tested, modified, adapted and extended by others [Rukhin, 2000a, b] [Richman and Moorman, 2000] [Chen, Wu and Yang, 2009]. The ApEn algorithm has been successfully applied in some important medical [Pincus and Viscarello, 1992] and other domains.

The other main approach to the determination of disorder in binary strings is the measurement of the length of algorithms used to generate such strings [Kolmogorov, 1965][Chaitin, 1966]. The BiEntropy algorithm is based upon the use of the binary derivatives of a string. The properties of binary derivatives have been studied by [Nathanson, 1971] and [Goka, 1970]. Binary derivatives have also been used in randomness tests [McNair, 1989] and in cryptographic applications [Carroll, 1989 and 1998] and attacks [Bruwer, 1995]. The use of Binary Derivatives in an algorithm for measuring order and disorder is thought to be unique.

3. Tacit Understanding of Binary Order and Disorder

Table 1 suggests our tacit understanding of the order and disorder of some example 8 bit binary strings.

Table 1 : Tacit understanding of binary order and disorder

Binary String	Description	Reason
....11111111....	Perfectly ordered	All 1's
....00000000....	Perfectly ordered	All 0's
....01010110...	Mostly ordered	Mostly 01's
....01010101...	Regular, not disordered	Repeating 01's
....11001100....	Regular, not disordered	Repeating 1100's
....01011010...	Mostly ordered	0101 then 1010
....01101011....	Somewhat disordered	No Apparent Pattern
....10110101...	Somewhat disordered	No Apparent Pattern

There do not appear to have been any previous attempts within the literature to characterise the relative order and disorder of the 256 possible 8-bit binary strings. There is a requirement for an algorithm which would rank binary strings of arbitrary length in terms of their relative order and disorder. We need to understand the order and disorder of a binary string in the same way as we need to understand its ordinality and whether it is odd, even, prime, semi-prime or composite *etc.*. Note that binary there are $256! \approx 8.578 \ldots \times 10^{506}$ different ways of ordering all 256 of the 8-bit binary strings.

4. Shannon Entropy and Binary Derivatives

4.1. *Shannon Entropy*

Shannon's Entropy of a binary string $s = s_1, \ldots, s_n$ where $P(s_i{=}1) = p$ (and $0 \log_2 0$ is defined to be 0) is:

$$H(p) \;=\; -p \log_2 p - (1{-}p) \log_2 (1{-}p)$$

For perfectly ordered strings which are all 1's or all 0's *i.e.* $p = 0$ or $p = 1$, $H(p)$ returns 0. Where $p = 0.5$, $H(p)$ returns 1, reflecting maximum variety. However, for a string such as 01010101, where $p = 0.5$, $H(p)$ also returns 1, ignoring completely the repetitive nature of the sequence.

4.2. *Binary Derivatives and Periodicity*

The first binary derivative of s, $d_1(s)$, is the binary string of length $n{-}1$ formed by XORing adjacent pairs of digits. We refer to the kth derivative of s, $d_k(s)$, as the binary derivative of $d_{k-1}(s)$. There are $n{-}1$ binary derivatives of s.

Some years ago [Nathanson, 1971], following the work of [Goka, 1970] defined the notions of period and eventual period within a binary string and outlined the related properties of the derivatives both individually and collectively. Amongst a number of useful results we find that (a) if a derivative of a binary string is eventually periodic with a period P then the binary sequence is also eventually periodic with a period P or $2P$, and (b) if a derivative is all zero's then the string has a period $2m$ for some m, $0 \leq m \leq n$.

For example, the first binary derivative of 01010101 (with period, $P = 2$) is 1111111 ($P = 1$), following which all the higher derivatives are all 0's. The third derivative of 00010001 ($P = 4$) is 11111, following which again all the higher derivatives are 0. The sixth derivative of 00011111 (with eventual period $P = 1$ from the fourth digit) is 10.

By calculating all the binary derivatives of s we can discover the existence of repetitive patterns in binary sequences of arbitrary length. If a binary sequence is periodic, or eventually periodic, all or some of the bits of a higher derivative eventually fall to zero. We assume that the derivatives of binary strings that exhibit more obvious periodicity fall to zero more rapidly than those of more disordered strings.

Although Nathanson's definitions of periodicity and eventual periodicity are useful, they are defined only for infinite right reading strings. We deal with finite strings, which can be both right reading and left reading. We rely solely upon the binary derivatives of a finite string to resolve the issue of the periodicity within the string.

5. BiEntropy

BiEntropy, or BiEn for short, is a weighted average of the Shannon Binary Entropies of the string and the first $n-2$ binary derivatives of the string using a simple power law. This polynomial version of BiEntropy is suitable for shorter binary strings where $n \leq 32$ approximately.

$$\text{BiEn}(s) = \left(1/(2^{n-1} - 1)\right) \times$$
$$\left(\sum_{k=0}^{n-2} \left(-p(k) \log_2 p(k) - (1-p(k)) \log_2 (1-p(k))\right) 2^k\right)$$

The final derivative d_{n-1} is not used as there is no variation in the contribution to the total entropy in either of its two binary states. The highest weight is assigned to the highest derivative d_{n-2}. If the higher derivatives of an arbitrarily long binary string are periodic, then the whole sequence exhibits periodicity. For strings where the latter derivatives are not periodic, or for all strings in any case, we can use a second version of BiEntropy, which uses a Logarithmic weighting, to evaluate the complete set of a long series of binary derivatives.

$$\text{Tres BiEn}(s) = \left(1/\sum_{k=0}^{n-2} \log_2 (k+2)\right) \times$$
$$\left(\sum_{k=0}^{n-2} \left(-p(k) \log_2 p(k) - (1-p(k)) \log_2 (1-p(k))\right) \log_2 (k+2)\right)$$

The logarithmic weighting or (TBiEn for short) gives greater weight to the higher derivatives. Depending upon the application, other weightings could be used. See [Croll, 2013] for the motivations behind the development of the BiEntropy algorithm and its variations.

6. BiEntropy of the 2, 4, and 8-bit Strings

6.1. *The 2-Bit Strings*

The BiEntropy of a 2-bit string is given in Table 2.

Table 2 : The BiEntropy of a 2-bit string

	0	**1**
0	0	1
1	1	0

The above table depicts the XOR operation and the computation of the binary derivatives of the 2-bit strings. There are two BiEntropy states of a 2-bit string. The two states are perfectly ordered (0) or perfectly disordered (1).

6.2. *The 4-Bit Strings*

We show in Table 3 the sequence of calculations required to compute the BiEn of a 4-bit string. The calculation shows at the left hand side the extraction of the binary derivatives from the string s = 1011. We tabulate the BiEntropy of all the 4-bit strings in Table 4.

Table 3 : Calculating the BiEn of a 4-Bit string

s	1's	n	p	(1-p)	-p.log(p)	- (1-p).log(1-p)	H(p)	k	2^k	H(p).2^k
1011	3	4	0.75	0.25	0.31	0.50	0.81	0	1	0.81
110	2	3	0.67	0.33	0.39	0.53	0.92	1	2	1.84
01	1	2	0.50	0.50	0.50	0.50	1.00	2	4	4.00
									7.00	6.65

BiEn(s) 0.95

Table 4 : The BiEntropy of the 4-bit strings

	0 0	1 1	0 1	1 0
00	0.00	0.41	0.95	0.95
11	0.41	0.00	0.95	0.95
01	0.95	0.95	0.14	0.41
10	0.95	0.95	0.41	0.14

Thus there are two perfectly ordered strings 0000 and 1111, two nearly ordered, periodic strings 0101 and 1010, four intermediately disordered strings where the left two bits are the 1's complement of the right two bits and eight disordered (aperiodic) strings where either a single 1 or a single 0 transits a four bit field. Note that these aperiodic strings would be classified under a generalisation of Nathanson's schema as either eventually periodic right reading or eventually periodic left reading, but not both. We have shaded the periodic and aperiodic strings for clarity. Strings that are unshaded are neither periodic nor aperiodic. Note the general XOR structure of the table. The BiEntropy algorithm classifies the strings so that those strings that have a non zero last derivative are accorded the highest weight. The logarithmic version of BiEntropy, TBien, is calculated in a similar way to BiEn but using logarithmic weights (with similar results) which we omit to avoid duplication. Mean BiEn for the 4-Bit strings is 0.594, standard deviation 0.3894. Mean TBiEn is 0.644, standard deviation 0.3554.

6.3. *The 8-Bit Strings*

Table 5 : The BiEn of the 8-bit strings

	0000	1111	0101	1010	0011	0110	1001	1100	0001	0010	0100	0111	1000	1011	1101	1110
0000	0.00	0.11	0.23	0.23	0.45	0.45	0.47	0.45	0.92	0.94	0.95	0.94	0.93	0.95	0.95	0.95
1111	0.11	0.00	0.23	0.23	0.45	0.47	0.45	0.45	0.95	0.95	0.95	0.93	0.94	0.95	0.94	0.92
0101	0.23	0.23	0.01	0.11	0.46	0.45	0.45	0.47	0.95	0.94	0.92	0.94	0.95	0.95	0.93	0.95
1010	0.23	0.23	0.11	0.01	0.47	0.45	0.45	0.46	0.95	0.93	0.95	0.95	0.94	0.92	0.94	0.95
0011	0.45	0.45	0.46	0.47	0.02	0.23	0.23	0.11	0.94	0.93	0.95	0.95	0.95	0.93	0.95	0.95
0110	0.45	0.47	0.45	0.45	0.23	0.02	0.11	0.23	0.95	0.95	0.94	0.93	0.95	0.95	0.95	0.93
1001	0.47	0.45	0.45	0.45	0.23	0.11	0.02	0.23	0.93	0.95	0.95	0.95	0.93	0.94	0.95	0.95
1100	0.45	0.45	0.47	0.46	0.11	0.23	0.23	0.02	0.95	0.95	0.93	0.95	0.95	0.95	0.93	0.94
0001	0.93	0.94	0.94	0.95	0.95	0.95	0.93	0.95	0.05	0.46	0.24	0.46	0.47	0.24	0.46	0.11
0010	0.95	0.95	0.95	0.92	0.93	0.94	0.95	0.95	0.46	0.05	0.45	0.24	0.24	0.47	0.11	0.46
0100	0.94	0.95	0.93	0.94	0.95	0.95	0.95	0.93	0.24	0.45	0.05	0.46	0.46	0.11	0.47	0.24
0111	0.95	0.92	0.95	0.95	0.94	0.93	0.95	0.95	0.46	0.24	0.46	0.05	0.11	0.46	0.24	0.47
1000	0.92	0.95	0.95	0.95	0.95	0.95	0.93	0.94	0.47	0.24	0.46	0.11	0.05	0.46	0.24	0.46
1011	0.95	0.94	0.94	0.93	0.93	0.95	0.95	0.95	0.24	0.47	0.11	0.46	0.46	0.05	0.45	0.24
1101	0.95	0.95	0.92	0.95	0.95	0.95	0.94	0.93	0.46	0.11	0.47	0.24	0.24	0.45	0.05	0.46
1110	0.94	0.93	0.95	0.94	0.95	0.93	0.95	0.95	0.11	0.46	0.24	0.47	0.46	0.24	0.46	0.05

8421

Note that Table 5 also has the general configuration of the XOR function. The equivalent 8 bit TBiEn table is similar in XOR structure, though the values of TBiEn differ and the sort order of 8 bit strings which are neither periodic nor aperiodic varies slightly. Mean BiEn for the 8-Bit strings is 0.625, standard deviation 0.340. Mean TBiEn is 0.747, standard deviation 0.209. The BiEn and TBiEn of the 8 bit strings are strongly correlated (Adjusted $R^2 = 0.85$). BiEntropy is fractal from the self-similarity exhibited in Tables 2, 4 and 5.

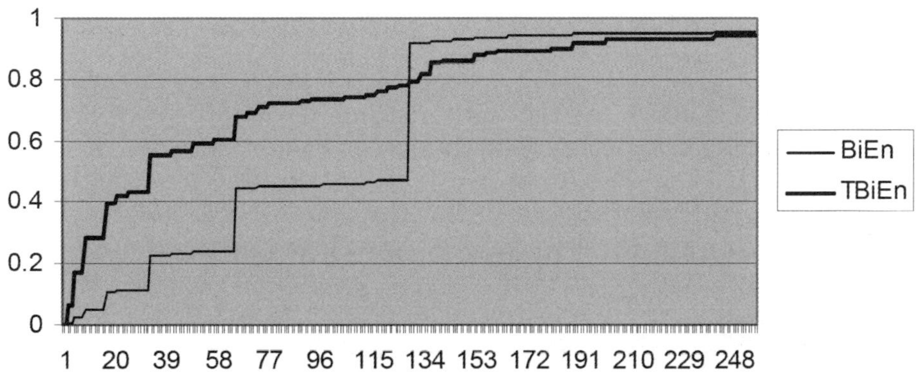

Figure 1 : 8 bit binary sequences in ascending BiEntropy order
Values of BiEn and TBiEn --- sorted

We show in Figure 1 above the sorted values of BiEn and TBien for the 8-bit strings. BiEn and TBiEn do not reach 1.0. Were they to do so, the $p(k)$ for all $k \leq n-2$ would have to be exactly 0.5, which is impossible as k is odd at least once for all $n \geq 3$. Note that in the absence of a closed form solution for determining the BiEntropy of finite binary string, BiEntropy has to be determined empirically.

We show in Table 5 the BiEn of all 256 8-Bit strings. They are again shaded such that the periodic strings (on the diagonal) and the aperiodic strings are shaded grey. Unshaded entries are neither periodic nor aperiodic. The diagram is structured such that the X and Y axes show the 4 bit strings of which each 8 bit string is comprised. The X and Y axes are sorted so that low BiEn or more ordered 4 bit strings appear towards the top and left of the table and high BiEn or more disordered 4-bit strings appear to the bottom and right. The Y-Axis corresponds to the first four bits of the string.

7. Empirical Testing of BiEntropy

Although BiEntropy was designed to answer a long standing question in the information sciences regarding the relative order and disorder of binary strings, the intention was and always has been to provide a function that has some practical utility. We have therefore tested the polynomial (BiEn) and logarithmic (TBiEn) versions of BiEntropy in a number of extremely diverse application areas.

7.1. *The Prime Numbers*

Consider the q natural numbers starting from 2 :

$$2, 3, 4, 5, 6, 7, 8, 9, 10, 11, 12, 13, 14, 15, 16, 17, 18, 19, \ldots$$

We can encode them in a binary string B of length n $(1 \leq n \leq q-1)$ such that the primes are encoded as 1 and the composites as 0. B_i is the i-th digit of B $(i \geq 1)$:

$$1, 1, 0, 1, 0, 1, 0, 0, 0, 1, 0, 1, 0, 0, 0, 1, 0, 1, \ldots$$

Thus B_1, corresponding to the natural number 2 is 1 and B_4, corresponding to the natural number 5 is also 1. We can easily compute the logarithmic BiEntropy of strings B for all n $(2 \leq n \leq q-1)$ which we show in Figure 2 for $q = 512$.

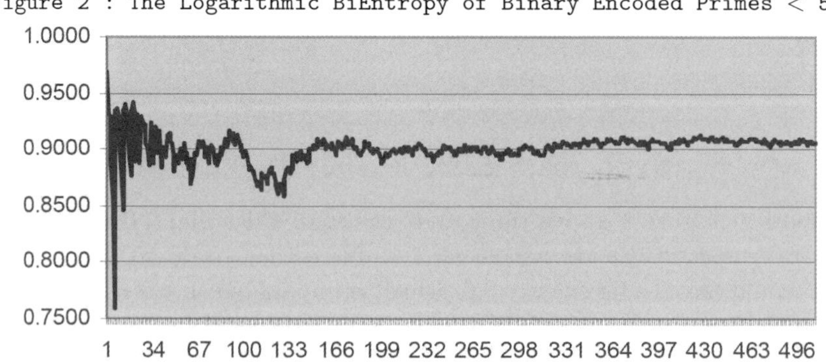

We can see that the logarithmic BiEntropy of the binary sequences corresponding to the primes < 512 are disordered, *i.e.* close to 1.0. They all sit within the aperiodic XOR partitions of the logarithmic equivalent of Table 5. The dip around 114–136 corresponds to a long sequence of composites broken by only two primes. This aperiodicity is implicitly due to the Prime Number Theorem. There is however a simple and explicit corollary that the primes are aperiodic, or more exactly, not periodic, which follows directly from Nathanson's definition of periodicity and the definition of B.

COROLLARY ONE — THE SEQUENCE OF PRIME NUMBERS IS NOT PERIODIC.

Consider a binary string B of even length n ($n \geq 4$) containing the binary encoding of the primes as above starting from 2. B_i is the i-th digit of B ($i > 0$). The binary sequence B is periodic if, for some positive integer p, $B_{i+p} = B_i$ for all $1 \leq i \leq n$.

This is impossible for $p = 1$ because the even numbers are composite. This is also impossible for $p \geq 2$ because:

(a) both $B_1 = 1$ and $B_2 = 1$ (because both 2 and 3 are prime) and

(b) no further pairs of natural numbers which are adjacent primes can occur because the even numbers are composite.

Hence the binary sequence B corresponding to the primes is not periodic for all n ($n \geq 4$) for all $p \geq 1$. The binary sequence B of length $n = 2$ corresponding to the first two primes 2 and 3 is periodic with $p = 1$. We choose not to compare sequences of unequal length.

7.2. *Human Vision*

We obtained the 7×5 dot matrix patterns of an ISO character set [Mitchell, 2008]. We converted each character into two 35 bit binary sequences using both horizontal and vertical raster scans and computed the mean (horizontal and vertical raster scan) polynomial BiEntropy for each character. We also did this for a Braille [Jiménez *et al.*, 2009], character set (BRAILLE) by superimposing the 3×2 bits of the Braille character set upon the central 5×3 bits of a 7×5 array. Using the insight gained from these exercises, we computed mean BiEntropy for a randomly generated 7×5 character set (RAND) and two further character sets designed specifically to be high BiEntropy (HECS) and low BiEntropy (LECS). We show 12 members of each character set and their corresponding mean BiEntropy etc. in Figure 3 below.

The mean polynomial BiEntropies of the five character sets are distinguished from each other at the $p < 0.01$ significance level. Note that any binary sequence with a length of 35 bits or more may have a member of these character sets as its 35-th last derivative. That the Braille character set is a low entropy character set is previously unknown.

Figure 3 :The BiEntropy of Some 7×5 Dot Matrix Character Sets

HECS   RAND   ISO   BRAILLE   LECS

Mean BiEn	0.945	0.634	0.494	0.068	0.014
StDev	0.006	0.240	0.291	0.053	0.012
N	12	96	96	64	12

7.3. *Random Number Generation*

We obtained one million decimal digits of each of *pi* [Andersson, 2013], [Champernowne's, 1933] number, *e*, $\sqrt{2}$ and $\sqrt{3}$ [Nemiroff and Bonnell, 1994]]. We used the random number function in Excel 2003 and Excel 2010 spreadsheets [McCullough, 1998] to generate two further sets of one million pseudo random digits. Finally, we obtained one million random decimal digits from a document produced by the RAND corporation [Rand, 2001] over 60 years ago using electromechanical machinery.

Table 6 : The BiEntropy of Some Irrational, Random and Normal Numbers

Sample	Mean	Stdev	Mean	Stdev	Mean	Stdev
	N=1,000		N=10,000		N=30,000	
PI	0.6190	0.3367	0.6302	0.3346	0.6283	0.3361
EXCEL10	0.6260	0.3422	0.6286	0.3381	0.6281	0.3374
CHAMP	0.6707	0.3763	0.6180	0.3443	0.6263	0.3353
SQRT(2)	0.6171	0.3400	0.6267	0.3384	0.6254	0.3384
EXCEL03	0.6271	0.3310	0.6237	0.3364	0.6250	0.3370
SQRT(3)	0.6408	0.3321	0.6307	0.3356	0.6247	0.3375
E	0.6425	0.3341	0.6243	0.3388	0.6235	0.3376
RAND	0.6288	0.3382	0.6179	0.3385	0.6217	0.3374

We converted these decimal series to binary by encoding $0-4$ as 0 and $5-9$ as 1. We then computed the mean polynomial BiEntropy for consecutive 32 bit sections of each series. Firstly 1,000 32 bit sections at a time, then 10,000 32-bit sections and finally 30,000 32-bit sections. We computed the mean and standard deviation of polynomial BiEntropy and tabulated the results in Table 6 above.

7.4. *Cryptography*

We have studied the use and abuse of spreadsheets in business for many years [Croll, 2005, 2009]. One of the risks of spreadsheet use is the general failure to protect spreadsheets using the cryptographically secure "Password to Open" menu in Microsoft Excel [Grossman, 2008].

We obtained or created four different spreadsheets. The most complex was a financial model previously used to support a large construction project. The file size of this model was 175 Kilobytes. We modified an address database held in Excel (by removing address rows) so that its file size was reduced such that it matched the size of the financial model. We then created two control spreadsheets. One contained the number 123 in every cell. The second contained a random number in each cell, created using the =RAND() function, which was then cut and pasted so that only the decimal values remained.

We encrypted each of the four files using three methods of encryption — no encryption, plain vanilla encryption (believed to be only XOR based) available in Excel 97 and finally strong 128 bit AES encryption.

We extracted 1,000 consecutive sequences of 128 bytes (1024 bits) from each of the 12 files and computed the logarithmic BiEntropy on each 1024 bit sequence. We show in Table 7 the results of this exercise.

Table 7 : The Logarithmic BiEntropy of Some Plain and Encrypted Spreadsheets

Spreadsheet	Encryption	TBiEn (1024 bit) N=1,000	
		Mean	Stdev
Numbers (all cells = 123)	None	0.8980	0.0732
Numbers (all cells = 123)	Office 97	0.9913	0.0545
Numbers (all cells = 123)	AES	0.9913	0.0545
Random Numbers	None	0.9857	0.0559
Random Numbers	Office 97	0.9913	0.0545
Random Numbers	AES	0.9914	0.0545
Address Database	None	0.9428	0.1770
Address Database	Office 97	0.9913	0.0545
Address Database	AES	0.9913	0.0545
Financial Model	None	0.9450	0.1736
Financial Model	Office 97	0.9912	0.0545
Financial Model	AES	0.9911	0.0545

BiEntropy simply and easily revealed significant differences ($p < 0.01$) between the encrypted and unencrypted binary files of each spreadsheet. Encryption clearly contributed to the disorder in the binary digits of each file. BiEntropy was able to distinguish between the more and less complex types of file in their unencrypted forms, but BiEntropy was unable to distinguish between the two methods of encryption, despite one of them being spectacularly insecure. BiEntropy has, of course, no knowledge of the Excel file structure. Table 5 implies that only half of all binary strings are aperiodic, which may have implications in the cryptographic security of key bit strings.

8. Algebras of BiEntropy

8.1. *Bit-String Algebra*

The construction of a bit string physics may require the performance of simple string and algebraic operations upon bit strings. These operations will naturally alter the order and disorder of the resultant strings in diverse ways, which may be reflected in physically relevant or even observable ways. We can easily investigate the algebras of simple operations on short strings empirically. In order to make some small progress in this direction, we have decided to classify the 2, 4 and 8 bit strings under the polynomial BiEntropy measure into three states: $-1 = $ periodic; $+1 = $ aperiodic; $0 = $ neither, following Table 5. We illustrate these states for the 2, 4 and 8 bit strings in Tables 8, 9 and 10 below.

Table 8 : The BiEntropy States of the 2-bit strings

	0	1
0	−1	+1
1	+1	−1

Table 9 : The BiEntropy States of the 4-bit strings

		−1 0 0	−1 1 1	+1 0 1	+1 1 0
−1	00	−1	0	+1	+1
−1	11	0	−1	+1	+1
+1	01	+1	+1	−1	0
+1	10	+1	+1	0	−1

Table 10 : The BiEntropy States of the 8-bit strings

					-1	-1	-1	-1	0	0	0	0	+1	+1	+1	+1	+1	+1	+1	+1
					0	1	0	1	0	0	1	1	0	0	0	0	1	1	1	1
					0	1	1	0	0	1	0	1	0	0	1	1	0	0	1	1
					0	1	0	1	1	1	0	0	0	1	0	1	0	1	0	1
					0	1	1	0	1	0	1	0	1	0	0	1	0	1	1	0
-1	0	0	0	0	-1	0	0	0	0	0	0	0	+1	+1	+1	+1	+1	+1	+1	+1
-1	1	1	1	1	0	-1	0	0	0	0	0	0	+1	+1	+1	+1	+1	+1	+1	+1
-1	0	1	0	1	0	0	-1	0	0	0	0	0	+1	+1	+1	+1	+1	+1	+1	+1
-1	1	0	1	0	0	0	0	-1	0	0	0	0	+1	+1	+1	+1	+1	+1	+1	+1
0	0	0	1	1	0	0	0	0	-1	0	0	0	+1	+1	+1	+1	+1	+1	+1	+1
0	0	1	1	0	0	0	0	0	0	-1	0	0	+1	+1	+1	+1	+1	+1	+1	+1
0	1	0	0	1	0	0	0	0	0	0	-1	0	+1	+1	+1	+1	+1	+1	+1	+1
0	1	1	0	0	0	0	0	0	0	0	0	-1	+1	+1	+1	+1	+1	+1	+1	+1
+1	0	0	0	1	+1	+1	+1	+1	+1	+1	+1	+1	-1	0	0	0	0	0	0	0
+1	0	0	1	0	+1	+1	+1	+1	+1	+1	+1	+1	0	-1	0	0	0	0	0	0
+1	0	1	0	0	+1	+1	+1	+1	+1	+1	+1	+1	0	0	-1	0	0	0	0	0
+1	0	1	1	1	+1	+1	+1	+1	+1	+1	+1	+1	0	0	0	-1	0	0	0	0
+1	1	0	0	0	+1	+1	+1	+1	+1	+1	+1	+1	0	0	0	0	-1	0	0	0
+1	1	0	1	1	+1	+1	+1	+1	+1	+1	+1	+1	0	0	0	0	0	-1	0	0
+1	1	1	0	1	+1	+1	+1	+1	+1	+1	+1	+1	0	0	0	0	0	0	-1	0
+1	1	1	1	0	+1	+1	+1	+1	+1	+1	+1	+1	0	0	0	0	0	0	0	-1

The arithmetic sum of the BiEntropy states of the 2, 4 and 8 bit strings is 0, 4 and 112 (128 aperiodic states (+1), 16 periodic states (-1), and 112 neither (0), respectively.

8.2. *Concatenation*

By reference to Tables 9 and 10 above, concatenation (&) of two 2-bit strings A and B to make a 4-bit string or two 4-bit strings A and B to make an 8-bit string follows these rules:

$$\begin{aligned}
&\text{if } (A = B) &&\text{then} &&A \& B = -1 \text{ else}\\
&\text{if } (A = +1) \text{ XOR } (B = +1) &&\text{then} &&A \& B = +1 \text{ else}\\
& &&&&A \& B = 0
\end{aligned}$$

That is, if A = B then the concatenated string is periodic, otherwise, if either A or B are aperiodic, then the concatenated string is also aperiodic, else the concatenated string is neither periodic nor aperiodic.

Concatenation is commutative, *i.e.* A & B = B & A.

8.3. *Negation*

Flipping all the bits of a string does not alter the BiEntropy i.e. A = NOT (A).

8.4. Addition

We can combine two 4-bit strings by adding them together or XORing their respective bits :

Table 11 : The Addition (XORing) of two 4-bit strings

		-1	-1	-1	-1	0	0	0	0	+1	+1	+1	+1	+1	+1	+1	+1
		0	1	0	1	0	0	1	1	0	0	0	0	1	1	1	1
		0	1	1	0	0	1	0	1	0	0	1	1	0	0	1	1
		0	1	0	1	1	1	0	0	0	1	0	1	0	1	0	1
		0	1	1	0	1	0	1	0	1	0	0	1	0	1	1	0
-1	0 0 0 0	-1	-1	0	0	0	0	0	0	+1	+1	+1	+1	+1	+1	+1	+1
-1	1 1 1 1	-1	-1	0	0	0	0	0	0	+1	+1	+1	+1	+1	+1	+1	+1
-1	0 1 0 1	0	0	-1	-1	0	0	0	0	+1	+1	+1	+1	+1	+1	+1	+1
-1	1 0 1 0	0	0	-1	-1	0	0	0	0	+1	+1	+1	+1	+1	+1	+1	+1
0	0 0 1 1	0	0	0	0	-1	0	0	-1	+1	+1	+1	+1	+1	+1	+1	+1
0	0 1 1 0	0	0	0	0	0	-1	-1	0	+1	+1	+1	+1	+1	+1	+1	+1
0	1 0 0 1	0	0	0	0	0	-1	-1	0	+1	+1	+1	+1	+1	+1	+1	+1
0	1 1 0 0	0	0	0	0	-1	0	0	-1	+1	+1	+1	+1	+1	+1	+1	+1
+1	0 0 0 1	+1	+1	+1	+1	+1	+1	+1	+1	-1	0	0	0	0	0	0	-1
+1	0 0 1 0	+1	+1	+1	+1	+1	+1	+1	+1	0	-1	0	0	0	0	-1	0
+1	0 1 0 0	+1	+1	+1	+1	+1	+1	+1	+1	0	0	-1	0	0	-1	0	0
+1	0 1 1 1	+1	+1	+1	+1	+1	+1	+1	+1	0	0	0	-1	-1	0	0	0
+1	1 0 0 0	+1	+1	+1	+1	+1	+1	+1	+1	0	0	0	-1	-1	0	0	0
+1	1 0 1 1	+1	+1	+1	+1	+1	+1	+1	+1	0	0	-1	0	0	-1	0	0
+1	1 1 0 1	+1	+1	+1	+1	+1	+1	+1	+1	0	-1	0	0	0	0	-1	0
+1	1 1 1 0	+1	+1	+1	+1	+1	+1	+1	+1	-1	0	0	0	0	0	0	-1

Adding (XORing) two 4-bit strings A and B to make another 4-bit string follows these rules :

$$
\begin{array}{lll}
\text{if } (A = B) & \text{then} & A + B = -1 \text{ else} \\
\text{if } (A = NOT(B)) & \text{then} & A + B = -1 \text{ else} \\
\text{if } ((A = +1) \text{ XOR } (B = +1)) & \text{then} & A + B = +1 \text{ else} \\
& & A + B = 0
\end{array}
$$

Adding strings together by XORing their bits reduces the BiEntropy on average as an additional 16 periodic (16×-1) states are created, leaving 128 aperiodic states ($+1$), 32 periodic states (-1) and 96 states that are neither (0).
Addition is also commutative, *i.e.* $A + B = B + A$.

8.5. Multiplication

We can combine two 4-bit strings by multiplying them together or ANDing their respective bits :

Table 12 : The Multiplication (ANDing) of two 4-bit strings

					-1	-1	-1	-1	0	0	0	0	+1	+1	+1	+1	+1	+1	+1	+1
					0	1	0	1	0	0	1	1	0	0	0	0	1	1	1	1
					0	1	1	0	0	1	0	1	0	0	1	1	0	0	1	1
					0	1	0	1	1	1	0	0	0	1	0	1	0	1	0	1
					0	1	1	0	1	0	1	0	1	0	0	1	0	1	1	0
-1	0	0	0	0	-1	-1	-1	-1	-1	-1	-1	-1	-1	-1	-1	-1	-1	-1	-1	-1
-1	1	1	1	1	-1	-1	0	0	0	0	0	0	+1	+1	+1	+1	+1	+1	+1	+1
-1	0	1	0	1	-1	0	0	-1	+1	+1	+1	+1	+1	-1	+1	0	-1	+1	0	+1
-1	1	0	1	0	-1	0	-1	0	+1	+1	+1	+1	-1	+1	-1	+1	+1	0	+1	0
0	0	0	1	1	-1	0	+1	+1	0	+1	+1	-1	+1	+1	-1	0	-1	0	+1	+1
0	0	1	1	0	-1	0	+1	+1	+1	0	-1	+1	-1	+1	+1	0	-1	+1	+1	0
0	1	0	0	1	-1	0	+1	+1	+1	-1	0	+1	+1	-1	-1	+1	+1	0	0	+1
0	1	1	0	0	-1	0	+1	+1	-1	+1	+1	0	-1	-1	+1	+1	+1	+1	0	0
+1	0	0	0	1	-1	+1	+1	-1	+1	-1	+1	-1	+1	-1	-1	+1	-1	+1	+1	-1
+1	0	0	1	0	-1	+1	-1	+1	+1	+1	-1	-1	-1	+1	-1	+1	-1	+1	-1	+1
+1	0	1	0	0	-1	+1	+1	-1	-1	+1	-1	+1	-1	-1	+1	+1	-1	-1	+1	+1
+1	0	1	1	1	-1	+1	0	+1	0	0	+1	+1	+1	+1	+1	+1	-1	0	0	0
+1	1	0	0	0	-1	+1	-1	+1	-1	-1	+1	+1	-1	-1	-1	-1	+1	+1	+1	+1
+1	1	0	1	1	-1	+1	+1	0	0	+1	0	+1	+1	+1	-1	0	+1	+1	0	0
+1	1	1	0	1	-1	+1	0	+1	+1	+1	0	0	+1	-1	+1	0	+1	0	+1	0
+1	1	1	1	0	-1	+1	+1	0	+1	0	+1	0	-1	+1	+1	0	+1	0	0	+1

Multiplying strings together by ANDing their bits significantly reduces BiEntropy such that there are 120 aperiodic states (+1), 82 periodic states (-1) and 54 states that are neither (0). There does not appear to be any obvious rule (other than ANDing the bits) to produce this state diagram. Multiplication is also commutative, *i.e.* A · B = B · A, which we can glean from the difficult to spot diagonal symmetry of Table 12.

We note that multiplication causes a reduction in BiEntropy, implying that maximal disorder may be a difficult state to achieve or maintain.

8.6. *Summary and Future Development of the Algebras*

We have shown how we are able to produce a series of simple BiEntropy Algebras for short strings using the concatenation, negation, addition and multiplication operators. These will be easy to generalise for longer strings. Due to the fact that BiEntropy is intimately connected to the binary world, it will be possible to develop a wide range of further algebras to meet a variety of circumstances in the physical sciences and elsewhere.

For example, each N-Dimensional BitSpace (*i.e.* set of BitStrings of length N, with XOR-addition as combination) has a Canonical Basis consisting of N primitive BitStrings such as 10000000, 01000000, 00100000 *etc.* from which every member of the space is constructible. We can generate every possible subspace from the various possible permutations of the Canonical Basis. We can then ask questions about the average BiEntropy (actual or statewise) of all the BitStrings in a subspace.

It is known that every such BitSpace is also a Projective Geometry [Amson, 1988] which will enable us to construct a BiEntropy of Projective Geometries. Every such BitSpace is also a Graph, which opens up the possibility of the BiEntropy of Graphs.

9. Summary

We have introduced a simple set of BiEntropy functions which can be used to measure the order and disorder of finite strings of arbitrary length. The functions are simply weighted averages of the Shannon Entropies of all but the last binary derivative of the string.

We have briefly covered the historical background regarding the development of tests and measures of order, disorder, randomness, irregularity and entropy. We show that binary derivatives have previously been used in the measurement of disorder and have also been used in cryptographic applications and attacks.

We have discussed our intuitive understanding of order and disorder, Shannon Entropy and Binary Derivatives and formally defined the polynomial (BiEn) and logarithmic (TBiEn) BiEntropy functions. We have successfully tested BiEntropy in four extremely diverse application areas of Prime Number Theory, Human Vision, Random Number Generation and Cryptography.

Two important results include a Corollary showing that the primes are not periodic, and evidence that the Braille characters set, extant for some 150 years, is a low entropy character set.

We developed a set of simple BiEntropy algebras and demonstrated that a BiEntropy of Projective Geometries and a BiEntropy of Graphs are each a future possibility, both of which may be of use in a bit string physics.

Acknowledgements

The author thanks: Qinqin Zhang, Math librarian at the Library of the University of Western Ontario for locating and scanning a copy of James Mc Nair's MSc Thesis report; Dr Mike Manthey and Dr John C. Amson for their reviews and detailed comments on various versions of this paper. The author first met Pierre Noyes at the 25-th ANPA meeting in Wesley College Cambridge in 2004, and thanks him for the interesting discussions at the time and his foundation and contribution to ANPA over three decades.

References

Amson, John. Discrimination Systems and Projective Geometries, in:
    Discrete and Combinatorial Physics, Proceedings of ANPA No.9, 158–189 (1988)
Andersson, E.A. (2013) *One Million Digits of Pi*
    http://www.eveandersson.com/pi/digits/1000000 Accessed 10th Feb 2013
Bruwer, C. S. (2005) *Correlation attacks on stream ciphers using convolutional codes*,   Masters Thesis, University of Pretoria

Carroll, J.M. (1989) The binary derivative test: noise filter, crypto aid, and random-number seed selector, *Simulation*, 53 pp129–135
http://sim.sagepub.com/content/53/3/129.full.pdf+html

Carroll, J.M., Sun, Y. (1998) *The Binary Derivative Test for the Appearance of Randomness and its use as a Noise Filter*
Tech Report No. 221 University of Western Ontario. Dept. of Computer Science

Chaitin, G. J. (1966) On the length of programs for computing finite binary sequences. *Journal of the ACM* (JACM), 13(4), 547–569.

Chen, W., Zhuang, J., Yu, W., Wang, Z. (2009) Measuring complexity using FuzzyEn, ApEn, and SampEn. *Medical Engineering & Physics*, 31(1), 61–68.

Champernowne, D. G. (1933) The construction of decimals normal in the scale of ten. *Journal of the London Mathematical Society*, 1(4), 254–260.
http://www.uea.ac.uk// h720/teaching/dynamicalsystems/champernowne1933.pdf

Croll, G. J. (2005) The importance and criticality of spreadsheets in the City of London. *Proc. EuSpRIG*. arXiv preprint arXiv:0709.4063.

Croll, G. J. (2009) Spreadsheets and the financial collapse.
Proc. EuSpRIG. arXiv preprint arXiv:0908.4420.

Croll, G. J. (2013) BiEntropy – The Approximate Entropy of a Finite Binary String. arXiv preprint arXiv:1305.0954.

Davies, N., Dawson, E., Gustafson, H., Pettitt, A.N., (1995) Testing for randomness in stream ciphers using the binary derivative.
Statistics and Computing, Vol 5, pp. 307–310

Gao, Y., Kontoyiannis, I., Bienenstock, E. (2008) Estimating the Entropy of Binary Time Series: Methodology, Some Theory and a Simulation Study, *Entropy*, 10, pp71–99

Goka, T. (1970) An operator on binary sequences, *SIAM Rev.*, 12 (1970), pp. 264–266

Grossman, T. A. (2008) Source code protection for applications written in Microsoft Excel and Google Spreadsheet. arXiv preprint arXiv:0801.4774.

Jiménez, J., Olea, J., Torres, J., Alonso, I., Harder, D., Fischer, K. (2009) Biography of Louis Braille and invention of the Braille alphabet.
Survey of Ophthalmology, 54(1), 142–149.

Kolmogorov, A. N. (1965) Three approaches to the quantitative definition of information. *Problems of information transmission*, 1(1), 1–7.

Marsaglia, G., Zaman, A. (1993) Monkey tests for random number generators.
Computers & mathematics with applications, 26(9), 1–10.

McCullough, B. D. (2008) Microsoft Excel's 'not the Wichmann–Hill' random number generators. *Computational Statistics & Data Analysis*, 52(10), 4587–4593.

McNair, J. (1989) *The Binary Derivative: a new method of testing the appearance of randomness in a sequence of bits.* MSc Thesis report, University of Western Ontario.

Mitchell, S. (2008) Open Source Urbanism
  `http://www.openobject.org/opensourceurbanism/Bike_POV_Beta_4`

Nathanson, M. B. (1971) Derivatives of binary sequences. *SIAM Journal of Applied Mathematics*, 21(3), 407–412.

Nemiroff, R., Bonnell, J. (1994) RJN's More Digits of Irrational Numbers Page, `http://apod.nasa.gov/htmltest/rjn_dig.html` Accessed 10 Feb 2013 13:40.14/14

Noyes, H. P. (1997) *A Short Introduction to Bit-String Physics.* arXiv preprint hep-th/9707020.

Pincus, S. (1991) Approximate entropy as a measure of system complexity. *Proc. Natn. Acad. Sci. USA* Vol 88 pp 2297– 2301

Pincus, S. & Kalman, R.E., (1997) Not all (possibly) "random" sequences are created equal. *Proc. Natn. Acad. Sci. USA* Vol 94 pp 3513–3518

Pincus, S. Singer, B. H. (1996) Randomness and Degrees of Irregularity, *Proc. Natl. Acad. Sci. USA.* March 5; 93(5): 2083–2088.

Pincus, S. M., Viscarello, R. R. (1992) Approximate entropy: a regularity measure for fetal heart rate analysis. *Obstetrics and Gynaecology*, 79(2), 249–255.

RAND (2001) *A Million Random Digits With 100,000 Normal Deviates*, `http://www.rand.org/pubs/monograph_reports/MR1418.html` Accessed 10th February 2013, 14:25

Richman, J. S., Moorman, J. R. (2000) Physiological time-series analysis using approximate entropy and sample entropy. *American Journal of Physiology – Heart and Circulatory Physiology*, 278(6), H2039–H2049.

Rukhin, A.L. (2000a) Approximate Entropy for Testing Randomness, *J. Appl. Prob.* Volume 37, Number 1, 88–100

Rukhin, A.L. (2000b) Testing Randomness: A Suite of Statistical Procedures, *Theory of Probability and its Applications*, 45:1, 111–132

Shannon, C. E. (1948) A Mathematical Theory of Communication, *Bell System Technical Journal*, 27 (3): 379–423

Brief Biography

Grenville obtained a Bachelors degree in computer science from Brunel University, London in 1980. He subsequently enjoyed a diverse 30–year career in the computer industry including 15 years running his own software publishing and management training companies. His research interests include end-user computing, cryptography and theoretical physics.

Constraints Theory Brief

Anthony Michael Deakin

1a Castle Rd, Rowlands Castle, Hants. PO9 6AP
E-mail: amdeak.deakin@btinternet.com

We find that the Schrödinger quantization of a rudimentary classical archetype produces operator equations that express most of the rules of classical mechanics. These rules include the Riemannian connectivity of physical space; but this seems to emerge only at the macrophysical scale. The field equations are fourth order PDEs. But they are satisfied by the conventional second order solutions both as regards gravity and EM; extra physics is therefore to be expected. The Maxwell equations, which are deducible from the Feynman-Dyson quantum mechanical argument, are hardly effected. Gravitation is governed by the fourth order Kilmister equation; in conjunction with the Einstein equation, this dictates the allowed distributions of matter-energy-momentum and, perhaps, offers an explanation of the Dark Energy effect. Preliminary machine calculations are encouraging.

1. Purpose

To examine the relationship between a slight generalisation of the Schrödinger method of quantisation and Classical Mechanics (CM); see [1]. In this case CM includes the Special and General theories of relativity (SR and GR). We use the primary characteristic of CM; that all the variables are continuous. This investigation is called 'Constraints Theory'.

2. The Archetype

Imagine a continuous space P, of infinite extent, with unknown dimension and unknown connectivity. The coordinates of the ath point in this space form the n_d tuple $q_a^1, q_a^2, \ldots, q_a^{n_d}$. Special points are defined, numbering n_p, called *particles*; individual particles can be anywhere in P. According to the axioms, that Schrödinger used, the axes of $q^1, q^2, \ldots, q^{n_d}$ are at right angles at the origin (any point); that is they are Cartesian. But this is to beg the question as to the geometry of P. More on this topic later.

Associated with the n_p particles is assumed to be one or more differentiable functions $\theta(\underline{q})$; here $\underline{q}$ denotes the aggregate of the coordinates of *all* the particles q_β^J in any convenient order: $\beta = 1, 2, \ldots, n_p$; $J = 1, 2, \ldots, n_d$. The functions $\theta(\underline{q})$ can be described as 'fields'. The particle coordinates may be differentiable functions of the continuous time t. This time needs further definition. A single observer is taken to observe all the particles; t is the time according to his/her consciousness and an adjacent, stationary clock.

3. The Identities

Given the above archetype we may write an infinite hierarchy of identities for $\dot{\theta}, \ddot{\theta}, \ldots$;

$$\dot{\theta} = \dot{q}^j \theta_{,j} \quad \theta_{,j} \equiv \frac{\partial \theta}{\partial q^j}; \quad \dot{\theta} \equiv \frac{d\theta}{dt}; \quad j = 1, 2, \ldots, n_c; \quad n_c = n_d n_p \tag{3.1}$$

$$\ddot{\theta} = \ddot{q}^j \theta_{,j} + \dot{q}^j \dot{q}^j \theta_{,jk}; \quad \theta_{,jk} \equiv \frac{\partial^2 \theta}{\partial q^j \partial q^k} \tag{3.2}$$

$$\dddot{\theta} = \dddot{q}^j \theta_{,j} + 3\ddot{q}^j \dot{q}^k \theta_{,jk} + \dot{q}^j \dot{q}^j \dot{q}^l \theta_{,jkl} \tag{3.3}$$

    Etc.

Here and in the sequel the Einstein summation convention is in force; and, unless otherwise stated, all indices lie in the range $[\,1,\, n_c\,]$.

4. Quantization:- The Rules

We quantize these identities, as if they are equations of motion with t as time, using the rules of Schrödinger. The usual choice of a mathematical model, from say classical physics (i.e. energy equation or a Lagrangian), is thereby avoided. The results are no longer identities; they are constraints. The constraints are operator equations in the coordinate operators $\underline{Q}$, the corresponding conjugate momentum operators $\underline{P}$, the operator $\Theta(\underline{Q})$ which represents the function $\theta(\underline{q})$ and the QM Hamiltonian $H(\underline{P}, \underline{Q})$. The rules of quantization are a generalisation of those used by Schrödinger in his famous hydrogen model. The constraints are taken not as inconsistencies but rather as laws of nature.

    In summary the rules are as follows:

(4.1)   Real observable a is replaced by an Hermitian symmetric operator A acting on an Hilbert space of 'state' functions;   notation $a \to A$.

(4.2)   Weighted sums: $\alpha\, a + \beta\, b \to \alpha\, A + \beta\, B$ where α, β are numerical and A, B may or may not commute.

(4.3)   Powers: $a^n \to A^n$ where n is an integer.

(4.4)   Products: Louis H. Kauffman deduces, from the previous rules,
$a_1 a_2 \cdots a_n \to \frac{1}{n!} \sum_{Perm} A_1 A_2 \cdots A_n \equiv \{A_1, A_2, \ldots, A_n\}$ whether or not the A_j mutually commute. Ref. [2] deduces this rule for $n = 2$ and asks, in an exercise, what is the result for $n = 3$?

(4.5)   Commutation of Position and Momentum (O and I are zero and unit operators): $[P_j, P_k] = O;$ $[Q^j, Q^k] = O;$ $[Q^j, P_k] = i\,\hbar\delta^j_k I;$ $j, k = 1, 2, \ldots, n_c \equiv n_d n_p;$ notation $[\,A,\, B\,] \equiv AB - BA.$

(4.6)   Nature of space P: P is flat and the axes of the $\underline{p}$ and the $\underline{q}$ are Cartesian. This assumption is to beg the question about the connectivity of P.

(4.7) Position Representation: In the *position* representation [3]
$$q^j \to Q^j \equiv q^j I; \quad a(\underline{q}) \to A(\underline{Q}) \equiv a(\underline{q})I; \quad -\infty < q^j < \infty.$$
$$p_k \to P_k \equiv -i\hbar \frac{\partial}{\partial q^k}; \quad b(\underline{p}) \to B(\underline{P}) \equiv b(\underline{P}).$$

(4.8) Momentum Representation: In the momentum representation [3]
$$p^j \to P^j \equiv p^j I; \quad b(\underline{p}) \to B(\underline{P}) \equiv b(\underline{p})I; \quad -\infty < p_j < \infty.$$
$$q_k \to Q_k \equiv i\hbar \frac{\partial}{\partial p^k}; \quad a(\underline{q}) \to A(\underline{Q}) \equiv a(\underline{Q}); \quad \text{P is flat.}$$

(4.9) Derivatives And Derivations:
$$\frac{\partial a}{\partial p^j} \equiv a^j \to \lfloor A, Q^j \rfloor \equiv \frac{i}{\hbar}\left[A, Q^j\right] \equiv A^{:j};$$
$$\frac{\partial a}{\partial q^j} \equiv a_{,j} \to \lfloor P^j, A \rfloor \equiv \frac{i}{\hbar}\left[P^j, A\right] \equiv A_{,j}$$
Kauffman calls commutators like the above *derivations*.

(4.10) Calculating With Derivatives And Derivations:

The quantities $\quad A_{,j} = \dfrac{\partial A}{\partial Q^j}, \quad A^{:k} = \dfrac{\partial A}{\partial P_k}$, where A is a multinomial in

the $\underline{P}$ and the $\underline{Q}$, are to be calculated as if A, $A_{,j}$, $A^{:k}$, Q^j, P_k are scalars providing that the order of non-commuting operators is preserved. If A is pure (either in the $\underline{P}$ or the $\underline{Q}$) then, it can be proved, that these results apply even when A is not a multinomial. It is necessary for the scalar partial derivatives $\partial a(\underline{p})/\partial p_k$ or $\partial a(\underline{q})/\partial q^i$, where $a \to A$, to exist. Notice that $A_{,j,k} = A_{,k,j}$; $A^{:j:k} = A^{:k:j}$ are identities for any operator A .

(4.11) Time: $\dfrac{da}{dt} \equiv \dot{a} \equiv \overset{1}{a} \to \lfloor H, A \rfloor;$
$$\frac{d^n a}{dt^n} \equiv \overset{n}{a} \to \lfloor H, \lfloor H, \lfloor H, \ldots \lfloor H, A \rfloor \ldots \rfloor \rfloor \rfloor.$$

Constant H is taken as the total energy of an *isolated system*.
$T \equiv t I$ commutes with all constant operators and functions of t. But if an operator A contains $\partial(\,.\,)/\partial t$ then T does not necessarily commute with A .

5. Quantization Of The Identities

The constraints limit the forms permitted for Θ and H . They are:

$$\{H^{:j}, \Theta_{,j}\} = \lfloor H, \Theta \rfloor \tag{5.1}$$

$$\{\lfloor H, H^{:j} \rfloor, \Theta_{,j}\} + \{H^{:j}, H^{:k}, \Theta_{,j,k}\} = \lfloor H, \lfloor H, \Theta \rfloor \rfloor \tag{5.2}$$

$$\{\lfloor H, \lfloor H, H^{:j} \rfloor \rfloor, \Theta_{,j}\} + 3\{\lfloor H, H^{:j} \rfloor, H^{:k}, \Theta_{,j}\} + \{H^{:j}, H^{:k}, H^{:l}, \Theta_{,j,k,l}\}$$
$$= \lfloor H, \lfloor H, \lfloor H, \Theta \rfloor \rfloor \rfloor \tag{5.3}$$

Etc.
It is a hypothesis that we only need the first two constraints to achieve our purpose.
Note, in particular, that the operators (expressed in the position representation),

$$\Theta(\underline{Q}) \equiv \theta(\underline{q})\, I; \quad \Theta_{,j} = \theta_{,j}\, I; \quad \Theta_{,j,k} = \theta_{,j,k}\, I; \quad \Theta_{,j,k,l} = \theta_{,j,k,l}\, I; \ etc. \tag{5.4}$$

are pure in the $\underline{Q} = q\,I$; the order of the suffices is immaterial.

6. When The First Constraint Holds

If $\theta(\underline{q})$ is arbitrary it can be shown that the first constraint (5.1) requires H to be quadratic in the $\underline{P}$

$$H = \{C^{uv}, P_u, P_v\} + \{E^j, P_j\} + U ; \quad C^{uv} = C^{vu} ; \tag{6.1}$$

$$u, v, j = 1, 2, \ldots, n_c \equiv n_d n_p;$$

Here the coefficient Hermitian operators $C^{uv}(\underline{Q})$, $E^j(\underline{Q})$, $U(\underline{Q})$ do not depend on the $\underline{P}$ and commute with the $\underline{Q}$. Thus the rate operator of the kth coordinate is

$$\dot{Q}^k \equiv \lfloor H, Q^k \rfloor = 2 \{C^{uv}, P_k\} + E^j; \qquad \text{See (4.5/4.12)} \tag{6.2}$$

It follows that if any of the C^{uv} or the E^k are represented, in the position representation, by non-scalar matrices (albeit infinite) then the spectrum of $\dot{Q}^k$ is not continuous; thus if, we require the spectrum of $\dot{Q}^k$ to be always continuous then, the C^{uv} , E^k must be scalars in the position representation. A form, consistent with this assumption, is

$$H \equiv \mathsf{K} \{G^{uv}, P_u, P_v\} + \{F^j, P_j\} + V ; \quad G^{uv} = G^{vu} ; \quad u, v, j = 1, 2, \ldots, n_c; \tag{6.3}$$

where K is a numerical constant. The constant K has physical dimensions mass^{-1} in order to ensure, as a convention, that the G^{uv} have no physical dimensions, the $\underline{P}$ have the physical dimension of momentum and H has that of energy. In the position representation, the coefficients G^{uv}, F^j, V are either scalar constants or scalar functions of the $\underline{q}$. Thus, in the position representation,

$$G^{uv}(\underline{Q}) = g^{uv}(\underline{q})\,I ; \quad F^j(\underline{Q}) = f^j(\underline{q})\,I ; \quad V(\underline{Q}) = v(\underline{q}\,I ; \quad Q^j = q^j\,I \tag{6.4}$$

7. When The First Two Constraints Hold

The operator equation, that is satisfied when *both* the first two constraints hold, is called the Operator Theta Equation (OTE). It can be shown that it is

$$-\frac{\hbar^2 \mathsf{K}^2}{3} G^{vj}(G^{uk}\Theta_{,j,k,u})_{,v} = 0 ; \ j, k, u, v = 1, 2, \ldots n_c; \ \theta(\underline{q}) \to \Theta(\underline{Q}); \ \text{OTE} \tag{7.1}$$

Notice that the OTE is independent of the coefficient operators F^j and V. In the position representation the OTE becomes the Scalar Theta Equation (STE):

$$g^{vj}(g^{uk}\theta_{,jku})_{,v} = 0 ; \ j, k, u, v = 1, 2, \ldots n_c; \qquad \text{STE See (6.4)} \tag{7.2}$$

This is a fourth order PDE with $\theta(\underline{q})$ as the dependent variable and the Schrödinger (Cartesian and flat) choice of coordinates as independent variables; see (4.10, 4.11).

If H is a complete descriptor of the system, as is usual, an *exhaustive list* of candidates for θ comprises the individual g^{uv}, the individual f^j and v; see (6.4).

Thus (7.2) is an *archetype field equation* that determines the various choices of θ in terms of the position coordinates. As far as is known, and we shall give examples, all the solutions of the customary second order field equations satisfy the STE; but not *vice versa*. Thus extra physics is to be anticipated.

8. Hamilton's Equations and the Space **C**

As a consequence of the Schrödinger definitions the $\underline{P}$, the $\underline{Q}$ and $H(\underline{P},\underline{Q})$ satisfy an operator form of Hamilton's equations. By inspection of the rules (4.9, 4.11) we see that

$$\dot{P}_j = -II_{,j} ; \quad \dot{Q}_k - -H^{:k} ; \quad \text{Hamilton's Equations in QM.} \tag{8.1}$$

These equations are the quantum analogue, in a flat space, of the classical Hamilton's equations

$$\dot{p}_j = -\frac{\partial h}{\partial q^j} ; \quad \dot{q}_k = -\frac{\partial h}{\partial p_k} ; \quad \text{Hamilton's Equations in CM.} \tag{8.2}$$

But keep in mind that the axioms of quantization require that P is flat and the coordinates are Cartesian. So the operator Hamilton's equations are restricted; their classical counterparts (p,q) are allowed to be curvilinear and are independent.

Equations (8.2) can be regarded as the *reverse quantizations* (every operator commutes with every other) of (8.1). Similarly, the reverse quantization of (6.3) is

$$h = \mathsf{K}g^{uv}(\underline{q})p_u p_v + f^j(\underline{q})p_j + \nu(\underline{q}) ; \quad u,v,j = 1,2,\ldots n_c ; \quad \text{Scalar Hamiltonian.} \tag{8.3}$$

If we eliminate the $\underline{p}$ between (8.2) and (8.3) we get

$$\ddot{q}^m + \Gamma^m_{kl}\dot{q}^k\dot{q}^l =$$
$$-g^{um}_{,v}\dot{q}^v f_u - g^{um}\left(f^j_{,u}(\dot{q}_j - f_j) - \frac{1}{2}g^{jk}_{,u}(\dot{q}_j f_k - f_j f_k) + 2\mathsf{K}\nu_{,u}\right) + f^m_{,v}\dot{q}^v \tag{8.4}$$

and if

$$f^j = 0; \quad v = 0 \implies h = \mathsf{K}g^{uv}(\underline{q})p_u p_v ; \quad \text{See (6.4)} \tag{8.5}$$

then

$$\ddot{q}^j + \Gamma^j_{kl}\dot{q}^k\dot{q}^l = 0; \quad \Gamma^l_{ij} \equiv \frac{1}{2}g^{lk}\left(g_{ik,j} + g_{jk,i} - g_{ij,k}\right) ; \quad \text{Geodesic Equation of C.} \tag{8.6}$$

which is the geodesic equation for a Riemannian space **C** with fundamental tensor g^{lk} , dimension n_c and distance element

$$ds^2 = c^2 dt^2 = g_{uv}dq^u dq^v ; \quad \text{First integral of (8.6).} \tag{8.7}$$

Note that we have to multiply (8.6) by c^{-2} to get the actual geodesic equation of C; further, C is flat (the Riemann-Christoffel tensor vanishes or, equivalently, the

Riemann curvature vanishes [4] [5]) and the coordinates $\underline{q}$ are Cartesian. If we have but one particle in P

$$n_p = 1 \implies \mathsf{P} \equiv \mathsf{C}. \tag{8.8}$$

Because we reverse quantized at (8.2, 8.3) results (8.4) to (8.7) apply only to *macrophysics*.

9. The Riemannian Space C'

We may define a Riemannian space C', which is of the same dimension and connectivity as C, but curved. The space C' is tangential to C at point X (in C') and at point Q (in C); and, if we use *Cartesian geodesic* coordinates in C' with pole X, all the definitions, rules and equations of Schrödinger QM apply at X and are approximated in its neighbourhood. At Q (in C) we have coordinates $\underline{q}$ and fundamental tensor $g^{uv}(\underline{q})$. At X (in C') we have coordinates $\underline{x}$ and fundamental tensor $g'^{uv}(\underline{x})$ with

$$g'^{uv}_{,w}(\underline{x}) = 0; \quad \underline{x} = \underline{q}; \quad g'^{uv}(\underline{x}) = g^{uv}(\underline{q}) \tag{9.1}$$

We may drop the primes on C etc. as long as we recognise that C may be curved. For clarity we denote the coordinates by $\underline{q}$ when the space is definitely flat; and by $\underline{x}$ otherwise.

10. Electromagnetism

The Feynman-Dyson argument [6] deduces classical electromagnetism from QM. Therefore we have only to show that Constraints Theory does not clash with classical EM. Taking EM alone C is supposed flat. Therefore the STE holds anywhere in C.

Taking the Minkowski metric for C

$$ds^2 = -(dq^1)^2 - (dq^2)^2 - (dq^3)^2 + (dq^4)^2 ; \quad n_c = 4 \tag{10.1}$$

where q^1, q^2, q^3 are the Cartesian spatial coordinates and q^4/c is the time coordinate, the STE becomes

$$\Box^2(\Box^2\theta) = 0; \quad \Box^2 \equiv -\nabla^2 + \frac{\partial^2}{\partial(q^4)^2} ; \quad q^4/c \quad \text{is coordinate time} \tag{10.2}$$

where

$$\theta \equiv f^j \text{ or } \theta \equiv v. \tag{10.3}$$

Notice that by choosing (10.1) we have assumed that there is a single particle in P.

A detailed analysis has shown that the only impact of Constraints Theory on Maxwell's equations, is that

$$\Box^2\underline{i} = \underline{0}; \quad \Box^2\rho = 0; \quad \text{See (10.2, 10.3)} \tag{10.4}$$

where $\underline{i}$ is the 3-vector of current and ρ is the charge density. The result (10.4) means that any disturbance of either the current or charge density is propagated at a rate c. Further, comparison of the motions of a single charge ϵ of mass m with (8.4), gives

$$f_{\mu,\lambda} - f_{\lambda,\mu} = \frac{\epsilon}{mc}(a_{\mu,\lambda} - a_{\lambda,\mu}); \quad f_{4,\lambda} - f_{\lambda,4} = -\frac{\epsilon}{mc}(\phi_{,\lambda} + a_{\lambda,4});$$

$$f^j_{,\lambda} f_j = 2\mathsf{K}\nu_{,\lambda}; \quad f^j_{,4} f_j = 2\mathsf{K}\nu_{,4} \tag{10.5}$$

where a_λ and ϕ are the components of 3-vector and scalar potentials. If we define

$$4\mathsf{K}\nu = f^j f_j \implies 4\mathsf{K}\nu_{,k} = f^j_{,k} f_j + f^j f_{j,k} = 2f^j_{,k} f_j \implies f^j_{,k} f_j = 2\mathsf{K}\nu_{,k} \tag{10.6}$$

then we get results consistent with (10.5). With this definition the scalar Hamiltonian can be written

$$h \equiv \mathsf{K}\left(p^u + \frac{f^u}{2\mathsf{K}}\right)\left(p_u + \frac{f_u}{2\mathsf{K}}\right) \qquad \text{See (6.3)} \tag{10.7}$$

This compares with an Hamiltonian which gives the usual equations of motion

$$h = \frac{1}{2m}\left(-\sum_{\mu=1}^{3}\left(p_\mu + \frac{\epsilon}{c}a_\mu\right)^2 + \left(p_4 + \frac{\epsilon}{c}\phi\right)^2\right);$$

$$\mathsf{K} \equiv \frac{1}{2m}; \quad f_\mu \equiv \frac{\epsilon}{mc}a_\mu; \quad f_4 \equiv -\frac{\epsilon}{mc}\phi; \tag{10.8}$$

Here Roman indices have a range 1 to 4 whereas Greek indices have a range 1 to 3.

11. Curvature of C, the GTE and the KE

Curvature of C is taken to be a symptom of gravity. Since C is Riemannian we may use it as an arena for Einsteinian gravity. Einstein's equations, by which he brought matter into an essentially geometric theory, are [4]

$$G^a_b + \chi T^a_b + \delta^a_b = 0; \ G^\mu_\nu \equiv R^\mu_\nu - \frac{1}{2}R\delta^\mu_\nu\Lambda = 0; \ R^a_b \ \text{is the mixed Ricci tensor} \tag{11.1}$$

where

$$\chi \equiv 8\pi\mathsf{G}/c^4 = 2.0761{\times}10^{-43}m^{-1}kg^{-1}s^2;$$

$$c = 2.99792458{\times}10^8 ms^{-1}; \tag{11.2}$$

$$\mathsf{G} = 6.672(59){\times}10^{-11} Nm^2 kg^{-2}$$

G is *Newton's constant*, Λ is the *cosmological constant* with T^a_b as the *matter-energy-momentum-stress tensor*. The tensor is subject to an identity [4]

$$G^a_{b;b} = 0 \implies T^u_{v,u} = 0. \quad \text{See (11.1)} \tag{11.3}$$

where ';' denotes covariant differentiation.

The STE is valid in the neighbourhood of the pole X (which may be any point of C) and its coordinates are Cartesian geodesic. If we choose

$$\theta \equiv g^{lm} \tag{11.4}$$

the STE becomes the Gravitational Theta Equation (GTE)

$$g^{vj}(g^{uk}g^{lm}_{,jku})_{,v} = 0; \quad \text{GTE} \tag{11.5}$$

having the same coordinates and the same range of validity. We may ask: What tensor equation reduces to the GTE when we use Cartesian geodesics with the pole X? The answer is the Kilmister Equation (KE) [11]

$$K_{ab} \equiv g^{ef}\left(R_{ab;ef} + \frac{2}{3}R_{ae}R_{fb}\right) = 0 \tag{11.6}$$

derived by the late Clive W. Kilmister. Three properties of the KE should be noted: Firstly,

$$R^a_b = 0 \implies K^a_b = 0 \tag{11.7}$$

Secondly, since

$$G \equiv -R \implies R^\mu_\nu = G^\mu_\nu - \frac{1}{2}G\delta^\mu_\nu \tag{11.8}$$

in the empty space between particles

$$T^a_b = 0 \implies G^a_b = -\Lambda\delta^a_b \implies R^a_b = \Lambda\delta^a_b \quad \text{See (11.1, 11.6)} \tag{11.9}$$

Therefore

$$K_{ab} = 0 \implies \Lambda = 0. \tag{11.10}$$

That is, if Einstein's law of gravity for the empty space between particles is to satisfy the K equation then the cosmological constant vanishes. Thirdly, because the K equation is purely geometrical, if the cosmological constant vanishes it determines T_{ab} (subject to boundary conditions) throughout the universe. That is

$$\Lambda = 0 \implies R_{ab} = -\chi\left(T_{ab} - \frac{1}{2}Tg_{ab}\right) \tag{11.11}$$

12. Newtonian Gravity, GR and Constraints Theory

The simplest discussion of *weak* gravity is of a single infinitesimal particle moving in a potential. Choose a perturbation of the Minkowski metric ($n_p = 1$, $n_c = 4$)

$$ds^2 = (1 + 2U)(dx^4)^2 + (-1 + 2U)(ds_0)^2; \quad |U| << 1; \quad |ds_0| << |dx^4|; \tag{12.1}$$

(see (10.1)) where the $\underline{x}$ are *quasi-Cartesian* coordinates and time as distance. The small, dimensionless potential U is a function of the spatial coordinates x^1, x^2, x^3 only. This is suitable to discuss Newtonian gravity (weak gravity at speeds low in comparison with c) in the context of GR; given (12.1) C is almost flat.

We assume that U increases without limit as a particle in P is approached; that is (12.1) is valid except in a closed neighbourhood that surrounds the particle. We restrict U so that it does not depend upon time because this ensures that the force, proportional to

$$\mathbf{F} \equiv -\nabla U; \quad \nabla \times \mathbf{F} = 0; \quad \text{Identity} \tag{12.2}$$

is *conservative*. It then turns out that U is approximately proportional to the New-tonian potential. It can be shown that, with the Newtonian approximations, the geodesic equations (8.6) reduces to

$$\frac{d^2 x^j}{dt^2} \approx -c^2 \Gamma_{44}^j \approx -c^2 U_{,J}\,; \quad J = 1, 2, 3\,; \quad |U| << 1 \tag{12.3}$$

The approximations, attached to (12.1), are [8]

$$R_{ab} \approx -\delta_{ab} \nabla^2 U\,; \quad G_4^4 \approx -2\nabla^2 U\,;$$
$$K_{ab} \approx g^{ef} R_{ab;ef} \approx \delta_{ab} \nabla^2(\nabla^2 U)\,; \quad |U| << 1 \tag{12.4}$$

So the Einstein/Newton law of gravity (between particles) is

$$R_{ab} = 0 \implies \nabla^2 U = 0\,; \quad |U| << 1 \tag{12.5}$$

The K equation reduces, in the Newtonian approximation, to

$$\nabla^2(\nabla^2 U) = 0\,; \quad |U| << 1 \quad \text{See (12.4)} \tag{12.6}$$

The Spherically Symmetric (SS) solution of (12.5) is

$$U = \frac{c_1}{r} + c_2\,; \quad c_1 = -\frac{m\mathsf{G}}{c^2}\,; \quad |U| << 1 \implies c_2 = 0 \tag{12.7}$$

where c_1, c_2 are arbitrary constants. The assignment of c_1 at (12.7) is Newtonian; it represents either a particle mass m at the origin or an SS distribution of matter, centre origin, of total mass m . By contrast the SS solution of (12.6) is

$$U = \frac{k_1}{r} + k_2 r^2 + k_3 r + k_4\,; \quad |U| << 1 \implies k_4 = 0 \tag{12.8}$$

where k_1, k_2, k_3, k_4 are constants. We set

$$k_1 = c_1 = -\frac{m\mathsf{G}}{c^2}\,; \quad k_4 = c_2 = 0 \tag{12.9}$$

to bring the two solutions together as far as possible.

But this still leaves *extra terms $k_2 r^2 + k_3 r$* at (12.8). To have escaped experiment these extra terms must be either very small (in the solar system) or zero. We assume, in what follows, that the extra terms are *non-zero*; but that they are negligible except at *cosmological* distances r . The extra terms may provide a crude explanation of at least one of two recently observed and mysterious phenomena [9] [11].

At great distances r the $k_2 r^2$ term dominates the $\dfrac{k_1}{r}$ and $k_3 r$ terms; in the Newtonian approximation, the respective radial accelerations are likewise

$$-c^2 \frac{\partial U}{\partial r} = -c^2 \left(-\frac{k_1}{r^2} + 2k_2 r + k_3 \right) \approx -2k_2 r c^2\,; \quad r \to \infty \quad \text{See (12.3)} \tag{12.10}$$

If the radial acceleration is positive, as observed [9], then $k_2 < 0$. This means that, according to the Newtonian approximations, any two infinitesimal test particles, at sufficient distance from the origin, will experience an acceleration away from each other in proportion to their distance apart. This relates to, and is almost an explanation of, the Dark Energy phenomena [9].

As far as the Voyager Anomaly [11] is concerned: The sun's $-c^2 k_{s3}$ term will be dominant over the $-2k_{s2}rc^2$ term, because the radius is so much less. Nevertheless it can be argued that the constant acceleration will be cancelled out by nearby stars. So, according to Constraints Theory, the Voyager Anomaly is *not* purely gravitational.

There is an argument which suggests that, in a *flat space,*

$$k_2 \propto m, \quad \text{and} \quad k_3 \propto m, \quad \text{See (12.9)} \tag{12.11}$$

So that, given (12.1), then (12.11) is, at the least, approximate. A more dubious argument, based on Newtonian mechanics, shows that for a *flat, single particle* model universe

$$k_{u1} = -5.9 \times 10^{24} m \,; \quad k_{u2} = -4 \times 10^{-53} m^{-2} \,; \quad k_{u3} = -5.9 \times 10^{-28} m^{-1} \,;$$
$$m_u = 7.9 \times 10^{51} kg \,; \tag{12.12}$$

where the mass has been chosen, arbitrarily, as twice the minimum allowed by the theory. A Newtonian theory of a *flat, fluid* model universe suggests that

$$k_{u1} = -2.5 \times 10^{25} m \,; \quad k_{u2} = -1.4 \times 10^{-52} m^{-2} \,; \quad k_{u3} = -4.9 \times 10^{-27} m^{-1} \,;$$
$$m_u = 3.4 \times 10^{52} kg \,; \tag{12.13}$$

The mass, in this case, is much higher because the density of the fluid has been given a high (closure) value. Note that both these theories, however dubious, give $k_{u2} < 0$ as is required.

13. The KE

The KE is still the subject of research; we can only give an example, a flavour, here. To compare the KE with observation we must have an *exact* solution with non-zero pressure and density of matter; this evades us! We have exact solutions without matter; and we have approximate solutions with matter.

The tensor K_b^a is so complicated that, for any serious manipulation, we are obliged to use a machine. The software we use is MAPLE 16; this is able to do many types of symbolic and numerical calculations.

Consider the general SS metric (in polars) [4]

$$ds^2 = -e^\lambda dr^2 - r^2 d\vartheta^2 - r^2 \sin^2 \vartheta \, d\varphi^2 + c^2 e^\nu d\tau^2 \,; \quad \tau = q^4/c \text{ is the time coordinate} \tag{13.1}$$

where λ and ν are both real functions of r. It can be proved that a necessary and sufficient condition for this metric to represent an Einstein space is

$$e^{-\lambda} = 1 + ar^2 + b/r \,; \quad \nu = f - \lambda = f + \ln(1 + ar^2 + b/r) \tag{13.2}$$

where a, b and f are arbitrary constants. An Einstein space is one that satisfies

$$R_b^a = \xi \delta_b^a \,; \quad \xi \text{ is a constant invariant; see (11.3)} \tag{13.3}$$

With (13.1, 13.2) the machine shows that R_v^u, G_v^u and K_v^u are all scalar 4×4 matrices with values $-3a$, $3a$ and $6a^2$ respectively. This is a 'proof', by calculation, that the metric (13.1, 13.2) is indeed an Einstein space. Further, if a is *small*, the result shows that the K equation is also satisfied *approximately* everywhere but the origin. If, we insist that $\Lambda = 0$ then, the Einstein equation reads

$$T_b^a = -\frac{3a\delta_b^a}{\chi} \; ; \quad T_b^a \text{ is a 4×4 scalar matrix} \tag{13.4}$$

So, given (13.1, 13.2), it seems that the K equation is satisfied *approximately* providing that the pressure and density are constant and low; if these quantities are to be positive then $a < 0$.

To explain the far field acceleration [9] we have to consider the motion of an infinitesimal test particle. Of the geodesic equations, given the metric (13.1, 13.2), the one to do with the polar angle ϑ shows that the motion is planar. The geodesic equation to do with the azimuth φ shows that we may choose straight line motion through the origin. The geodesic equation to do with the time is simplified if we set

$$b = 0 \tag{13.5}$$

so that there are no singularities in the gravity field or in the metric. The remaining geodesic equation, to do with the radius, is (previous results taken account of)

$$r'' = -\frac{ar(e^f - r'^2)}{1 + ar^2} \; ; \quad (\,.\,)' = \frac{\partial(\,.\,)}{\partial s} \; ; \quad ds = cd\tau \tag{13.6}$$

Therefore if the radial acceleration $c^2 r''$ is to be proportional to the distance r

$$c^2 r'' \propto r \implies \frac{e^f - r'^2}{1 + ar^2} \equiv \mu \quad \text{is constant} \tag{13.7}$$

Detailed analysis shows that it is desirable to choose

$$f = 0 \tag{13.8}$$

Further, the radius must not be too big ($1 \gg ar^2$); and, consistent with this condition, the radial speed cr' must also not be too big ($1 \gg r'^2$).

So the above argument does not take us much further than the SS Newtonian argument since that is contingent on the condition $|U| \ll 1$; see (12.8). However the constant acceleration has disappeared!

If the Hubble law holds out to the ranges we are concerned with then

$$\dot{r} = cr' = \mathsf{H}r \tag{13.9}$$

where $\mathsf{H} = 2.055 \times 10^{-18} \, s^{-1}$ is Hubble's constant. So that if we choose

$$a = -\left(\frac{\mathsf{H}}{c}\right)^2 \tag{13.10}$$

then $\mu = 1$ is truly constant.

Acknowledgements

I joined ANPA more than thirty years ago. I was proposed by Ted Bastin and interviewed by the then president H. Pierre Noyes. I made many friends and was able to make contributions to the proceedings. My contributions were, for the most part, naïve; but, nevertheless, I was stimulated to develop Constraints Theory along with Louis H. Kauffman and Clive W. Kilmister.

References

[1] C. W. Kilmister [Ed.], *Schrödinger: Centenary Celebration of a Polymath*, C.U.P. (1987)

[2] G. Temple. *The General Principles Of Quantum Theory*, Methuen (1945)

[3] J. Binney & D. Skinner. *The Physics Of Quantum Mechanics*, Cappella Archive, (2010)

[4] B. Spain. *Tensor Calculus*, Dover, (2003)

[5] J.L. Synge & A. Schild, *Tensor Calculus*, Dover, (1978)

[6] F. J. Dyson. 'Feynman's proof of the Maxwell Equations', *Am. J. Phys.*, pp 209–211 58 (3), March (1990)

[7] Louis H. Kauffman & H. Pierre Noyes, 'Discrete Physics and the Derivation of Electromagnetism from the formalism of Quantum Mechanics', *Proc.Roy.Soc.Lond. A*, **452**, pp. 81–95, (1996)

[8] A. S. Eddington, *The Mathematical Theory Of Relativity*, Chelsea, (1965)

[9] A. G. Riess, *et al.* 'Observational Evidence from Supernovae for an Accelerating Universe and a Cosmological Constant', *Astron.J.* 116 1009–1038, (1998)

[10] M. M. Nieto & S. G. Turyshev, 'Finding the Origin of the Pioneer Anomaly', *Classical and Quantum Gravity*, 21 (17) 4005–4024, (2004)

[11] Communicated by letter to AMD.

An Elegance First Approach
to Looking for the Universe in Finite Geometry
The Combinatorial Hierarchy Revisited

Thanks again Pierre! Happy 90 th !

Herb Doughty

herbdoughty@gmail.com

Abstract It is easy to imagine that in the near future someone will find a theory free from apparent contradiction which will account for all known physical phenomena; but which will soon be replaced by a succession of simpler, and hence more likely, alternatives. Inspired by Feynman, I propose a method for attempting to locate the simplest first:

1. Begin with the very simplest structure;
2. Find the most conspicuous way in which it is different from the universe;
3. In the most natural way, fix that problem, obtaining next candidate;
4. Find the most conspicuous way in which it is different from the universe;
5. Iterate the process.

We begin by examining a few mathematical seven day creation stories, each beginning with a day of rest. In the first story, *each succeeding day*, the Powerset operator is applied to the previous days results. Each *succeeding story* uses a more enriched version of the day to day operator. Our main focus is on the later stories which take us up the tree of double fields of Catalan-Mersenne irreducible trinomials, to the 39 digit Mersenne prime $L = 2^{127} - 1$.

We reach a tensor product of Cartesian products of disjoint unions of double fields each with $L+1$ elements, providing steps toward an elegant *rooted* pedigree for non-commutative, discrete, complex, projective, 3-spaces with coordinates in 2×2 matrices over the field $\mathbf{F}_{L^2}$ or over the ring $\mathbf{Z}_{L^2}$, which we aspire to soon explore with computer graphics.

This should shed light on finite M-theory, Penrose twistors, and Loop quantum cosmology, or so I dream.

Mathematical Structures

Mathematical structures are really not created or destroyed. Fancier structures have roots in much simpler structures that are more fundamental but not earlier in physical time. They are always there, some it seems to me, in a more robust sense than others.

This structural ancestry secondarily provides a means by which they may come to our attention, or the attention of any other minds. We speak of their structural ancestry as their creation, and discover them by finding such ways to point to them. Sometimes years later, by noticing an inconsistency, we realize that one of our attempts to point at a structure failed. Loosely, we say the structures is destroyed. It only seemed to exist. By a *rooted pedigree* of a structure, I mean a sequence of tame operators, deriving its structure from nothing.

A type of structure that to me seems particularly at risk, is bases for the vector space of real numbers over the field of rational numbers. There are respected proofs that these bases exist, and also proofs that none of them can be individually pointed at in any way.

Somewhat safer, I feel, are countably infinite structures, and safer still are the finite well founded lists we will deal with here, candidate structures, which moreover, don't even contain more than 2^{65536} items. In respect to these, I am totally a realist.

Sensory experience and Mathematical Structure

The feeling that the things we see, hear, feel, taste, and smell, are more real than things that are not part of the geometry we live in, is indeed about a very important distinction, but it is not about those objects having a different kind of existence, it is about them having a different relationship to us.

Surely observers in other geometries experience a similar distinction between the objects of their experience, and object outside the geometry they live in. Can objects of one's experience ever include things outside the geometry one lives in?

I think they can!

Graphics feedback

Through the wonders of modern computer graphics, I expect to soon be seeing high resolution views within various candidate geometries using software being developed using the powerful free math package *Sage*, to calculate high resolution images, spectra, and polarizations in views within a variety of geometries, including non-commutative discrete complex projective 3-spaces with coordinates in 2×2 matrices over fields including $\mathbf{F}_{L^2}$ and rings including $\mathbf{Z}_{L^2}$.

This feedback should provide much better insight to guide our quest. Maybe *you* will find better ways to tweek the operators, and reveal which geometry we live in.

First Story — Sets

On Sunday, nothing is created. This is not our universe, *there is nothing here!* The one Monday item is the empty set, $\varnothing$. *Something from nothing!* The two Tuesday items are $\{\varnothing\}$, and $\varnothing$. Two is not enough. The Wednesday items are $\{\{\varnothing\}\varnothing\}$, $\{\{\varnothing\}\}$, $\{\varnothing\}$, and $\varnothing$. Again not enough! The Thursday items are the 16 sets of Wednesday items. Still not enough. The Friday items are the 65536 sets of items created on Thursday, but 65536 items is still clearly not enough for this to be our Universe. The items created on Saturday are the 2^{65536} sets of Friday items. It is by no means clear that there are still too few items to make up our universe, the digits of their cardinality take up more than four pages! However, *this is not our universe*, it lacks the algebraic structure and implicit geometry that would provide a basis for Physics.

Structure — Finite Well formed Sets, Lists, Alphabets, Words and Vectors

Here, a **set** is a *finite* family of distinct, structurally previous items, that is of items each definable without reference to the set. The conventional language used to define well formed ness speaks of previously defined items, which taken literally, muddles the existence of these always present items with the human activity of discovering them.

A set with an order defined upon it is called an *alphabet*. Alphabets are lists without duplicates. The elements of an alphabet are called, *letters*. A list of letters from any alphabet with or without duplicates is called a *word* on that alphabet. If under some addition and multiplication, the letters of an alphabet form a field, $\mathbf{F}$, then the words of length n on that alphabet inherit place by place addition and a scalar multiplication, thereby forming an n–dimensional vector space, $\mathbf{V}^n\mathbf{F}$, over $\mathbf{F}$.

The space of all mappings from one structure to another

In Algebra in general, a very important way that a pair of simple structures give rise to a richer structure is by forming the alphabet of all mappings from one into the other which preserve all the relations and operations on some (possibly empty) list. Since spaces of this type provide the main links of the structural ancestries, they are significant here in two ways. First, they are how simple structures give rise to fancier structures. Second, they are an essential part of how we find out about them.

When we have discovered an elegant structural pedigree for a particular candidate structure, which might be the one we live in, we may also discover related, often vastly larger spaces of mappings which help us understand it, without being part of it. For example the space of mappings from the plane within the candidate structure, into itself.

These auxilliary structures have their own longer structural pedigrees, and like countless other structures, exist in the same way as the candidate structure.

Let's associate them with an eighth day of of creation ;–)

Cataloging all algebraic structures on a very short alphabet

Notice that since any binary operation on an alphabet, $\mathbf{A}$, of n letters is a mapping from $\mathbf{A} \times \mathbf{A} \to \mathbf{A}$, so there are inherently exactly $n^{n \times n}$ of them. *Usually* a given algebraic structure will involve only a very short list of these.

On an alphabet, $\mathbf{A} = [a]$, of one letter, a, the only binary operation is $[[(a, a) \mapsto a]]$ and so $[[a][[(a, a) \mapsto a]]]$ is the only structure that a group with just one element can have. If two people each bring you a group with just one element, then the mapping that maps the one element of the first to the one element of the second, preserves the operation and is therefore a group isomorphism. We say that the one element group, $\mathbf{G}_1$, is unique up to group isomorphism.

On an alphabet, $\mathbf{AB} = [a\, b]$, of two letters there are *inherently exactly* $2^{2 \times 2} = 16$ binary operations. Fifty years ago I encountered common names for each of them in introductory classes in logic and set theory. Although I have not taken such courses, I would expect that by now, all 16 *may* have common names that you might encounter in in a class in data structures and algorithms.

But aside from their common names, each of the 16 binary operations on $\mathbf{AB}$ has an elegant name, by which I suppose it is called in *The Book* that Paul Erdős spoke of. Each of these 16 elegant names is a 4 letter word on the alphabet $\mathbf{AB}$, with the four letters telling the output of applying that operation to (b, b), (b, a), (a, b), and (a, a).

We may define a BiOpStructure on $\mathbf{AB}$ to be an ordered pair consisting of $\mathbf{AB}$ together with one of the 65536 subsets of the set of 16 binary operations on $\mathbf{AB}$. We should consider the roles that each may have in the elegant pedigrees of structures beyond. Since I encountered names for all 16 of them 50 years ago, and since deleting operations from the list does not look promising as a way of adding richness, my guess would be that we should seriously examine the BiOpStructure ($\mathbf{AB}$, 1111111111111111) with all 16 binary operations on $\mathbf{AB}$ as the one most likely to be prominent in elegant pedigrees of structures beyond. However, in evaluating elegance, should we charge less for BiOpStructures with fewer operations? What pricing algorithm looks appropriate?

Just as we counted and named the binary operations on $\mathbf{AB}$, you can also count and name the relations on $\mathbf{AB}$, I'll let you do it. (A relation on $\mathbf{AB}$, is a subset of $\mathbf{AB} \times \mathbf{AB}$, the set of ordered pairs of elements of $\mathbf{AB}$). We should explore structures like RelBiOpSt, with binary operations and relations. Can you count and name them? So besides binary operations and relations, what do you recommend? Can you count and name them?

By examining the structures on alphabets of 2 letters, we have shed light on what is to be done for longer alphabets. Seeing what is to be done, we can program it. Through the wonders of modern information technology, we may explore all the structures on alphabets of 3 and even 4 letters, and it seems to me that doing so will really tell us how they best combine to give geometries rich enough to have inhabitants who can appreciate them.

From sets via bitstrings to vector spaces over $\mathbf{F}_2$

Attempting to remedy the problem with the first story, we slightly enrich the Powerset operator, assigning bitstring names to the items as we generate them. To facilitate this naming, we assign an order, reverse lexicographic or revlex for short, to the items created each day. A good way to think of this enriched day to day operator is as a PowerAlphabet mapping the Alphabet of items created yesterday into the revlex ordered Alphabet of mappings of from that Alphabet into $[0\,1]$. No duplicates occur.

If the number of items that were created one day is n, then the 2^n items created the next day will each be named by a bit string of length n, which we may think of as an attendance chart in which the kth bit tells whether the kth of the previous day's n items is present or absent in the new item being named. More precisely, let us define the kth bit of the name of the new item to represent the parity of the presence of the kth of the previous days items, in the new item being named. In this way we also see each of these bits as odd or even, rather than just present or absent.

$$
\begin{array}{c|cc}
+ & 0 & 1 \\
\hline
0 & 0 & 1 \\
1 & 1 & 0
\end{array}
\qquad
\begin{array}{c|cc}
\times & 0 & 1 \\
\hline
0 & 0 & 0 \\
1 & 0 & 1
\end{array}
$$

$$+ \;:\quad [[(1,1) \mapsto 0][(1,0) \mapsto 1][(0,1) \mapsto 1][(0,0) \mapsto 0]]$$
$$\times \;:\quad [[(1,1) \mapsto 1][(1,0) \mapsto 0][(0,1) \mapsto 0][(0,0) \mapsto 0]].$$

So defined, odd and even, usually written 1 and 0, together with their familiar addition and multiplication, provide an elegant pedigree for the smallest field, $\mathbf{F}_2$.

We then see our bitstrings as vectors over $\mathbf{F}_2$. And our new day–to–day operator as mapping lists, (vectors over $\mathbf{F}_2$, if nonempty), into longer vectors also over $\mathbf{F}_2$.

Second Story — Vector spaces over $\mathbf{F}_2$

On Sunday, the day of rest, no items are created. The alphabet of Sunday items, is the only item created Monday. It is the bitstring with one bit for each of the 0 items created Sunday, like any empty list, is also usually denoted Λ. The one letter alphabet, $[\Lambda]$, of Monday items, together with its *only* binary operation, $[[(\Lambda, \Lambda) \mapsto \Lambda]]$, is the simplest instance of the one element group structure $\mathbf{G}_1$. The Tuesday items have names, 1 and 0, defined to be the parity of the presence of Λ in the item named. The alphabet they form, $[1\,0]$, together with the operations defined above, form $\mathbf{F}_2$ itself, and also the vector space, $\mathbf{V}^1\mathbf{F}_2$, of dimension 1 over it. While retaining their auxiliary role as an attendance chart, the Wednesday items 11, 10, 01, and 00, with their inherited operations, form a 2-D vector space, $\mathbf{V}^2\mathbf{F}_2$. Thursday, 1111, 1110, 1101, 1100, 1011, 1010, 1001, 1000, 0111, 0110, 0101, 0100, 0011, 0010, 0001, and 0000, together with their inherited operations form $\mathbf{V}^4\mathbf{F}_2$. Friday, 65536 of length 16, with their inherited operations, form $\mathbf{V}^{16}\mathbf{F}_2$. Saturday's 2^{65536} vectors of length 65536, with their inherited operations, $\mathbf{V}^{65536}\mathbf{F}_2$.

From vectors to finite fields

Over a field $\mathbf{F}$, there is an obvious bijection between vectors of length at most n and polynomials of degree less than n in one variable,

$$a_{n-1} \cdots a_2\, a_1\, a_0 \;\longleftrightarrow\; \sum_{k=0}^{n-1} a_k\, x^k$$

which preserves vector addition and scalar multiplication, so the polynomials of degree less than n over $\mathbf{F}$ form a vector space of dimension n over $\mathbf{F}$. Polynomials can be multiplied, but under this multiplications, the polynomials of degree $< n$ are not closed.

This can be remedied by replacing polynomials of degree at less than n, with the residue classes of polynomials modulo an irreducible polynomial, φ, of degree n, producing a field $\mathbf{F}\varphi$ of residue classes of polynomials over $\mathbf{F}$, modulo φ. Fortunately, for every finite field, $\mathbf{F}$, and every $n > 1$, there is at least one irreducible polynomial of degree n over $\mathbf{F}$.

Actually, for every prime p and positive integer, n, there is *exactly one* field with p^n elements, that is, it is unique up to field isomorphism, meaning that if two people bring you fields with p^n elements that they have obtained in any way, such as perhaps by extended $\mathbf{F}_p$ using different combinations of irreducible polynomials with degrees summing to n, then there is a way for you to find a field isomorphism, that is to say a bijection preserving both field operations, between their fields with p^n elements.

Classifying finite fields by characteristic modulo 4

A subset of a field $\mathbf{F}$ which is a field *under the same operations as* $\mathbf{F}$, is called a subfield of $\mathbf{F}$. A field is always a vector space over any of its subfields, and hence, the number of elements in the field, if finite, is always a power of the number of elements in any of its subfields.

Given any field $\mathbf{F}$, there is always a unique smallest subfield of $\mathbf{F}$, which is either the infinite field of rational numbers $\mathbb{Q}$, or for some prime p, it is $\mathbf{F}_p$. In the latter case, the prime is called the *characteristic* of $\mathbf{F}$. If the smallest subfield is $\mathbb{Q}$, as is the case for the famous infinite fields $\mathbb{R}$, of real numbers, and $\mathbb{C}$, of complex numbers, then the characteristic of $\mathbf{F}$ is 0. All finite fields and many infinite fields are of prime characteristic, their cardinality, if finite, is always a power of their characteristic.

Later in a section on computer graphics, a local distance function will be defined that lets us speak of nearest neighbors on a line, to a point on that line. Let's look mod 4 at the characteristic Char of a field $\mathbf{F}$, and consider the quadratic extensions of $\mathbf{F}$, and nearest neighbors to points on a line over $\mathbf{F}$.

If $\mathrm{Char}(\mathbf{F}) \equiv 0 \mod 4$, then $\mathbf{F}$ contains $\mathbb{Q}$, and is infinite, and not relevant here.

If $\mathrm{Char}(\mathbf{F}) \equiv 1 \mod 4$, and $\mathbf{F}$ is finite, its cardinality is a power of a prime 1 more than a multiple of four, like the Fermat primes. In this case, $\mathbf{F}$ already contains a square root of -1, so $\mathbf{F}$ cannot be extended by adjoining i. However, it doesn't yet contain a square root of two. It can be extended by adjoining $\sqrt{2}$, a root of the quadratic polynomial $x^2 - 2$. The plane over such a field always contains a self-perpendicular line whose points have 4 nearest neighbors on it. In both ways $\mathbf{F}$ is very unlike $\mathbb{C}$.

If $\mathrm{Char}(\mathbf{F}) \equiv 2 \mod 4$, and $\mathbf{F}$ is finite, its cardinality is a power of 2. These include our bit string fields. But in a geometry over one of these, taking two steps in the same direction returns you to where you were, and no two lines are perpendicular. The only quadratic polynomial by which we may extend $\mathbf{F}_2$ is $x^2 + x + 1$ which we will call q. This extension adjoins a root of q, which is ω, a complex cube root of one. In several ways $\mathbf{F}$ is very unlike $\mathbb{C}$.

If $\mathrm{Char}(\mathbf{F}) = 3 \mod 4$, and $\mathbf{F}$ is finite, its cardinality is a power of a prime 1 less than a multiple of four, like the Mersenne primes. In this case, $\mathbf{F}$ already contains a square root of 2, so $\mathbf{F}$ can't be extended by adjoining one. However, it doesn't yet contain a square root of -1, so it can be extended by adjoining i, a root of the quadratic $x^2 + 1$. Over such a field, within any line, each of its points has just 2 nearest neighbors. In both ways $\mathbf{F}$ is very much like $\mathbb{C}$. Large fields of this last type are the only finite fields that an observer could easily mistake for the infinite field $\mathbb{C}$.

Our Numbers

Known to ANPA members as the Combinatorial Hierarchy,
the Catalan-Mersenne numbers are defined by:
for $n \in \mathbb{N}$, let Mersenne of n, $M_n = 2^n - 1$; and let $C_0 = 2$, $C_{n+1} = M_{C_n}$.
Then $C_0 = 2$, $C_1 = 3$, $C_2 = 7$, $C_3 = 127$, and
$C_4 = 170141183460469231731687303715884105727$ are each prime.
If for any n, C_n is not prime, then for $k > n$, C_k is also not prime.

There is another important sequence of primes closely related to ours.

Let $B_k = (C_k + 2)/3$. For $k = 2, 3$, and 4, B_k is also prime! Also

$$B_k = (2^{C_k - 1} + 1)/(2 + 1) \qquad \text{and} \quad C_k = (2^{C_k - 1} - 1)/(2 - 1);$$
$$3 \times B_k \times C_k = (2^{C_k - 1} - 1) \times (2^{C_k - 1} + 1) = 2^{C_k - 1 \times 2} - 1.$$

Have fun relating these to products of Fermat numbers, and to products of Mersenne numbers of Fermat numbers!

My conjecture

In 1986 I gave a short talk *Finite Double Fields* at The International Congress of Mathematicians in Berkeley, concluding with the conjecture that if p is any prime, then over $\mathbf{F}_2$ the trinomial $x^p + x + 1$ is irreducible if and only if $p = C_k$ for some k.

It is a theorem that if a prime $p = C_k$, then even if its beyond L, the trinomial is irreducible. At least up to 30000, if p is a non-hierarchy prime, the trinomial can be factored.

The number of directions, Professor DeVogelaere, and Trig functions

In a plane in any vector space over field $\mathbf{F}_m$, with m elements, the number of directions is twice the number of points in the projective line $\mathrm{P}^1\mathbf{F}_m$ over $\mathbf{F}_m$. $\mathrm{P}^1\mathbf{F}_m$ has $m{+}1$ points, so the number of directions in the plane over $\mathbf{F}_m$ is $2 \times (m + 1)$. Interestingly, if m is a Mersenne prime $2^n - 1$, then the number of directions in the plane is 2^{n+1}, and moreover, if for some $k > 1$, $m = C_k$, then the number of directions is $2^{2^{C_{k-2}}} = 12{\times}B_k{\times}C_k + 4$.

Associated to each of these directions (except up and down, because we can't divide by zero) is a slope, which is a ratio of two integers mod C_k.

When I arrived in Berkeley in 1967, Rene DeVogelaere, a student of Georges Lemaître at Louvain, was a Berkeley Math Professor widely known for his work on differential equations, but mainly interested in finite geometry as a basis for Cosmology. When he died, Rene was writing a large book on Euclidean and non-Euclidean Finite Spaces. I sure hope that someone is finishing his book! I remember him telling me shortly before he died about functions related to any finite field $\mathbf{F}_q$, which like trig functions, were their own 4th derivatives. Their periods were always either $q{+}1$, or $q{-}1$, so I expect that they were defined on either the projective line $\mathrm{P}^1\mathbf{F}_q$, or on the multiplicative group $\mathbf{F}_q^\times$. I wonder whether our slopes are ratios of two of them.

Generators

If there is any element g in a multiplicative group $\mathbf{G}$, such that every element of $\mathbf{G}$ is a power of g, then $\mathbf{G}$ is cyclic, and g is a generator. For any other element h of $\mathbf{G}$ there is a k such that, $h = g^k$. Groups of prime order, and the multiplicative groups of finite fields, are always cyclic, and hence commutative. In the groups of prime order each element is either the identity, or a generator. The additive group of a field, $\mathbf{F}_{p^k}$, is not cyclic if $k > 1$.

In many contexts it is necessary to designate a particular generator, in some by a criterion depending upon the context. For our hierarchy prime case, for $k > 1$, in the field $\mathbf{F}_{C_k}$, the element   mod (C_{k-1}, C_k) is a generator for both the cyclic groups $\mathbf{F}_{C_k}^+$ of order C_k, and $\mathbf{F}_{C_k}^\times$ of order C_{k-1}.   For $\mathbf{F}_{L^2}$ I use $g = 1 + 8{\times}I$.

Our Polynomials

Because of a bijection between double fields with 2^d elements and irreducible polynomials of degree d over $\mathbf{F}_2$, so we should expect relationships between double fields to reflect relationships between irreducible polynomials over $\mathbf{F}_2$. Of the mappings between polynomials over $\mathbf{F}_2$ which preserve irreducibility. Two are well known, and also preserve degree, they are:

1. Flip the order of the coefficients, to get the reciprocal roots polynomial.
2. Teichmuller : Substitute $x + 1$ for x (not used here).

More significant here, the third mapping, and its composition with flip, each preserve the number of terms, but not the degree. Over $\mathbf{F}_2$:

$$\text{If } A = \sum_{k=0}^{d} a_k\, x^k, \quad \text{let } A_0 = \sum_{k=0}^{d} a_k\, x^{M_k},$$

$$\text{and flipping it,} \quad \text{let } A_1 = \sum_{k=0}^{d} a_{d-k}\, x^{M_k}\,.$$

It is a theorem that if A is irreducible and M_d is prime, then A_0 and A_1 are irreducible. So, over $\mathbf{F}_2$, from q, the only irreducible quadratic, we get the Catalan-Mersenne irreducible trinomials :

$$q = x^2+x+1$$

$q_{00} = x^7+x+1$	$q_{000} = x^{127}+x+1$	$q_{100} = x^{127}+x^7+1$
$q_{01} = x^7+x^6+1$	$q_{001} = x^{127}+x^{126}+1$	$q_{101} = x^{127}+x^{120}+1$
$q_{10} = x^7+x^3+1$	$q_{010} = x^{127}+x^{63}+1$	$q_{110} = x^{127}+x^{15}+1$
$q_{11} = x^7+x^4+1$	$q_{011} = x^{127}+x^{64}+1$	$q_{111} = x^{127}+x^{112}+1$

with also $q_0 = x^3+x+1$ and $q_1 = x^3+x^2+1$.

With no pedigree that I yet know of, there are, over $\mathbf{F}_2$, exactly two other degree 127 irreducile trinomials, $x^{127} + x^{30} + 1$ and its flip, $x^{127} + x^{97} + 1$. For some kinds of string theory, ten dimensions are needed, so these might be useful.

L is prime, so we *can* apply operator *Awe* defined below to get 20 more irreducible trinomials each of degree L.   Then :

$$\text{if} \quad C_5 = M_L \sim 5.4543129 \times 10^{51217599719369681875006054625051616349} \quad \textit{is prime,}$$

each of those trinomials is associated to a double field with 2^L elements, and C_5 can be the characteristic of the coordinate field for the geometry of some universe. *If so*, I don't think it will be ours, because in such a universe the ratio of large and small distances would be approximately C_5, *way too big* to be ours.

The Tree of Catalan–Mersenne Trinomials

Modulo irreducible trinomials of degree d there are 2^d residue classes, so the fields of their double field have orders 2^d and 2^d-1.

d	2^d	2^d-1		
2	4	3		q
3	8	7		$q_0 \quad\quad q_1$
7	128	127		$q_{00} \quad q_{01} \quad q_{10} \quad q_{11}$
127	2^{127}	$2^{127}-1$		$q_{000}\ q_{001}\ q_{010}\ q_{011}\ q_{100}\ q_{!01}\ q_{110}\ q_{111}$

The branches are labelled 0 (left) and 1 (right): q branches to q_0 and q_1; q_0 to q_{00}, q_{01} and q_1 to q_{10}, q_{11}; and each q_{ij} to q_{ij0}, q_{ij1}.

Awe

We define for the next two stories the day-to-day mapping *Awe*.

I chose the name because this mapping which I found around 1999 to me seemed *Awe*some, or at least *Awe*spicious. This name will also provide for the benefit of those holding other opinions, either now or after more data has come to light, the opportunity to claim '*Awe* spicious', or to say that it is '*Awe* full'.

Intuitively the intention is that each day's structure is a list of mapping from yesterday's structure into $\mathbf{F}_2$, together with a way to inherit the appropriate operations, in this way preserving the cardinalities from our first story with Powerset. *Via* hierarchy irreducible trinomials, we first climb their tree of algebraic extensions of $\mathbf{F}_2$; and then similarly we will climb their tree of associated double fields.

Let $\mathbf{F}_2 q_\alpha$ is the field of residue classes mod $q_\alpha = x^d + x^r + 1$;
let $\mathbf{D}_\alpha$ be the double field combining $\mathbf{F}_2 q_\alpha$ with the prime field $\mathbf{F}_{M_d}$.

Let $\mathbf{F}_2 q_{\alpha_0}$ be the field of residue classes mod $q_{\alpha_0} = x^{M_d} + x^{M_r} + 1$;
let $\mathbf{D}_{\alpha_0}$ be the double field combining $\mathbf{F}_2 q_{\alpha_0}$ with prime field $\mathbf{F}_{MM_d}$.

Let $\mathbf{F}_2 q_{\alpha_1}$ be the field of residue classes mod $q_{\alpha_1} = x^{M_d} + x^{M_d - M_r} + 1$;
let $\mathbf{D}_{\alpha_1}$ be the double field combining $\mathbf{F}_2 q_{\alpha_1}$ with the prime field $\mathbf{F}_{MM_d}$.

Let $\mathbf{F}_2 q_\alpha \, \textit{Awe} = \mathbf{F}_2 q_{\alpha_0} {<}0{>} \mathbf{F}_2 q_{\alpha_1}$, and let $\mathbf{D}_\alpha \, \textit{Awe} = \mathbf{D}_{\alpha_0} {<}0{>} \mathbf{D}_{\alpha_1}$.

We then define *Awe* for any two same type structures, α and β, built of hierarchy fields or double fields by $(\alpha {<}n{>} \beta) \textit{Awe} = \alpha \textit{Awe} {<}n{+}1{>} \beta \textit{Awe}$.

In traditional contexts

> ${<}0{>}$ is a disjoint union, sometimes denoted $\sqcup$,
>
> ${<}1{>}$ is a Cartesian product denoted by $\times$, and
>
> ${<}2{>}$ is a tensor product denoted by $\otimes$.

A Cartesian product of disjoint unions of items of the same type gives rise to a $2{\times}2$ matrix of Cartesian products of such items.

Third Story — The tree of even fields
modulo Catalan-Mersenne trinomials

Sunday	0	Λ
		$\Downarrow$
Monday	1	$\mathbf{G}_1$
		$\Downarrow$
Tuesday	2	$\mathbf{F}_2$
		$\Downarrow$
Wednesday	4	$\mathbf{F}_2 q$
		$\Downarrow Awe$
Thursday	16	$\mathbf{F}_2 q_0 {<} 0 {>} \mathbf{F}_2 q_1$
		$\Downarrow Awe$
Friday	65536	$(\mathbf{F}_2 q_{00} {<} 0 {>} \mathbf{F}_2 q_{01}) {<} 1 {>} (\mathbf{F}_2 q_{10} {<} 0 {>} \mathbf{F}_2 q_{11})$
		$\Downarrow Awe$
Saturday	2^{65536}	$(\mathbf{F}_2 q_{000} {<} 0 {>} \mathbf{F}_2 q_{001}) {<} 1 {>} (\mathbf{F}_2 q_{010} {<} 0 {>} \mathbf{F}_2 q_{011})$
		${<} 2 {>}$
		$(\mathbf{F}_2 q_{100} {<} 0 {>} \mathbf{F}_2 q_{101}) {<} 1 {>} (\mathbf{F}_2 q_{110} {<} 0 {>} \mathbf{F}_2 q_{111})$

Double fields

Around 1960, I had conjectured that physical spacetime was some particular instance of a discrete mathematical structure analogous to a manifold, which was embedded in a finite dimensional vector space over the field of integers modulo some large prime, but that puzzlingly the information at each point in it was Boolean, and involved a field of two power order. In February of 1968, I thought of *double fields*.

Wedderburn's theorem tells us that for *any* finite field $\mathbf{F}_q$, with q elements, its multiplicative group $\mathbf{F}_{q^\times}$, is the *cyclic group* with $q-1$ elements, hence it is isomorphic to the additive group of the ring $\mathbf{Z}_{q-1}$ of integers modulo $q-1$. If $q-1$ is prime, then $\mathbf{Z}_{q-1}$ is a field $\mathbf{F}_{q-1}$; and together, $\mathbf{F}_q$ and $\mathbf{F}_{q-1}$ form a double field $\mathbf{D}$ with q elements.

A double field, $\mathbf{D}$, is a set together with three commutative associative binary operations, Δ, $+$, and $\times$, with identity elements ∞, 0, and 1, respectively, such that $\mathbf{D}$ under Δ and $+$, and $\mathbf{D}$ without ∞ under $+$, and $\times$, are both fields. It inherits distributive laws and inverses from its two fields. ∞ is an annihlator for $\times$. There are infinite double fields which we will ignore.

I chose to use as names for the second and third operations of a double field, their operation names from its field of large characteristic which is needed for our computer graphics. This way, we are also helped by associations with the operations we were first taught in school. The addition of its characteristic two field is symmetric difference, Δ.

Given a finite double field with q elements, $\mathbf{D}$ under Δ; $\mathbf{D}$ without ∞ under $+$; and $\mathbf{D}$ with neither ∞ nor 0 under $\times$; are Abelian groups with q, $q-1$, and $q-2$

elements, respectively. The middle group with $q-1$ elements is the multiplicative group of the field with q elements, so Wedderburn's Theorem tells us that it is cyclic. It is also the additive group of the field with $q-1$ elements. The additive group of a field can only be cyclic if it is of prime order. And q must be either a prime or a power of a prime.

One of q and $q-1$ must be even. So $q-1$ is either 2 or it is a Mersenne Prime. If $q-1$ is 2, then **D** has 3 elements, and is the only odd double field.

Together the double fields with 4 and three elements form the only triple field, **T**. Maybe someday we will see how the entire triple field is involved in our hierarchy, but for now we will use the double field with 4 elements, and ignore the one with 3.

Very briefly in February of 1968, I mistakenly supposed that as with fields, two double fields with the same number of elements would be isomorphic. But I quickly found that the *field isomorphisms* which preserve two of the three operations of the double field don't preserve the third so they are not *double field isomorphisms*.

In any double field **D** with 2^d elements, each element is a residue classes of polynomials modulo an irreducible binary polynomial φ of degree d. Polynomial addition and multiplication modulo φ are represented by Δ, and $+$, each residue class contains a polynomial of least degree, whose bitstring coefficients, padded if necessary with leading zeros to bring its length up to d, will be referred to as the *polynomial name* of the class. It tells the role of this class in respect to the field $\mathbf{F}_{2^d}$.

The residue class also contains either a monomial of least degree, x^a, with a being called the *numeric name* of the class, or will contain the polynomial zero and no monomials. The *numeric name* of that class is ∞. It will be mapped to the point at ∞ on the projective line $\mathrm{P}^1\mathbf{F}_{\mathbf{M}_d}$. The numeric name tells the role of this class in the field $\mathbf{F}_{\mathbf{M}_d}$.

Since $x^a \times x^b = x^{a+b} \bmod \varphi$, polynomial multiplication $(+)$ of two classes may be accomplished by adding the numeric names of the two classes. The third operation, $\times$, is multiplication of the numeric names of the two classes. The mapping from the field $\mathbf{F}_{2^n}$ to the projective line over the field with $2^n - 1$ elements is $Log_x(\ )$.

Given the tables for all three operations of any double field **D** you can sort the powers of x into equivalent piles and find the least nonzero polynomial in the same class with zero; it is the unique irreducible polynomial φ which divides the difference between any two polynomials in the same class. Hence, in **D** the 2^n element field is $\mathbf{F}_2\varphi$. Hence, the bijection between 2^n element double fields and irreducible polynomials of degree n over $\mathbf{F}_2$, which by the Möbius inversion formula is known to number $(2^n - 2)/n$.

Operation tables for the double field D_1 of the Catalan-Mersenne trinomial $q_1 = x^3 + x^2 + 1$

$\triangle$	∞ 000	0 001	1 010	2 100	3 101	4 111	5 011	6 110
∞ 000	∞ 000	0 001	1 010	2 100	3 101	4 111	5 011	6 110
0 001	0 001	∞ 000	5 011	3 101	2 100	6 110	1 010	4 111
1 010	1 010	5 011	∞ 000	6 110	4 111	3 101	0 001	2 100
2 100	2 100	3 101	6 110	∞ 000	0 001	5 011	4 111	1 010
3 101	3 101	2 100	4 111	0 001	∞ 000	1 010	6 110	5 011
4 111	4 111	6 110	3 101	5 011	1 010	∞ 000	2 100	0 001
5 011	5 011	1 010	0 001	4 111	6 110	2 100	∞ 000	3 101
6 110	6 110	4 111	2 100	1 010	5 011	0 001	3 101	∞ 000

$+$	∞ 000	0 001	1 010	2 100	3 101	4 111	5 011	6 110
∞ 000	∞ 000	∞ 000	∞ 000	∞ 000	∞ 000	∞ 000	∞ 000	∞ 000
0 001	∞ 000	0 001	1 010	2 100	3 101	4 111	5 011	6 110
1 010	∞ 000	1 010	2 100	3 101	4 111	5 011	6 110	0 001
2 100	∞ 000	2 100	3 101	4 111	5 011	6 110	0 001	1 010
3 101	∞ 000	3 101	4 111	5 011	6 110	0 001	1 010	2 100
4 111	∞ 000	4 111	5 011	6 110	0 001	1 010	2 100	3 101
5 011	∞ 000	5 011	6 110	0 001	1 010	2 100	3 101	4 111
6 110	∞ 000	6 110	0 001	1 010	2 100	3 101	4 111	5 011

$\times$	∞ 000	0 001	1 010	2 100	3 101	4 111	5 011	6 110
∞ 000	∞ 000	∞ 000	∞ 000	∞ 000	∞ 000	∞ 000	∞ 000	∞ 000
0 001	∞ 000	0 001	0 001	0 001	0 001	0 001	0 001	0 001
1 010	∞ 000	0 001	1 010	2 100	3 101	4 111	5 011	6 110
2 100	∞ 000	0 001	2 100	4 111	6 110	1 010	3 101	5 011
3 101	∞ 000	0 001	3 101	6 110	2 100	5 011	1 010	4 111
4 111	∞ 000	0 001	4 111	1 010	5 011	2 100	6 110	3 101
5 011	∞ 000	0 001	5 011	3 101	1 010	6 110	4 111	2 100
6 110	∞ 000	0 001	6 110	5 011	4 111	3 101	2 100	1 010

Fourth Story — The Tree of Double Fields
of the Catalan-Mersenne trinomials

Sunday	0	Λ
		$\Downarrow$
Monday	1	$\mathbf{G}_1$
		$\Downarrow$
Tuesday	2	$\mathbf{F}_2$
		$\Downarrow$
Wednesday	4	$\mathbf{D}$
		$\Downarrow Awe$
Thursday	16	$\mathbf{D}_0{<}0{>}\mathbf{D}_1$
		$\Downarrow Awe$
Friday	65536	$(\mathbf{D}_{00}{<}0{>}\mathbf{D}_{01}){<}1{>}(\mathbf{D}_{10}{<}0{>}\mathbf{D}_{11})$
		$\Downarrow Awe$
Saturday	2^{65536}	$(\mathbf{D}_{000}{<}0{>}\mathbf{D}_{001}){<}1{>}(\mathbf{D}_{010}{<}0{>}\mathbf{D}_{011})$

$$<2>$$

$$(\mathbf{D}_{100}{<}0{>}\mathbf{D}_{101}){<}1{>}(\mathbf{D}_{110}{<}0{>}\mathbf{D}_{111})$$

Fifth Story — Tree with projective lines over odd fields
of the Catalan-Mersenne trinomials,
with traditional names for operators

Same as 4th Story through Tuesday.   So we start here with Wednesday

$$\Downarrow$$

Wednesday	4	$\mathrm{P}^1\mathbf{F}_3$
		$\Downarrow$
Thursday	16	$\mathrm{P}^1\mathbf{F}_7{}_0 \uplus \mathrm{P}^1\mathbf{F}_7{}_1$
		$\Downarrow$
Friday	65536	$(\mathrm{P}^1\mathbf{F}_{127}{}_{00} \uplus \mathrm{P}^1\mathbf{F}_{127}{}_{01}) \times (\mathrm{P}^1\mathbf{F}_{127}{}_{10} \uplus \mathrm{P}^1\mathbf{F}_{127}{}_{11})$
		$\Downarrow$
Saturday	2^{65536}	$(\mathrm{P}^1\mathbf{F}_{L}{}_{000} \uplus \mathrm{P}^1\mathbf{F}_{L}{}_{001}) \times (\mathrm{P}^1\mathbf{F}_{L}{}_{010} \uplus \mathrm{P}^1\mathbf{F}_{L}{}_{011})$

$$\otimes$$

$$(\mathrm{P}^1\mathbf{F}_{L}{}_{100} \uplus \mathrm{P}^1\mathbf{F}_{L}{}_{101}) \times (\mathrm{P}^1\mathbf{F}_{L}{}_{110} \uplus \mathrm{P}^1\mathbf{F}_{L}{}_{111})$$

Sixth Story — With matrices of cartesian products of projective lines over the odd fields of the trinomials, with standard names for operators

Same as 5th Story through Wednesday.   So we start here with Thursday

Thursday
$$\underset{0}{\mathrm{P}^1\mathbf{F}_7} \;\bigcup\; \underset{1}{\mathrm{P}^1\mathbf{F}_7}$$

$$\Downarrow$$

Friday
$$\begin{bmatrix} \underset{00}{\mathrm{P}^1\mathbf{F}_{127}} \times \underset{10}{\mathrm{P}^1\mathbf{F}_{127}} & \underset{00}{\mathrm{P}^1\mathbf{F}_{127}} \times \underset{11}{\mathrm{P}^1\mathbf{F}_{127}} \\[2ex] \underset{01}{\mathrm{P}^1\mathbf{F}_{127}} \times \underset{10}{\mathrm{P}^1\mathbf{F}_{127}} & \underset{01}{\mathrm{P}^1\mathbf{F}_{127}} \times \underset{11}{\mathrm{P}^1\mathbf{F}_{127}} \end{bmatrix}$$

$$\Downarrow$$

Saturday
$$\begin{bmatrix} \underset{000}{\mathrm{P}^1\mathbf{F}_{L}} \times \underset{010}{\mathrm{P}^1\mathbf{F}_{L}} & \underset{000}{\mathrm{P}^1\mathbf{F}_{L}} \times \underset{011}{\mathrm{P}^1\mathbf{F}_{L}} \\[2ex] \underset{001}{\mathrm{P}^1\mathbf{F}_{L}} \times \underset{010}{\mathrm{P}^1\mathbf{F}_{L}} & \underset{001}{\mathrm{P}^1\mathbf{F}_{L}} \times \underset{011}{\mathrm{P}^1\mathbf{F}_{L}} \end{bmatrix} \otimes \begin{bmatrix} \underset{100}{\mathrm{P}^1\mathbf{F}_{L}} \times \underset{110}{\mathrm{P}^1\mathbf{F}_{L}} & \underset{100}{\mathrm{P}^1\mathbf{F}_{L}} \times \underset{111}{\mathrm{P}^1\mathbf{F}_{L}} \\[2ex] \underset{101}{\mathrm{P}^1\mathbf{F}_{L}} \times \underset{110}{\mathrm{P}^1\mathbf{F}_{L}} & \underset{101}{\mathrm{P}^1\mathbf{F}_{L}} \times \underset{111}{\mathrm{P}^1\mathbf{F}_{L}} \end{bmatrix}$$

Seventh Story — With matrices of projective lines over $\mathbf{F}_{127^2}$ and $\mathbf{F}_{L^2}$

Same as 5th Story through Wednesday. So we start here with Thursday.

Thursday
$$\mathrm{P}^1\mathbf{F}_{7 \atop 0} \;\underline{\cup}\; \mathrm{P}^1\mathbf{F}_{7 \atop 1}$$

$$\Downarrow$$

Friday
$$\begin{bmatrix} \mathrm{P}^1\mathbf{F}_{127^2} \atop {0\,0 \atop 1\,0} & \mathrm{P}^1\mathbf{F}_{127^2} \atop {0\,1 \atop 1\,0} \\[2em] \mathrm{P}^1\mathbf{F}_{127^2} \atop {0\,0 \atop 1\,1} & \mathrm{P}^1\mathbf{F}_{127^2} \atop {0\,1 \atop 1\,1} \end{bmatrix}$$

$$\Downarrow$$

Saturday
$$\begin{bmatrix} \mathrm{P}^1\mathbf{F}_{L^2} \atop {0\,0\,0 \atop 0\,1\,0} & \mathrm{P}^1\mathbf{F}_{L^2} \atop {0\,0\,1 \atop 0\,1\,0} \\[2em] \mathrm{P}^1\mathbf{F}_{L^2} \atop {0\,0\,0 \atop 0\,1\,1} & \mathrm{P}^1\mathbf{F}_{L^2} \atop {0\,0\,1 \atop 0\,1\,1} \end{bmatrix} \;\otimes\; \begin{bmatrix} \mathrm{P}^1\mathbf{F}_{L^2} \atop {1\,0\,0 \atop 0\,1\,0} & \mathrm{P}^1\mathbf{F}_{L^2} \atop {1\,0\,1 \atop 0\,1\,0} \\[2em] \mathrm{P}^1\mathbf{F}_{L^2} \atop {1\,0\,0 \atop 0\,1\,1} & \mathrm{P}^1\mathbf{F}_{L^2} \atop {1\,0\,1 \atop 0\,1\,1} \end{bmatrix}$$

$$\Downarrow$$

$$\begin{bmatrix}
\mathrm{P}^1\mathbf{F}_{L^2} \atop {0\,0\,0 \atop 0\,1\,0} \times \mathrm{P}^1\mathbf{F}_{L^2} \atop {1\,0\,0 \atop 1\,1\,0} + \mathrm{P}^1\mathbf{F}_{L^2} \atop {0\,0\,1 \atop 0\,1\,0} \times \mathrm{P}^1\mathbf{F}_{L^2} \atop {1\,0\,0 \atop 1\,1\,1} & \mathrm{P}^1\mathbf{F}_{L^2} \atop {0\,0\,0 \atop 0\,1\,0} \times \mathrm{P}^1\mathbf{F}_{L^2} \atop {1\,0\,1 \atop 1\,1\,0} + \mathrm{P}^1\mathbf{F}_{L^2} \atop {0\,0\,1 \atop 0\,1\,0} \times \mathrm{P}^1\mathbf{F}_{L^2} \atop {1\,0\,1 \atop 1\,1\,1} \\[2em]
\mathrm{P}^1\mathbf{F}_{L^2} \atop {0\,0\,0 \atop 0\,1\,1} \times \mathrm{P}^1\mathbf{F}_{L^2} \atop {1\,0\,0 \atop 1\,1\,0} + \mathrm{P}^1\mathbf{F}_{L^2} \atop {0\,0\,1 \atop 0\,1\,1} \times \mathrm{P}^1\mathbf{F}_{L^2} \atop {1\,0\,0 \atop 1\,1\,1} & \mathrm{P}^1\mathbf{F}_{L^2} \atop {0\,0\,0 \atop 0\,1\,0} \times \mathrm{P}^1\mathbf{F}_{L^2} \atop {1\,0\,1 \atop 1\,1\,0} + \mathrm{P}^1\mathbf{F}_{L^2} \atop {0\,0\,1 \atop 0\,1\,1} \times \mathrm{P}^1\mathbf{F}_{L^2} \atop {1\,0\,1 \atop 1\,1\,1}
\end{bmatrix}$$

Calculating views in finite spaces

Considering structural pedigrees, there is an obvious chicken and egg question: *Which came first the spaces or their coordinates?* In pedigrees that I am aware of, coordinates come first; then vector spaces; and then projective spaces, *etc.*. If you find elegant ones of another kind, please tell me!

Although interested in finite geometry since 1950, thinking of it in a Euclidian way, without coordinates or distinguished points, it had not occured to me that they might contain anything of interest to plot. With vector spaces plotting looked interesting.

Finally, around 1970 I drew by hand an image the plane over $\mathbf{F}_{31}$. In the 1980s, I was a programmer developing gridding software at Dynamic Graphics Inc. then in Berkeley, now in Alameda. The office minicomputer was connected to a large flat bed plotter, which I arranged to sometimes use after work. I plotted images of planes similar to the one I had done by hand, but now up to $\mathbf{F}_{127}$.

In all these plots, the entire plane was represented. The origin, $(0, 0)$, was at the center, the x and y coordinates plotted were centerlifts, least absolute value integers representing their residue class.

Each point was represented by an unframed square traversed by a straight line segment pointing toward its nearest neighbors in the line determined by the given point and the origin.

For a given amount of paper, the squares representing points in the plane were getting smaller as the primes got larger. Pondering this as I went to sleep, I dreamed of color plots with one pixel per point, and points on the same line assigned the same color.

I designed, and without an adequate computer to run it on, clumsily began writing a program to display such views, but had no hope of finishing it in time for ANPA West 7 in February of 1991.

My friend Edward Blair an outstanding programmer who had just graduated in Mathematics from UC Santa Cruz, and was not yet employed, moved into the rooming house where I lived, took an interest in my project, and rescued me.

At ANPA West 7 we presented *Blobs in Galois Fields*. My dream via his program Blobs, running on his 386. Both Edward Blair's program Blobs and the later MAC version ported by Carl Hanson, *Events in small planes* enabled calculating color movies of portions of the plane over any small prime field, $\mathbf{F}_p$, or integer ring $\mathbf{Z}_n$, up to n or $p = 2^{31} - 1$.

The following year Ed was working, I had a Mac with Mathematica II, and Carl and I presented *3-D Lucas Zooms*. With flawed, slowly calculated, greyscale stills over L. I had really goofed in respect to the third dimension. Think of taking a snapshot of a sphere, and then viewing the snapshot at an angle. The spheres looked like ellipsoids.

Finite Vector Spaces and Projective Spaces

In general, each finite vector space is $\mathbf{V}^d\mathbf{F}_q$, for prime p and some positive integers d and n, with $q = p^n$.

Given any nonorigin point $\mathbf{v}$ in $\mathbf{V}^d\mathbf{F}_q$ together with the origin $\mathbf{o}$, $\mathbf{v}$ determines a unique line, with exactly q points, the origin $\mathbf{o}$, and $q-1$ others. Each of the others is $s\times\mathbf{v}$ for some nonzero scalar s in the multiplicative group $\mathbf{F}_p^\times$ of the field $\mathbf{F}_q$. We say they are *projectively equivalent* to $\mathbf{v}$.

Each such line through the origin in $\mathbf{V}^d\mathbf{F}_q$ is a projective point in $\mathbf{P}^{d-1}\mathbf{F}_q$. Any two of *these* lines intersect only in $\mathbf{o}$. So, the q^d points of $\mathbf{V}^d\mathbf{F}_q$ are *partitioned* into $\{\mathbf{o}\}$ and $(q^d - 1)/(q - 1)$ lines with $q - 1$ nonorigin points each. Hence there are $(q^d - 1)/(q - 1) = \sum_{j=0}^{d-1} q^j$ projective points in $\mathbf{P}^{d-1}\mathbf{F}_q$.

Both for calculating views, and I believe for physics, the lines through the origin are much more significant, than other lines not through the origin.

Distance, wavelength, energy and photons

Because of lines wrapping around we lose transitivity, and don't have a general purpose distance function. We do have local distance functions, adequate to find for a nonorigin point $\mathbf{v}$ in $\mathbf{V}^d\mathbf{F}_q$, the nearest projectively equivalent neighbors of the origin on that line; and to assign their absolute value as a distance between consecutive points on that line, we may call it a *step* along that line. It seems surprisingly like a wavelength!

So we call it wavelength, and in our graphics, it is the basis for color, for spectra, and for assigning, inversely proportional to wavelength, an energy to each point.

If our vector space is over a prime field $\mathbf{F}_p$ then we may traverse the line additively.

In this case, if $\mathbf{v}' = [\,\mathbf{v}'_1, \ldots, \mathbf{v}'_d\,]$ is a nearest neighbor to the origin on the line through $\mathbf{v}$, then the wavelength assigned to that line and to each of its nonorigin points is defined by

$$w(\mathbf{v}) \;=\; \sqrt{\sum\nolimits_{k=0}^{d} \operatorname{centerlift}(\,\mathbf{v}'_k\,)^2}\;.$$

We may then speak of $t{\cdot}\mathbf{v}'$ as being t steps from $\mathbf{o}$. In our movies, at time t only points $t{\times}\mathbf{v}'$ and $-t{\times}\mathbf{v}'$ are shown, and representing one point per pixel, the other points on that line were black. In zoomed plots, each pixel represents a large number of points, on seperate lines through the origin, with different wave lengths. The pixel has a different spectrum at each time t. For plotting purposes considering the points on the three shortest wavelength lines through the pixel which are visible at time t should suffice. Fortunately, Sage is fast.

Of course although we can do all this over $\mathbf{F}_p$, these lines through the origin are not photons. For photons we need electromagnetism, Maxwell's equations, and a discrete version of $\mathbb{C}$, either $\mathbf{F}_{p^2}$, or for twistors, $\mathbf{Z}_{p^2}$.

It is here that we restrict ourselves to primes $p \equiv 3 \mod 4$, where the elements of $\mathbf{F}_{p^2}$ are of the form $a + b{\times}I$ with a and b in $\mathbf{F}_p$ and $I{\times}I = -1$.

Over extensions of $\mathbf{F}_p$

Although $\mathbf{F}_{p^2}$, is a particularly important case, several remarks are in order about the more general case of $\mathbf{F}_{p^k}$, which of course subsumes the $\mathbf{F}_{p^2}$ case.

In any extended field $\mathbf{F}_{p^k}$ and hence in any vector space $\mathbf{V}^d\mathbf{F}_{p^k}$, over it we cannot traverse the line additively, because repeating the same step p times returns you to where you were, with the $p^k - p$ other points on the line unvisited.

It is possible to traverse the nonorigin points of the line multiplicatively; we may do this in the generic (no zeros) case by designating a particular generator g of $\mathbf{F}_{p^k}$ (for $\mathbf{V}^4\mathbf{F}_{L^2}$, I use $g = 1 + 8{\cdot}I$).

For example let

$$\mathbf{v}' = [\,\mathbf{v}'_1,\, \mathbf{v}'_2,\, \mathbf{v}'_3,\, \mathbf{v}'_4\,] = [\,g^{b'_1},\, g^{b'_2},\, g^{b'_3},\, g^{b'_4}\,]\,;$$

and then for time t,

$$\mathbf{v}_t = [\,\mathbf{v}'^{\,t}_1,\, \mathbf{v}'^{\,t}_2,\, \mathbf{v}'^{\,t}_3,\, \mathbf{v}'^{\,t}_4\,] = [\,g^{t\cdot b'_1},\, g^{t\cdot b'_2},\, g^{t\cdot b'_3},\, g^{t\cdot b'_4}\,],$$

which is projectively equivalent to

$$[\,g^{t\cdot(b'_1-b'_4)},\, g^{t\cdot(b'_2-b'_4)},\, g^{t\cdot(b'_3-b'_4)},\, g^{t\cdot(b'_4-b'_4)}\,]\,.$$

We also need to treat nongeneric lines with some coordinates zero.

For Penrose twistors we instead use modules over the ring $\mathbf{Z}_{p^k}$. Then after p steps we are once around in one dimension but increment the next dimension.

Representing the integer mods of $\mathbf{Z}_{p^k}$ expressed as a number written base p with d digits, we now see that the difference is that in $\mathbf{Z}_{p^k}$ you carry, and in $\mathbf{F}_{p^k}$, you don't!

So, over $\mathbf{Z}_{p^k}$ you can traverse the line additively, and I think you can use the same type of definition of wavelength as over $\mathbf{F}_p$.

Another interesting ring to try is $\mathbf{Z}_3 \cdot B_k \cdot C_k$.

I am still too ignorant to say much about the noncommutative cases with 2×2 matrices over any of these rings as coordinates.

If the universe does not involve any infinite sets:

Then, for some prime $p \equiv 3 \mod 4$, of the order of $\sim 10^{38}$, $\mathbf{F}_p$ will take in Physics, the place of $\mathbb{R}$ must be taken by $\mathbf{F}_p$; and the place of $\mathbb{C}$ by the corresponding quadratic extension, $\mathbf{F}_{p^2}$, or for twistors, $\mathbf{Z}_{p^2}$. You know that I expect that that prime is L.

Although there cannot be a finite noncommutative division ring, because of zero divisors, I expect that the space of 2×2 matrices over the quadratic extension, will provide an adequate form of finite noncomutative geometry.

Here, the space of mappings from one space to another is of finite dimension, rather than of uncountably infinite dimension as with the corresponding space over $\mathbb{R}$. But they should do the same job here. So, I expect that the relationship between conservation laws and symmetries, should be unchanged. Speaking *ex cathedra* from the office of ignorant amateur, I expect that General Relativity, Quantum Electrodynamics, Discrete String Theory, Loop Quantum Cosmology, and perhaps Penrose Twistors will be living peacefully together here, and be well approximated by Standard Model QM.

The anticipated computer graphics should let us look inside and find out what is *really* going on in the candidate geometry. After looking around inside, we should again look for the most obvious difference between that candidate and our universe; tweek, and try again.

A Challenge

The 2^{65536} bitstrings of length 65536, created Saturday in the second story can form vector spaces over 17 different fields! —

$$\mathbf{V}^{65536}\mathbf{F}_2 \qquad \mathbf{V}^{32768}\mathbf{F}_{2^2} \qquad \mathbf{V}^{16384}\mathbf{F}_{2^4} \qquad \mathbf{V}^{8192}\mathbf{F}_{2^8}$$
$$\mathbf{V}^{4096}\mathbf{F}_{2^{16}} \qquad \mathbf{V}^{2048}\mathbf{F}_{2^{32}} \qquad \mathbf{V}^{1024}\mathbf{F}_{2^{64}} \qquad \mathbf{V}^{512}\mathbf{F}_{2^{128}}$$
$$\mathbf{V}^{256}\mathbf{F}_{2^{256}} \qquad \mathbf{V}^{128}\mathbf{F}_{2^{512}} \qquad \mathbf{V}^{64}\mathbf{F}_{2^{1024}} \qquad \mathbf{V}^{32}\mathbf{F}_{2^{2048}}$$
$$\mathbf{V}^{16}\mathbf{F}_{2^{4096}} \qquad \mathbf{V}^{8}\mathbf{F}_{2^{8192}} \qquad \mathbf{V}^{4}\mathbf{F}_{2^{16384}} \qquad \mathbf{V}^{2}\mathbf{F}_{2^{32768}}$$
$$\mathbf{V}^{1}\mathbf{F}_{2^{65536}}$$

Some or all of these may be involved in the pedigree of the structure we live in in ways not discussed here. *Think about fancy multifield spaces.*

In our later stories, the mappings of today's items into the Tuesday items of $\mathbf{F}_2$, become tomorrow's items. You might look at other spaces of mappings between available structures that might lead more directly toward auspicious structures. Can you hear them purring as they mutate?

For the scale ratios found by Physicists to appear, a field with characteristic $\sim L$, must be involved. Since in the later days of each story the scale changes dramatically, care is needed to not jump past the relevant characteristic. Thinking again of *The Sand Reckoner*, the number of rooted pedigrees that produce candidate structures involving such a field is not infinite, in fact it is not even very large. Perhaps you can count them.

What can we do toward building rooted pedigrees of our tame operators? We should expect that a candidate structure may have multiple structural pedigrees, and that they may be connected by a network of commutative diagrams proving their resulting structures to be the same. At least, progress in Commutative Algebra and Algebraic Geometry encourages this amateur to so dream.

Since I may not be around long enough to do the job, it may be up to you to write and maintain as a sequel to this paper — *The Catalogue of Rooted Pedigrees of Cosmic Geometries.* And it may be up to you to keep *Sage* updated with ever better tools with which to explore them.

Cosmic implications of many diverse explorers viewing a variety of geometries

Amid the more than $\sim 10^{11}$ other planets in our galaxy, and similarly the $\sim 10^{11}$ planets of each of the $\sim 10^{11}$ other galaxies in our light cone, and the far more planets, in the many other light cones of this geometry; and likewise throughout countless other adequate geometries; *many diverse explorers* have been calculating high resolution views within a variety of geometries, including theirs, and also ours.

Some of these explorers calculate with technology, others use highly evolved brains, still others with Parallel Distributed Processing in a biosphere of microbes or by means still harder for us to imagine.

Those explorers with adequate candidate geometries certainly find galaxies, stars, and planets, some of the planets have life or technology. By scanning future-ward along the world lines of such planets, the explorer observes the fate of the life or technology on that planet. In most cases, it will eventually be wiped out by some kind of natural disaster; sometimes it spreads to nearby planets; in rare cases it roars out into interstellar space at near light speed.

I expect that planets which eventually have their own explorers calculating high resolution views within a variety of geometries are far more common than those which launch an interstellar culture, but they may be harder for the viewer to detect.

However, any explorer discovering one of these, will be able to examine the discoveries of the explorers on the observed planet, and discover the ever growing tree of explorers over whose shoulders they are looking. Unlike intersteller travel, it is an *un-menacing way* to find and learn from far greater minds!

We may join that club too, possibly within your lifetime!

I certainly expect that most of the major insights that mankind has discovered *or will ever discover* are widely known elsewhere in this *and other* geometries, most being independently discovered many times, but I also expect that some of our insights become known elsewhere mainly by *looking over our shoulders*. I especially expect these to include insights that we are getting, as we read the recipes, methods, and data structures in the genomes of Earth's diverse life forms.

You may be surprised that I do not expect that we will
ever discover ourselves in the geometries we explore!

Great Entity

Herb Doughty 1968

To that great entity to be,
whose embryo is our technology;

Astronaut, Bioengineer,
worthy child of our Biosphere;

Reconfiguring your self,
and life;

Throughout the planets,
without strife.

Someday you will others find,
who elsewhere have been of your kind.

Finally you, your place will gain,
a neuron, of the cosmic brain.

We never know who else found it.

Richard G. E. Pinch, while an Oxford undergraduate in the 1970s, independently rediscovered, and properly wrote up double fields, the triple field, and the bijection between even double fields and irreducible polynomials of Mersenne exponent degree.

Over a Japanese meal in Berkeley around 1990 he asked about my adventures. As I told him about double fields his face lit up, and he said "Now that's a *nice* idea. When I was a kiddie, I thought of that too!" When he got home he kindly sent me a copy, which unfortunately, I have mislaid at the moment. It was either in Eureka! or in Kvant. He used a nice system of names :

> commutative groups were called 1-farms.
> fields were called 2-farms.
> double fields were called 3-farms.
> and the triple field was called the 4-farm.

My February 1968 discovery of these, while independent, was not the first. News spread rapidly by word of mouth through Lehmer's students and friends, especially John Selfridge. Around 1970, playing Go in my kitchen I learned that our new Go player John Hathaway Lindsay was an outstanding Mathematician. I asked him what area of Math was his favorite.

He said finite algebraic structures with more than two operations. I told him about double fields, he said it is a great idea, but it is not new. I asked for a reference. He said he could not think of any. I asked how he found out about them. He said that John Selfridge had told him about them in Dekalb IL the previous year. I was relieved, double fields had seemed new to Selfridge, when I told him about them in Berkeley two years earlier. My illusion of possibly being first was rescued . . .

. . . until 1986, when a colleague showed me an *excellent* introduction to Galois theory for high school kids written in 1959 by W. W. Sawyer, *A Concrete Approach to Abstract Algebra*, in it the first two even double fields were presented, without giving a name to the concept. He also discussed using $\mathbf{F}_{p^2}$, in place of $\mathbb{C}$. It is now a free PDF!

Derrick Henry Lehmer in his 1947 Scripta Math. paper, *The Tarry-Escott problem*, referred to the triple field as a ternary field. I had enjoyed the luxury of getting to bounce my ideas off of him since February of 1968.

In spring of 1986, I mentioned to him that since in the triple field each element was the identity for one of its 4 operations, one could use the name of that element as the symbol for that binary operation, and view the four binary operations as a single ternary operation.

He opened his file cabinet and gave me a copy of his 1947 paper. Then he began making unreasonably flattering remarks about my observations, including guessing that I was the first to think of this bijection; harnessing my silly vanity to save my life, he concluded the conversation with "And just think, if you quit smoking that pipe, you will be able to contribute such insights twenty years longer!" I quit smoking May 31, 1986.

What I have learned about double fields, was likely known to Frobenius 100 years ago and perhaps to Gauss 200 years ago. I expect that it was discovered by lots of people about whom I know nothing, *and by many other diverse explorers.*

Aknowlegements

Currently as in the past, I am indebted to Geoffrey Chew and Evan O'Dorney for very helpful conversations and emails. I am also especially grateful to my wife Hafida and to my friends at Berkeley Go Club, both for suggestions, and for freeing me from other concerns, so I could write this.

For inspiration and insight I am grateful to my ANPA friends, especially, Lou Kauffman, Geoffrey Chew, John Amson, Jean Burns, Peter Gaposhkin, Vladimir Karmanov, David McGoveran, and of course Pierre; and to Math friends Earl Lhamon, Dan Giesy, Roy Meyers, Jack Tull, Arno Cronheim, Arnold Ross, Hans Zassenhaus, Dick and Emma Lehmer, John Selfridge, John Baez, Rene DeVogelaere, Joe Gerver, Royce Wolff, Nancy Blachman, Peter Lawrence Montgomery, Elwyn Berlekamp, David Eisenbud, Ruchira Datta, Zvezda Stankova, and Tom Rike; to Bob Gaskins, Felix Lev, Clive Hayzelden, and Nick Kersting.

I am grateful to my colleagues, programmers Edward Blair and Carl Hanson; and to those making the best mathematical software available to all of us; especially Henri Cohen of PARI and William Stein of Sage.

I am especially grateful to three people for giving me very special lessons in the art of exploring: to my father, Herbert C. Doughty Jr., who by my second birthday had given me a deeply ingrained habit of responding to any interesting idea, with a search for diverse examples; to Richard Feynman for six very inspiring months in Physics-X at Caltech; and to Minoru Tazima, who discovering my Go interest as he arrived for a three month meeting, moved into the house where I lived to be my teacher, and gave me two hours a night of excellent, one–on–one lessons in exploring while introducing me to Go.

My Quest

In 1950 simultaneously contemplating Cosmology, Ackermann's function, and Archimedes' *The Sand Reckoner*, I began to suspect that the geometry we live in has only finitely many points, and involves no infinite sets. My ever growing suspicion got a huge boost in 1956–7 when, briefly at Caltech, I heard from fellow students about finite fields and Gauss's interest in them, and heard from Feynman, about various exploring techniques including frequent rethinking from scratch.

In 1960 after 15 months as a computer at SAO attending weekly Harvard Astronomy Colloquia, my suspicion seemed to be shared by the best Mathematicians, but not by Astronomers, so returning to Ohio State, I changed my major from

Astronomy to Math. For the next eight years, my attention was flipping between fields of two power order and fields of large prime characteristic. In February of 1968, working late one night as a programmer at UC Berkeley, I noticed that both can occur within the same structure, and found the bijection between even double fields and irreducible binary polynomials of Mersenne exponent degree.

In 1986, I gave a talk on them at The International Congress of Mathematicians. In 1988, I heard of ANPA and quickly met Pierre and our ANPA West friends. On my behalf, David McGoveran kindly presented my paper Finite Double Fields at the 1988 Cambridge, UK, meeting. The next few Februaries, I presented results at ANPA West.

About me

At 75, I am the proud father of the 61 year old Lima Astronomical Society, and the 45 year old Berkeley Go Club where I am again President. Without a PhD, I was a programmer on diverse research projects. In my retirement, with a view towards Cosmology, I explore finite geometry with computer graphics and some newer algebra, while amateurishly speculating with friends young and old on the nature of cosmos, life, mind, and the role we each have as explorer and participant in a situation that we do not understand at all **yet**.

Favorite Sources

Physics
```
http://front.math.ucdavis.edu/categories/physics
```

Mathematics
```
http://front.math.ucdavis.edu/categories/math
http://front.math.ucdavis.edu/categories/nlin
http://front.math.ucdavis.edu/journals
http://www.msri.org/web/msri
https://www.msri.org/web/msri/online-videos
http://cims.nyu.edu/ kiryl/Algebra/Section_4.1--Elementary_Basic_Concepts.pdf
http://www.sciencedirect.com/science/article/pii/0012365X9400375S
http://www.alainconnes.org/docs/bookwebfinal.pdf (for noncommutative geometry)
http://link.springer.com/chapter/10.1007%2F978-94-009-6487-7_29#page-1
http://www-history.mcs.st-andrews.ac.uk/Biographies/Frobenius.html
http://www-history.mcs.st-andrews.ac.uk/history/
http://archive.org/details/AConcreteApproachToAbstractAlgebra
```

Computing
```
http://front.math.ucdavis.edu/categories/cs
http://www.sagemath.org/
https://github.com/sagemath
https://github.com/sagemath
https://github.com/bhamrick/multitwitch
http://en.wikipedia.org/wiki/Class_%28computer_programming%29
```

Boolean Geometry and Non-boolean Change

Thomas Etter (1928-2013)

Boundary Institute
P.O. Box 10336 San Jose, CA 95157 USA

The following paper was the basis for my talk at ANPA 16 (1994). At the time I believed its basic idea to be original, a belief supported by a number of knowledgeable readers. However, I subsequently learned that this idea, which I called Boolean geometry, was actually around as early as the 1930s. Having never quite made it into the mainstream of logic, it was reinvented not only by me but by several others, including Gordon Pask and, I believe, Gian-Carlo Rota — the full history here remains to be uncovered. However, the connection I made to negative quantum amplitudes does appear to be new, so my plan became to revise the paper into a larger work in which this connection is developed in detail. This larger work has indeed become larger! What was to be only an introductory section on link states turned into my 90-page IJGS paper, followed by a series of shorter papers on the same topic, and I'm afraid the grand synthesis of link theory and Boolean geometry is still only a sketch. Thus it seems like a good idea to go ahead and release this paper in its present unfinished state. Like the DOC paper, it's a historical record, and also I believe that it's not a bad introduction to its subject, which could well be of interest to other investigators of "strange" logics.

0. Introduction

Von Neumann showed that quantum observables with eigenvalues 0 and 1 can be interpreted as propositions about the outcome of measurement. When two such observables commute, their product as operators is their conjunction as propositions, *i.e.* PQ means (P AND Q). However, since (P AND Q) = (Q AND P), this cannot be true if P and Q don't commute. For such propositions, von Neumann defined AND in a new way, which led to new non- Boolean laws for AND, OR and NOT; the resulting non-Boolean "logic" was called quantum logic.

Quantum logic was a dismal flop, and is all but forgotten today.

But the logical issues raised by von Neumann's deep insight into the meaning of eigenvalues 0 and 1 are as alive today as ever. What are we to make of propositions that don't commute? To put it another way, why is it that sometimes we can't say "P and Q", taking "and" in its usual sense? Though these question arose in physics, they don't belong to physics; they're not about matter in motion but about propositions.

Non-commuting quantum propositions are always asserted from different "viewpoints", *i.e.* their eigenvectors belong to different bases in Hilbert space. When we can't say "P and Q", it's because P is true or false here, while Q can only be true or false true after we move to there. Taken at face value, this tells us not that Boolean logic is wrong but that it is relative. "AND" only jumps the Boolean track when we move our logical vantage point. The adjective "non- Boolean" is misapplied to logic — what it really applies to is change!

What, if anything, is invariant under non- Boolean change? In the present paper I explore the thesis that, however we may experience this constancy, mathematically speaking it is the connectivity of the Boolean lattice stripped of its arrows. This mathematical structure, which I call *Boole space*, is isomorphic to the undirected edge graph of the Euclidean hypercube. Given Pascal's logical definition of probability as the number of favorable cases divided by the total number of cases, this weakening of Boolean algebra to Boolean geometry turns out to be mathematically equivalent to generalizing probability so that it can go negative as well as positive. The result is a hidden variable theory that specializes to quantum mechanics as a simple large-number case. Of course the hidden variables here are highly non-classical; their invisibility is not just de-facto but logical, and they are not only hidden from the classical observer but from each other!

1. Illogic, Pre-logic and Logic

Early in the twentieth century, at a time when scientific idols were toppling right and left, even logic itself began to totter. "If Euclidean geometry has fallen, which according to Kant is so built into human reason that it's humanly impossible to rationally doubt it, then why is Boolean logic still standing?" So asked the spirit of the times, and so asked von Neumann when in the 1930s, in trying to clarify Bohr's notion of complementarity, he proposed his so-called quantum logic.

Quantum logic at first attracted an enthusiastic and distinguished following. But, some 60 lackluster years later, even its best-known advocate has declared it to be a flop. Why did quantum logic fail? There were no mistakes in von Neumann's mathematics, but it turned out to be strangely sterile. It produced not a single empirical prediction, and wasn't even helpful in making calculations. Not surprisingly, most physicists have turned away from logic altogether.

This is unfortunate. The powerful winds of change that were felt by pioneers like Bohr and Pauli were not just about calculating the radiation spectra of atoms. The really new thing in quantum mechanics is a very general idea, superposition, and superposition makes sense in any context whatsoever that presents us with a range of alternatives, never mind alternatives for what. Physicists, though they may claim to disdain logic, make unabashed use of this generality. For instance, they assume without hesitation that it makes sense to speak of the superposition of alternative topologies. But what would you call the theory that tells them they can do that? I don't know its name, but I'd certainly call it some kind of logic.

Among those of us who still do call it some kind of logic, the common wisdom is that we must continue the search for alternatives to Boolean logic. I believe that this is wrong and that the common wisdom here has overlooked the obvious.

There's nothing at all odd or non-Boolean about logic in quantum mechanics so long as we stay within a context where one measurement doesn't disturb another, *i.e.* within a single set of commuting observables. We can also freely move our ordinary Boolean logic to any other such context. It's only when we try to combine statements from different contexts that problems arise. Context sensitivity is of course very familiar in everyday life — think of statements about left and right. What the quantum logicians seem to have overlooked is the possibility that it's the context sensitivity of logic that is causing all the trouble. Why do we need a new kind of logic? Perhaps ordinary logic is not wrong, but just relative!

How do we explore this hypothesis? The approach that works well for spatial relativity is to start with relative descriptions and extract from them a certain absolute or invariant part. For instance, you can start with "X is to the left of Y and Y is to the left of Z" and extract the weaker invariant statement "Y is between X and Z", a statement that remains true under reversal of left and right. That's close to the sort of thing we're going to do here. However, in one respect our enterprise has no precedent. In all prequantum theories of relativity the job has been to find a system of invariant statements. But we are looking for something a little different, which is an invariant way to logically combine statements.

Actually, the same thing could be said about von Neumann's quantum logic. How the present approach diverges from quantum logic, to put it in a nutshell, is that it does not seek a competitor to Boolean logic with different laws, but an objectification of Boolean logic with weaker laws. I shall call the first *illogic*, the second *pre-logic*.

The particular pre-logic that I shall describe here is what I call *Boolean geometry*. Von Neumann compared his non-Boolean logic to non-Euclidean geometry. By way of contrast, the present approach can be more accurately compared to Euclidean geometry. To see what this means, let's indulge in a fantasy.

Imagine that it was not Euclid but Descartes who in 300 BC invented the definitive mathematics of space. Since Descartes was an egocentric fellow, Cartesian geometry was centered on his own person, and all points of space were designated by their distance from his navel taken in three directions: left/right, front/ back, and above/below. This was a wonderfully practical system when Descartes was sedentary, but it got a bit confusing when he was moving about. The problem seemed to be solved by his death, but grave robbers kept reviving it. Nevertheless, the system persisted for almost two thousand years until Euclid came up with his great theory of relativity. This theory solved once and for all the problem of Descartes' peregrinations by completely doing away with the idea of a Cartesian center, putting a Cartesian clone at every point of space! Note that Euclid's is not an alternative geometry. It doesn't contradict the viewpoint of the wandering Descartes, but weakens it by abstracting only what it shares with its clones.

For Descartes substitute Boole. Our new logical geometry is to Boolean algebra what Euclidean geometry is to Cartesian geometry. Indeed, as we shall see, a Boolean logic is literally a coordinate system on a Boolean geometry. Starting with

Boolean logic, we can define Boolean geometry as the invariant structure under a new group of transformations that translate the Boolean origin (Boolean 0). In geometry a translation is a congruence transformation that takes lines into parallel lines. We shall define the concept of parallel in Boolean algebra by saying that the line x,y is parallel to the line x',y' iff (x XOR y) = (x' XOR y'), where XOR is exclusive OR. Given this definition, we can move the concept of geometric translation into logic word-forword! All this will become clear in chapters 3 and 4; for now I just want to stress again how different Boolean geometry is from quantum logic, whose symmetry group bears an entirely different relationship to Boolean algebra.

To summarize: The laws of logic are not wrong. Indeed, without these laws, these Boolean laws, it would make no sense to speak of laws of any kind. The problem is that we have been using Boolean logic too egocentrically. It is only after we become aware that there is more than one Boolean viewpoint and begin to study transformations of viewpoint that the fundamental meaning of quantum superposition becomes clear.

This may sound like a soothing message: there's a technical fix for all this weird logic stuff that turns it into science as usual. However, let me end here with a word of warning: When you mess around with logic, you mess around with how you think! Lobotomy may be a routine procedure for a brain surgeon, but not when he performs it on himself!

2. Boolean Logic

2.1. *Boolean algebras*

In this paper I will be treating Boolean geometry as something abstracted from Boolean algebra; as mentioned, this is like treating Euclidean geometry as something abstracted from a system of Cartesian coordinates. For space it seems more natural to go the other way, since our everyday experience of space involves a constantly moving origin. But for logic, I don't see how we can go the other way. What is our everyday experience of a moving Boolean origin, of a moving nothingness?

This is actually a fascinating question, and one that can start you reflecting about all sorts of things: creation and annihilation, virtuality, Bergson's duration, the Hegelian dialectic between being and not-being, Jacob Boehme's un-ground, Heraclitus vs. Parmenides — there's a call of the wild here that echoes down through the ages.

Unfortunately we haven't yet found the concepts that can connect this kind of speculative reflection to empirical science, and so I believe it's better to begin with concepts that have securely made this connection, namely with AND, OR and NOT. In this chapter I shall set the stage for Boolean geometry by reviewing some familiar aspects of Boolean algebra and introducing some others that may not be so familiar.

A Boolean algebra is a collection of objects, its elements, on which there are certain operators AND, OR, NOT etc. that satisfy certain rules. It's usual to introduce these rules by means of certain axioms. That would be irrelevant here, though, since we are only concerned with finite Boolean algebras, and the mathematical structure of a finite Boolean algebra is quite transparent to common sense; it is simply the structure of the set of all subsets of a finite set. We will use this fact to freely go back and forth between logical and set-theoretic terminology, which will make it easier to visualize some of our new concepts. Let's remind ourselves of how the two kinds of terminology correspond:

The INTERSECTION of x and y is the subset of all things that are both in x AND in y. We'll abbreviate INTERSECTION and AND by "&".

The UNION of x and y is the set of all things that are either in x OR in y. We'll abbreviate UNION / OR by "∨".

The COMPLEMENT of x is the set of all things that are NOT in x. We'll abbreviate COMPLEMENT by "∼".

Most of the time we'll be dealing with Boolean algebras whose elements are subsets of some explicitly given finite set S. The members of S are called the atoms of that Boolean algebra, and the number of these atoms is called the dimension of the Boolean algebra. Here, as in everyday life, we won't distinguish a set with one member from that member itself, so atoms, as one-member subsets of S, are also elements of the Boolean algebra. S itself is of course an element; it's called the universal element and is abbreviated 1. The null set is also an element, abbreviated 0. The context allows us to distinguish them from numerical 0 and 1.

2.2. *Boolean logics*

A Boolean logic, as I shall be using the term, is defined as the Boolean algebra of subsets of a set S of mutually exclusive possibilities, *i.e.* of cases. Think of S as a menu. It's usual in the restaurant business to number the items in a menu so we can point to them by their indices: "Hey, one #5 special coming up!" Let i be a menu index that ranges, say, from 1 to 10. Let S be the set of all items on the menu. Then, roughly speaking, the Boolean logic of S is the set of all the things we can say about i, for instance "i > 3".

This last statement needs to be carefully qualified. Taken out of context, "i > 3" could mean almost anything. It could be about the ratio of the circumference to the diameter of a circle, or about the number of dimensions of space-time, or it could tell us that the customer is not ordering Hamburger Supreme or Heavenly Chicken or Fisherman's Delight. To discriminate among these possible messages, we must first know the answers to two questions: What values does i take? and what alternatives do these values point to? Given this essential background, it makes sense to ask what information does "i > 3" supply?

The elements of a Boolean logic are often represented by sentences or sets of indices, but what they are is items of information, taking that term in Shannon's sense as the narrowing of a range of possibilities. My use of the word "logic", which is also von Neumann's, is rather different from what you find in most textbooks on logic, which are largely concerned with formalizing the science of inference. Logic in this textbook sense has almost nothing to do with our present subject matter, and a case could be made for our using some other term such as "Boolean informational structure". However, "logic" is a pretty flexible word, spanning the gamut from Hegel's dialectic to logic gates, and since von Neumann's use of it has already taken root in the foundations of physics community, we'll stick with "Boolean logics".

To recapitulate: An item of information is defined as a selected subset of a fixed set C of cases. A boolean logic is defined as the set of all items of information on C structured by the set operators INTERSECTION, UNION and COMPLEMENT etc.

2.3. *Projections*

Information has another aspect which is not captured by Shannon's definition: It accumulates, and as it accumulates, the act of acquiring each new item of information alters all the others. Consider our example of the item i > 3. Taken by itself, this places i in the range 4,5,6,7,8,9,10. But suppose we had already learned that i < 8. Then in Shannon's terms, i > 3 is a different item of information, since now it confines i to the range 4,5,6,7 against a background of an S that includes only the first eight members of the original S.

To put it another way, an item of information is not just a subset, it is an operator on its Boolean logic. The item of information p, as an operator, takes every element x into p&x. We'll adopt the convention of using capital letters for operators and small letters for their operands, so we can write this as $P(x) = p\&x$. P is a projection operator in the sense that it is idempotent ($PP = P$) and it preserves a certain algebraic structure - we'll see just what this structure is later. This projective character of information is important to us here for two interrelated reasons: First, it tells us something about why propositions are represented by projections in quantum mechanics, and second, it makes it possible to extend the definition of information so that it still makes sense in Boolean geometry, as we'll see in Section 3.

In everything we do we make in one way or another an important distinction between information of two kinds: first there is information that applies only to a particular occasion or event, and then there is information that applies to a whole group of occasions, or is a fixed aspect of a changing object. There are many different words for this contrast. In natural science we speak of data vs. regularities, of boundary conditions vs. laws. In computer science it's input and memory vs. program. In everyday life we have events vs. rules, activity vs. condition or status, accident vs. order etc. The concept of "state" straddles the boundary; a state is something that tends to persist unless it is "forced" or "induced" to change by something outside of itself.

The concept of projection most naturally applies to the second kind of information. Typically when we encounter an object we imagine a large range of nominal possibilities for it, most of which are not realistic. Narrowing down this nominal set S to a realistic set C is the kind of information that we call a principle. Since a new item of data is most commonly expressed in terms of S, it can be very useful to understand how a principle projects data onto C, since this tells us something about how changing data bears on the state of the object as a whole. Indeed, understanding such a projection is just what we ordinarily mean when we say we understand the principle of the thing. We'll pick up this train of though again in section 2.7.

2.4. *The Boolean lattice*

There are a number of ways to characterize the structure of a finite Boolean algebra. We've done it here by giving that structure a canonical representation, so-to-speak, as the set of all subsets of a finite set. A useful variation on this is to identify the Boolean elements with bit strings which represent the characteristic functions on these subsets. There is a third quite different kind of representation, though, which is important for studying Boolean geometry, and that is the Boolean lattice.

Thinking of the Boolean elements as sets, the Boolean lattice is the partial ordering of these sets by inclusion. We'll write this relation as x < y, meaning that every element of x is an element of y. In logic, if all x's are y's we say that x implies y, so the logical name for < is implication. We must be careful not to confuse this relation of implication with the Boolean operator $\sim$x$\lor$y which is sometimes called material implication. The two are closely related, though, since x < y iff $\sim$x$\lor$y = 1. Also we must not confuse it with the concept of deducibility, which is a metaconcept that occurs in the study of formal languages.

Here is a picture of the relation > for a 3-atom Boolean logic; x < y if you can follow the arrows from x to y.

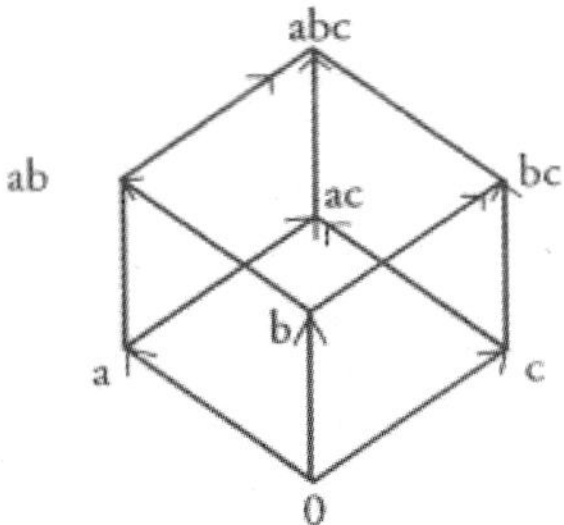

Figure 2.4.1.   The Three-atom Boolean Lattice

Note that we can define inclusion in terms of &: x is included in y means that x is the intersection of x and y. We can also go backwards from < to &: The intersection of x and y is the largest set included in both x and y, *i.e.* the g.l.b. (greatest lower bound) of x and y under the partial ordering < .

A less obvious fact is that we can define **NOT** in terms of < . First we define 0 as the element that is included in all others; 0 is of course the null set. Next, we define "x and y are disjoint" to mean that x&y = 0 (recall that we have already seen how to define & in terms of <). Finally we define $\sim$x to be the element that is disjoint from x and that includes every other element that is disjoint from x; clearly $\sim$x is the complement of x. (This last step only works for Boolean lattices. In non-Boolean lattices like quantum logic, there is more than one maximal disjoint element, and to define negation requires additional structure.)

Here's something that may come as a surprise. We've seen that we can define > in terms of AND, and NOT in terms of < , but this means that we can define NOT in terms of AND! Sounds impossible? That's because we're used to algebraically generating the other Boolean operators from AND and NOT, and we know it can't be done with AND alone. What we've done above, though, is something quite different, which is to define NOT as an aspect of the whole structure of the AND operator.

To summarize: The Boolean lattice is a partial ordering of the elements of a Boolean algebra in terms of which all of the Boolean operators can be defined, and conversely, which itself can be defined in terms of the Boolean operators; it is thus equivalent to Boolean algebra in the sense that it characterizes the same abstract structure.

2.5. *The Boolean graph*

In Fig. 2.4.1 the Boolean lattice is pictured as an oriented graph, where x < y means you can follow a path of arrows from x to y. Let's now look at this graph as a relation in its own right, which we'll call *arrow*. We'll write x → y to mean that there's an arrow from x to y. In the language of sets, x → y says that y is the result of adding another member to x. We can define x → y in terms of < to mean that x < y and there's nothing in between, *i.e.* for any z such that x < z and z < y, either z=x or z=y. Conversely, we can define < as the ancestral of arrow, *i.e.* the smallest transitive relation containing arrow — just what this fancy language means becomes pretty obvious when you look at the picture.

Since we can use it to define the Boolean lattice, the Boolean graph is another way to give the structure of a Boolean algebra. But far more important, it immediately leads to a simple definition of the basic object in Boolean geometry:

Boole space: The un-oriented graph that results from stripping a Boolean graph of its arrows.

We'll leave the arrows in place for the rest of this chapter, but we'll concentrate on those concepts that still remain important after we remove them. One of these is the concept of a path, which is a series of steps in the Boolean graph.

Step: An ordered pair of elements x,y such that either x → y or y → x; in the first case we call it a step up or a positive step, in the second, a step down or a negative step.

Path: A sequence of elements whose adjacent pairs are steps. A path from x to y will be notated x..y. Here's a path in the three-atom graph, shown by the dotted line (see Fig. 2.5.1).

By the *length* of a path we mean the number of steps in it. A *geodesic* is defined as a path from x to y of minimum length, and this length is called the Boolean *distance* between x and y. It's easy to see that Boolean distance is a metric in the sense that $D(x,x) = 0$, $D(x,y) = D(y,x)$, and it satisfies the triangle inequality. Giving this metric is another way to define Boole space, and in fact that's how we will start off defining it in Section 3. Here is an important theorem: The Boolean metric $D(x,y)$ together with the Boolean origin 0 determine the structure of Boolean algebra.   This will be proved in Section 4.

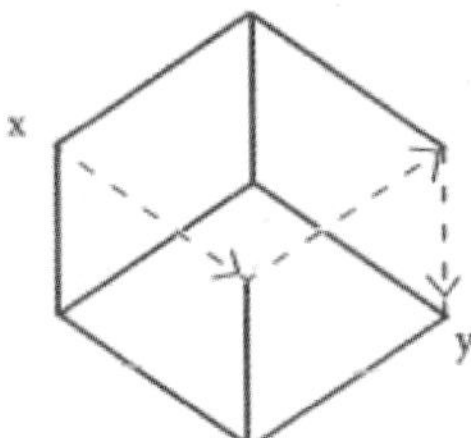

Figure 2.5.1.   Path from x to y

A step up adds a new member to an element, while a step down removes a member. Thus if a path starts from 0, we can determine the cardinality of its endpoint by taking the number of steps up and subtracting the number of steps down. In the language of finance, the net gain for someone traversing a certain path is his income along the way minus his outgo. It may help us remember what we are doing to use such words as technical terms:

Income: The number of positive steps in a path.

Outgo: The number of negative steps in a path.

Net gain: Income minus Outgo, abbreviated $\text{net}(x..y)$

Theorem :   The net gain of any path $x..y$ from x to y is the cardinality of y minus the cardinality of x. More briefly,

$$\text{net}(x..y) = \text{card}(y)\text{-card}(x)$$

Proof: We've seen that

$$\text{net}(0..x) = \text{card}(x) \text{ and } \text{net}(0..y) = \text{card}(y).$$

Now consider a path $0..y$ from 0 to y which results from splicing $x..y$ onto the end of any path $0..x$ from 0 to x. Clearly the net of a path is the sum of the nets of its consecutive parts, so

$$\text{net}(0..y) = \text{net}(0..x) + \text{net}(x..y)$$

and hence

$$\text{card}(y) = \text{card}(x) + \text{net}(x..y),$$

i.e. $\text{net}(x..y) = \text{card}(y) - \text{card}(x)$. QED.

Notation: This theorem shows that we can write $\text{net}(x,y)$ for $\text{net}(x..y)$.

We'll say that a path is ascending if all its steps are positive, descending if they are all negative. A path that is either ascending or descending will be called monotone. A monotone path is always a geodesic. A geodesic needn't be monotone, however, unless it starts from 0. The geodesics from 0 have a special place in Boolean algebra, since their lengths are the cardinalities of their upper endpoints.

2.6. *Probability*

The above discussion of paths has been part of our preparation for a definition of amplitude modeled on Pascalian probability. Recall that Pascal defined probability as the number of favorable cases divided by the total number of cases. In our current terminology, the probability of x is card(x)/card(1). In the language of graphs it's the net gain from 0 to x over the net gain from 0 to 1, and it's also the distance from 0 to x over the distance from 0 to 1. This last statement is purely geometric except for singling out a point called 0, so let's adopt it as the definition of probability.

Probability: prob(x) $=$ D(x,0) / Dmax, where Dmax is the greatest distance between any two points. (note that Dmax is D(1,0), which shows that 1 is the unique antipode of 0. Dmax is also what we called the dimension of the Boolean algebra, and it will turn out literally to be the dimension of the corresponding Boole space.

Pascal's definition of probability is a purely logical definition, and in fact it has its uses in logic proper. Perhaps the most important of these is to give us an easy way to define what it means for two items of information to be logically separate or independent: ' x is *independent* of y ' means that

$$\mathrm{prob}(x\,\&\,y) = \mathrm{prob}(x)\,\mathrm{prob}(y).$$

We'll see in a little while how to define independence in terms of Boolean factorization. The concept of independence is of course one of the cornerstones of science as we know it. Will it still make sense after we geometrize logic? Will science as we know it still be possible?

2.7. *Sublogics, factors and independence*

Generally speaking, a sub-structure of a mathematical structure is a subset of its elements together with the operators and relations that define the structure of the whole. Here is the basic Boolean sub-structure:

Subalgebra: A subset of a Boolean algebra closed under AND and NOT.

Theorem 2.7.1 : The atoms of a subalgebra form a partition of the atoms of the algebra.

Proof: Let B1 be a subalgebra of B. If B1 includes x and y, it includes -x and x&y. Therefore it includes 0 = x&-x and 1 = x∨-x and x-y = x&-y. Suppose x and y are atoms of B1. Then x and y must be disjoint, for otherwise one of x&y or x-y or y-x would be smaller than either x or y, contradicting the assumption that x and y are atoms. A similar train of reasoning shows that every element of B1 is a union of atoms of B1. But since 1 is an element of B1, we conclude that every atom of B is in some atom of B1. QED.

Sublogic: A subalgebra of a logic.

If the state of a (classical) physical object is described by several variables, each of these variables ranges over the case set of a sublogic of the state logic. More generally, any exhaustive set of mutually exclusive properties of the state defines a sublogic. The relationship of sublogic to logic captures much of what in everyday life we think of as the relationship of part to whole.

Since Boolean lattices, Boolean graphs, Boolean rings, etc. are all equivalent to Boolean algebras, it would be reasonable to suppose that sub-lattices, sub-rings, sub-graphs etc. are also subalgebras. Oddly enough, this is not true! Every subalgebra is a sub-ring (defined as a set closed under AND and exclusive OR) and every subring is a sub-lattice. However, not every sub-lattice is a sub-ring, nor is every sub-ring a subalgebra. An important substructure that is not a subalgebra is the set S' of elements of the form p&x that results from a projection P. S' is closed under OR (union) so, unless P=1, it does not include the negations (complements) of its elements, since x $\vee$ -x = 1, while P&1 = P. S' is a subring, however, namely that generated by OR from all the atoms of S'.

How do we characterize the logical relationship between two objects that have nothing to do with each other? In order to say anything at all about it, we must conceptually bring these objects into the same universe, which means that we must see them as sublogics of the same logic. Then what? There is a clear way to proceed if they together generate all the elements of this common logic:

Generation: We say that a set A of elements generates a Boolean algebra B if we can obtain every element of B by applying AND and NOT to the members of A. For instance S, the set of atoms, generates its Boolean algebra.

Spanning: Suppose that B1 and B2 are sublogics of B such that all of their elements together generate B; we then say that they span B. If B1 and B2 are any two sublogics of B there is a sublogic of B which well call B1&B2 that is spanned by B1 and B2.

Factors: We say that B1 and B2 are factors of B if they span B and if s1&s2 is never 0, where s1 is an atom of B1 and s2 is an atom of B2. More generally, sublogics B1, B2 ..Bi.. are factors of B if for any Bi, Bi and B' are factors, where B' is the sublogic spanned by all the other sublogics.

The concept of factoring captures what it means to break something into a "heap" of parts, assuming that this is possible. The converse operation, bringing parts together into a heap, will be known as:

Boolean multiplication: Given Boolean logics B1 and B2 over case sets S1 and S2, we define their product B1.B2 to be the Boolean logic over the Cartesian product of S1 and S2.

Theorem 2.7.2 : Given a product B1.B2, there are natural isomorphisms of B1 and B2 onto factors B1 and B2 of the product.

Proof: The atoms of the product are the ordered pairs $\langle xi, yj \rangle$, where xi is any atom of B1 and yj is any atom of B2. Define B1$'$ as the sublogic whose atoms xi$'$ are of the form $\langle xi, y1 \rangle \vee \langle xi, y2 \rangle \vee \langle xi, y3 \rangle \ldots$, and similarly B2$'$ with atoms yi$'$. Clearly the atoms of B1$'$ and B2$'$ correspond to those of B1 and B2, all these atoms together span B, and xi & yj is never 0, so B1$'$ and B2$'$ are factors. QED.

Because of these obvious natural isomorphisms, we'll regard multiplying and factoring as inverse operations, and we'll speak of a logic as the product of its factors. The dot notation B1.B2 is a close relative of the dot notation in computer science for separating fields in labels or addresses (think of an Internet address). If B1 and B2 are factors of B, we'll allow ourselves to write B = B1.B2; in such a case we have B1.B2 = B1&B2.

Now we come to a crucial theorem. Recall that we defined independence to mean prob(x & y) = prob(x)prob(y).

Theorem 2.7.3 : x and y are *independent* in B if and only if B can be factored into two subalgebras one of which contains x, the other y.

Proof: It's easy to show that elements in different factors are independent; indeed, this is an immediate corollary of the fact that the cardinality of a Cartesian product is the product of the cardinality of its factors. (Proving the converse is harder, though, and since the details of the proof are not relevant to what we're doing here, we won't go into them.) QED.

If the elements of one sublogic are independent of the elements of another, we say that the sublogics are independent. A necessary and sufficient condition for this is that the atoms of one are independent of the atoms of the other. This is the logical meaning of independent variables. That is, to say that two variables are independent means that they range over the case sets of independent sublogics. Independent sublogics are factors of the sublogic that they span, but they need not be factors of the whole logic. For a sublogic A to qualify as a factor of a logic B, the atoms of A must be equiprobable in B. A sublogic that meets this condition will be called separable. Note that if A is separable, the probabilities defined in B for the elements of A are the same as those defined in A alone, and if A is all we care about, we can leave B out of the picture. If A is not separable, however, we cannot use Pascal's definition on A alone but must also take B into account. One way to do this without mentioning B explicitly is to describe the effect of B in terms of a probability measure.

A *measure* on a Boolean algebra will be defined as a numerical function on its elements that adds for mutually exclusive cases. Some definitions of measure also includes the requirement that the numbers be non-negative, but we'll omit this requirement since we'll soon be studying amplitude measures that can go negative and complex.

A *probability measure* is a real non-negative measure that is 1 for Boolean 1. Pascal's definition of probability applied to B defines a probability measure on each of its sublogics.

The relative frequency interpretation of probability is often presented as a competitor of Pascalian probability. However, it can be brought under the Pascalian umbrella by treating a series of trials as a set of possibilities for this trial. The series then becomes the set of cases for the logic B of all statements about which trial is this one, and relative frequency becomes a Pascalian measure on a sublogic A whose cases are the possible outcomes of the trials. If the outcomes are equally frequent, as with a fair coin toss, then A is separable. It may seem that all this business about factors is laboring the obvious. After all, we all know what it means for things to have nothing to do with each other — why all the fuss? At this point I can only ask the reader to be patient a little longer. We really do need a rather abstract approach to familiar ideas like separateness, connectedness and equality, since our common sense intuition about such things turns out to be a poor guide in the strange new realm of Boolean geometry.

3. Boolean Geometry

3.1. *Boole space*

In section 2.6 we defined Boole space in terms of the Boolean graph; here we'll start with the Boolean metric.

Metric space: A set of points on which there is a distance function $D(x,y)$ satisfying three axioms: First, $D(x,x) = 0$, second $D(x,y) = D(y,x)$ and third, $D(x,y)+D(y,z)$ is greater than or equal to $D(x,z)$ (the triangle inequality.)

Boolean Distance: The distance $D(x,y)$ between two elements x and y of a Boolean algebra is defined as $\text{card}(x+y)$, *i.e.* the cardinality of $x+y$, where $+$ means exclusive OR.

Exclusive OR, or XOR for short, is central to Boolean geometry, so it's important to be clear about its basic properties.   First, let's be clear about its definition: x XOR y, abbreviated $x+y$, means $(x \lor y)$ & $\sim(x \& y)$ Unlike inclusive OR, exclusive OR is a group operator. Applied to bit strings, it is bit-by-bit addition mod 2, which justifies, or at least excuses, the symbol "$+$" for it. Boolean 0 is the identity of the XOR group, and every element is its own inverse; the latter property alone characterizes a group as isomorphic to a XOR group. If we define scalar multiplication by 0 and 1 by the rules $0x=0$ and $1x = x$, the XOR group becomes a vector space over the binary field; we'll call this vector space XOR space. AND distributes through XOR, so AND and XOR together form a ring, the so-called Boolean ring. The dual of XOR is IFF, defined as $(x \& y) \lor (-x \& -y)$.

Theorem 3.1.1 : Boolean distance is a metric.

Proof: Since $x+x = 0$, $D(x,x) = 0$. Since $x+y = y+x$, $D(x,y) = D(y,x)$. To prove the triangle inequality, first note that $D(x,y)$ is a maximum when x and y are mutually exclusive, in which case it is $\text{card}(x)+\text{card}(y)$. Then note that since $y+y = 0$, we have $x+z = (x+y)+(y+z)$, showing that $D(x,z) = \text{card}(x+z)$ is at most $\text{card}(x+y)+\text{card}(y+x) = D(x,y)+D(y,z)$, which is the triangle inequality. There are two basic theorems that launch Boolean geometry. Let's now turn to the first, which is that that Boole space is homogeneous. To understand what this means and how to prove it, we need the concepts of congruence, symmetry and displacement.

Congruence: A congruence between two subsets of a metric space is a 1-1 correspondence between them that preserves distance.

Symmetry: A self-congruence of a subset is called a symmetry.

Homogeneity: We say that a metric space is homogeneous if for any two points x and y there is a symmetry of the whole space taking x into y.

Displacement: A transformation on Boole space which for some fixed element d takes every element x into d+x. Note that this resembles a vector displacement in ordinary space, and is in fact literally a vector displacement of the linear algebra we called XOR space.

Theorem 3.1.2 : Displacements are symmetries of Boole space.

Proof: D(d+x,d+y) = card(x+d+d+y) = card(x+y) = D(x,y)

Theorem 3.1.3 : Boole space is homogeneous.

Proof: Given any two elements d and e, we have (d+e)+d = e, so the displacement by d+e takes d into e. QED.

The second big theorem that starts things off is that you can get from the Boole space of a Boolean algebra back to the Boolean algebra itself simply by saying which point is 0. Since Boole space is homogeneous, you can choose any other point to be 0 and get another Boolean algebra. The relativity of Boolean logic with respect to the choice of origin is the essential novelty of Boolean geometry, and in future papers we'll see how it leads to the relativity of logic in quantum mechanics.

At this point it will be helpful to consider Boole space in terms of the Boolean graph as we did in Section 2.5. Recall that we defined Boole space there to be the Boolean graph stripped of its arrows. More formally, it is the structure defined by the set of all unordered pairs of neighboring points in the graph. Referring to Fig. 2.4.1, we see that such neighbors are the endpoints of the straight lines in the Boolean cube. This is no accident, and it will turn out to be a good move to define straight lines in our geometry to be neighboring pairs. The definition now will be algebraic:

Line: A straight line, or simply a line, in a Boolean algebra is a two-element set x, y such that x+y is an atom.

We can see that this is clearly the case for the lines in Fig. 2.4.1. The logical meaning there of an arrow x $\rightarrow$ y is that y results from adding one more member to x. To remove the arrowhead means asserting the weaker relationship x $\rightarrow$ y or y $\rightarrow$ x, which doesn't say which of x and y is included in the other but merely that they differ by one atom, *i.e.*, that x+y is an atom. Thus we see that:

Theorem 3.1.4 : The two points on a line are neighboring points on the graph, which means that the structure of the unoriented graph is given by specifying the set of lines.

Theorem 3.1.5 : A line is a pair of points whose distance apart is 1.

Proof: If x+y is an atom then D(x,y) = card(x+y) = 1. QED.

The point of stating this obvious theorem is to make it clear that the concept of line is purely geometric, *i.e.* it depends only on the metric and not on other features of the Boolean algebra. In non-Euclidean geometry the role of straight lines is taken over by geodesics. Now we have both straight lines and geodesics, which in section 2.5 were defined as edge paths containing a minimal number of steps. We must now see what this means algebraically:

Path: A sequence of connected lines, *i.e.* a sequence points xi such that xi,xi+1 is a line.

Path length is the number of points minus one.

Geodesic: A shortest path between two points.

Theorem 3.1.6 : The length of a geodesic between x and y is D(x,y).

Proof: Consider first the case where x=0. Since 0+y = y, D(0,y) = card(y). We saw in 2.5 that card(y) is the length of any geodesic from 0 to y, so the theorem is true for x=0. But now suppose we apply a displacement d, turning our interval into d,d+y. We saw (theorem 3.1.2) that displacements are geometric symmetries, so D(d,d+y) = D(0,y). But we also saw that lines and hence paths are geometric concepts, so the shortest path between d and d+y has the same length as the shortest path between 0 and y, We can choose d arbitrarily. By letting y = d+z for arbitrary z, we can also choose d | y = z arbitrarily. Thus the theorem holds for any x and y. QED.

We now see that our two ways of defining Boole space agree. If we know the metric D(x,y) we know the geodesic distance on the graph and hence the structure of the (unoriented) graph itself. Conversely, if we know the graph, we know its graph distance and hence D(x,y).

Theorem 3.1.7 : Given any point x, and any path, every step forward in that path either increases the distance from x by 1 or decreases the distance from x by 1, *i.e.* there are no side steps.

Proof: The theorem is true for x=0, since distance from 0 is cardinality. It is true of any other point since the metric is homogeneous. QED.

Theorem 3.1.8 : The structure of a Boolean algebra is given by its metric together with its origin (Boolean 0).

Proof: The metric defines the unoriented graph while the origin defines the arrow as the direction of increasing distance from 0. As mentioned, it is theorems 3.1.3 and 3.1.8 that really get us going. These theorems tell us that a Boolean algebra is a Boole space on which we have singled out an origin, and that the points all look alike so any point will do. Theorem 3.1.7 is also very important, since, as we shall see, it plays a crucial part in our definition of amplitude. QED.

We started this chapter by defining the Boole space metric as the number of atoms in x+y. This of course presupposes a Boolean algebra in which atoms and + are defined. We then found that by applying a geometric symmetry we can displace the origin to any other element and obtain a new Boolean algebra in which this new origin is 0. Since the atoms are the neighbors of 0, this new 0 defines a new set of atoms, different from those we counted in defining the distance between x and y. In fact, + in the new Boolean algebra is also a different operator from + in the old one. Nevertheless, since the origin was shifted by a symmetry of the metric, we know that the new atom count is the same as the old count. Shifting the origin not only relativizes atoms and +, it relativizes all the familiar Boolean operators except NOT, and it also relativizes inclusion and mutual exclusion. Nevertheless, a surprisingly large part of Boolean structure remains invariant, and describing this invariant structure will be our next job.

3.2. *Geometric invariants*

What is the full symmetry group of Boole space? It includes the displacement group, and it of course includes the symmetry group of Boolean algebra, which I shall call the *logic group*.

Theorem 3.2.1 : The logic group of a finite Boolean algebra consists of all transformations that result from permuting the atoms.

Proof: A finite Boolean algebra is isomorphic to the set of all subsets of its atoms, whose intersections and complements obviously don't depend on how the atoms are arranged, so any permutation of the atoms will generate a symmetry. Since a logical symmetry must preserve the Boolean lattice, it must map neighbors of 0 into neighbors of 0, so the symmetries generated by permutations are the only logical symmetries. QED.

Theorem 3.2.2 : The geometric group, *i.e.* the full symmetry group of Boole space, is generated by the logic group and the displacement group in the sense that every geometric symmetry can be written in the form DL, where D is a displacement and L a logical symmetry.

Proof: Let T be any geometric symmetry of a Boolean algebra B. Let d = T(0). Then by theorems 3.1.3 and 3.1.8, T must map B isomorphically onto the Boolean algebra B$'$ whose origin is d. But the displacement D(x) = d+x also maps B isomorphically onto B$'$. Thus the symmetry L = DT is a logical symmetry on B (recall that D is its own inverse). We thus have T = DL, the product of a displacement and a logical symmetry. QED.

Theorem 3.2.3 : The displacement group is a geometric invariant, *i.e.* it is a normal subgroup of the geometric group.

Proof: We must show that for any symmetry T and any displacement D, $T^{-1}DT$ is a displacement. We have seen that we can write T in the form EL, where E is a displacement and L is logical. Since $(EL)^{-1} = L^{-1}E$, we have $T^{-1}DT = L^{-1}EDEL = L^{-1}DL$, so the problem reduces to showing that $L^{-1}DL$ is a displacement. We have $L^{-1}DL(x) = L^{-1}D(L(x)) = L^{-1}(d+L(x))$. But since L^{-1} is a logical symmetry, it preserves + and we have $L^{-1}(d+L(x)) = L^{-1}(d)+x$. Thus $T^{-1}DT$ is a displacement by $L^{-1}(d)$. QED.

Our aim in this section is to see how much of logic carries over into Boolean geometry, *i.e.* to find the important logical concepts that are geometric invariants. Now of course logical concepts are already invariant under the logic group — they don't depend on any particular arrangement of the atoms. Thus by theorem 3.2.2, to show that a logical concept is geometrically invariant it is sufficient to show that it is invariant under displacement. This reasoning is essentially what underlies the proof of theorem 3.2.3; displacement is a logical concept and the displacement group is invariant under itself. The most familiar geometric invariant: is NOT. NOT is in fact displacement by 1, *i.e.* -x = 1+x, which shows that it is invariant. The negation of an element x is, geometrically speaking, its antipode, *i.e.* the point furthest away from x. To see why this is so, note that the longest possible geodesic is of length n, where n is dimension, since each step in a geodesic must add or take away a different atom. But a geodesic of length n starting at x must take away all the atoms in x and add all the atoms not in x, since there are no other atoms to add or take away; the result is NOT x. Incidentally, note that the dimension n is a geometric invariant, since n is the number of neighbors of 0, and by homogeneity every point has the same number of neighbors. Let's now move on to some important unfamiliar geometric concepts.

Parallel: Two lines {x, y} and {z, w} are called parallel if x+y = z+w.

Theorem 3.2.4 : Parallel is a geometric concept.

Proof: Since + is a logical concept, it is sufficient to notice that a displacement D of two parallel lines leaves them parallel, *i.e.* if x+y = z+w, then d+x+y = d+z+w.

Recall that a line is an edge in the Boolean cube. The arrow on an edge represents adding an atom to a set. Note that parallel arrows add the same atom, which agrees with our definition above. This is no coincidence. It can be shown that the edge graph of the Euclidean n-cube is the Boolean graph of an n-atom Boolean algebra, and that the Boolean metric is the square of the Euclidean metric on the vertices of the n-cube. The Boolean n-cube is very useful for visualizing definitions, theorems and proofs in Boolean geometry, and its Euclidean geometry can be rigorously incorporate into the mathematics of Boole space, though we won't do so here. However, here is a quick intuitive account of why the two structures coincide: The set of parallel edges in a certain direction connects together two squares to make up the cube (see fig 3.2.1).

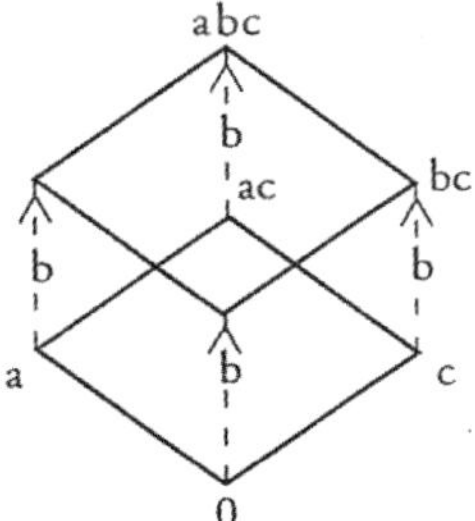

Figure 3.2.1   Adding Atom b to the Lower Square

The lower square represents the Boolean algebra of all subsets of a two member S, while the upper square represents the new subsets that result from adding a third member to S. If we add a fourth member to S, we connect up our cube to another cube by eight parallel lines in a new dimension get a four-dimensional hypercube. We can keep doing this, which shows that an n-dimensional hypercube is a natural representation for an n-atom Boolean algebra. Note that lines in the cube that are not parallel are orthogonal. This is true also in the n-cube, so it makes sense to define orthogonality as not being parallel. It's easy to show that dimension is the maximum number of orthogonal lines.

After this brief side excursion, let's get back to parallel lines. What parallel lines have in common is a particular atom, the atom x+y. Though parallelism is an invariant geometric concept, this statement depends on two non-invariant algebraic concepts: atom and +. Let's now try to think more geometrically. Suppose we choose a particular origin, call it 0. Then an atom can be defined geometrically as a neighbor of that origin.

Theorem 3.2.5 : For any point k, every line is parallel to a line of the form {k, x}, and no two lines of this form are parallel.

Proof: Given a line z,w, we have z+w = k+(k+z+w), so {k, k+z+w} is a parallel line. If x is unequal to y, then k+x is unequal to k+y by the group property of +. QED.

We see from this theorem that for any point k, an equivalence class of parallel lines has a canonical representative as a line connecting k with a neighbor. This is a geometric way of stating that what parallel lines have in common is a *unit displacement*, something that follows immediately from the definition of parallel. Looking at fig 3.2.1, we see that a unit displacement along the dotted lines reverses the two solid squares, keeping the dotted lines fixed. On the n-cube, a unit displacement reverses two n-1 cubes, keeping their connecting lines fixed. The unit displacement is such an important concept that we will give it a special name:

Step: A displacement of the form $A(x) = a+x$, where a is an atom.

Theorem 3.2.6 :   Every displacement is a product of steps in which no step occurs more than once, and the set of steps in that product is unique.

Proof: Any element d is a sum of its atoms, so d+x = a1+a2 ..+ai..+x. Thus D(x) = A1(a2...+x) = A1(A2(a3.. +x)) = etc. = A1(A2..(Ai(x))..). Clearly the ai must include the atoms in d, and any other atom would have to occur at least twice in it. QED.

Steps: Define the set steps (D) to be the unique set of steps that are the factors of displacement D according to theorem 3.2.6. Theorem 3.2.6 tells us that the displacement group is in natural 1-1 correspondence with the set of all sets of steps. As the set of all subsets of a set, it is a Boolean algebra under intersection and complement. What makes this Boolean algebra so important is that it is geometrically invariant! It will be useful to think of this Boolean algebra as a ring, *i.e.* as a set closed under **XOR** and **AND**. Recall that a ring is equivalent to a Boolean algebra, but a subring need not be a subalgebra.

Displacement ring: The displacement group together with an **AND** operator defined as follows:

D&E is the displacement such that steps(D&E) is the intersection of steps(D) and steps(E).

Theorem 3.2.7 : The displacement ring of a Boolean algebra is isomorphic to the ring of that algebra itself under the mapping $D \rightarrow d$, where $D(x) = d+x$.

Proof: $D \rightarrow d$ maps steps 1-1 onto atoms, so & is obviously a corresponding operation on the two rings. DE(x) = D(e+x) = d+e+x, which means that DE = d+e. QED.

Incidentally, this also shows that the so-called displacement ring actually is a Boolean ring, with + defined as the group product. With the geometrically invariant displacement ring we have arrived at a way to automatically transform any Boolean algebraic concept into a Boolean geometric concept without having to check it for invariance. What we have done is almost exactly analogous to pulling the normed linear algebra of Euclidean displacements out of Cartesian geometry. In each case we begin by thinking of the points as vectors and define a displacement as the addition of a vector to all points, the only difference being that Boolean vectors are over the binary field rather than over the real field. In each case we get to the full geometric group by combining the displacements with the rotations and reflections. In the Boolean case, the rotations (which only rotate by multiples of 90 degrees) and reflections are the permutations of coordinate axes, *i.e.* the logic group. Lifting a Boolean atom from the status of a mere object d to the status of a displacement D is to give it an essentially dynamic quality. A step is an object that is either joining or leaving a set. Which set? That doesn't matter; it's only the bare fact that the object is coming or going that creates a step. Which object? That depends on what other steps have been taken! In the face of a moving 0, only the fusion of the mercurial object with the ephemeral set is something fixed, invariant, objective.

I now see the logic of the classical observer in quantum mechanics as a displacement ring on an underlying "hidden" Boolean underworld of incompatible Boolean logics which can only be combined geometrically. It appears that only by dealing directly with logics that change can we find a logic that doesn't change, a logic that everyone can agree on and that will serve for the writing of history and the accumulation of facts.

3.3. *Independence*

In Section 2 we asked whether Boolean geometry is compatible with science as we know it. The issue is whether, given non-logical geometric change, it still makes sense to isolate parts of the world from the general flux. In our present terms, the question is whether we can find a geometrically invariant concept of separation that does the same job for some new science that logical separation, *i.e.* independence, does for present day science. Logics A and B are independent if taken together they form a product logic A.B whose atoms are ordered pairs of atoms from A and B.

A good way to visualize this is to think of the atoms of A being arranged horizontally and those of B vertically so that their ordered pairs form a rectangular set.

Every region of this rectangle is an element of A.B. However, the elements of A and the elements of B are regions of a special kind. An element of A is a vertical stripe, or more exactly, a set of atoms that can be turned into a vertical stripe by rearranging the columns. An element of B is a horizontal stripe in the same sense.

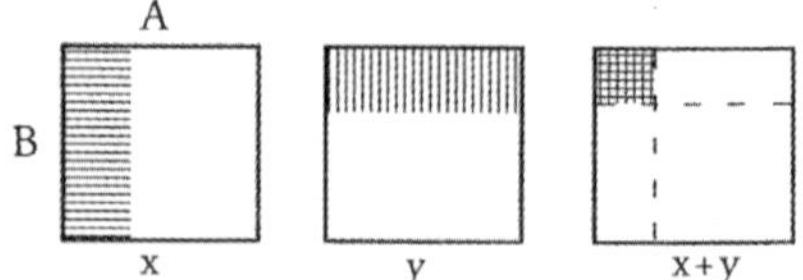

Figure 3.3.1   The Separable Element x&y

An element of the form x&y with x in A and y in B is a rectangular region, *i.e.*
a region that can be turned into a rectangle by rearranging rows and columns; in
particular, 1&1 is the whole rectangle. If an element of A.B is not a rectangle, *i.e.*
if it cannot be separated into an element of A and an element of B, we'll borrow a
term from quantum mechanics and speak of it as "entangled".

Let's now consider what happens when we apply a geometric transformation
G(x) to A.B. First let's assume that G is logical, meaning that it permutes the joint
atoms in the rectangle. To say that G only applies to A means that it only permutes
the columns. This has no effect on the set of atoms in a given row, so the atoms
of B are unaffected by G. More generally, a necessary and sufficient condition for
G to preserve the separability of separable elements is that it only rearranges rows
and columns, *i.e.* that it is a product of a logical transformation on A and a logical
transformation on B. This is just what common sense would expect: if two things
are independent, changing them separately can't make them correlated.

But now suppose that G is a displacement. If G applies to only one of the
components, we shouldn't expect it to affect the other, right? Let G(x) = a+x,
where a is in A. If we look only at what G does to the elements of A, *i.e.* to the
vertical stripes, then it displaces them just as if B didn't exist. B could be on
another planet, in another universe, as far as A is concerned. Thus it seems to be
OK to isolate a Boole space from the rest of the world as an object of study. So
far, Boolean geometry is still just another step in the cheerful march of scientific
progress. But wait just a minute, stop the band! A messenger has just arrived from
planet B reporting that something very odd has happened to the elements on his
planet, which is that they have all become entangled with the elements on ours!
What has happened? Our geometric displacement G, which seemed to be confined
to our A logic alone, has actually zapped his planet and turned every element y into
the inseparable element y+a!

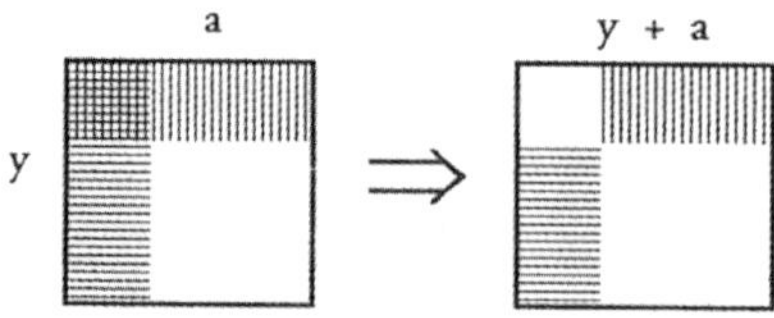

Figure 3.3.2   Displacemnt of a Applied to y

There is no way to analyze a Boole space into separate geometric parts. Separation is simply not a geometric concept. Unlike the action of a logical transformation

or the action of a projection, the action of a displacement cannot be broken down into distinct actions on independent factors. The reason for this is that any "rectangular" separation of possibilities still leaves all the parts with a common null set, a common 0, so a shift in the 0 of one part is a shift in 0 of all!

Bur before you take off for a Tibetan monastery or banish Boolean geometry to some mystical netherworld, remember the displacement algebra. Though it's a geometric invariant, it's also a Boolean *algebra* and as such is factorable like any other Boolean algebra. Even though the underlying geometry has no objective parts, there is at least an objective way to take apart an aspect of this geometry. Furthermore, if the geometry acquires a 0, the factorization of the displacement algebra can be as-it-were projected onto it. All is not lost to science after all.

Of course if the geometry acquires a fixed origin it simply turns into logic, and the displacement algebra as such becomes irrelevant. What we're really interested in is the laws governing a moving origin, which is the likely source of of non-Boolean change in quantum mechanics. The conceptual framework within which we shall formulate and study these laws is the displacement algebra, and our main tool in this study will be the concept of *amplitude*, which will be the focus of another paper.

. . . .

Brief Biography

Thomas Etter's educational background includes broad study in mathematics, philosophy, and computing. He holds several early patents on integrated circuits, and has conducted research *via* grants from the State of New York and the University of Minnesota. In the 1990s, Etter was Senior Software Architect at the E-Speak division of Hewlett-Packard, and later worked at Interval Research Corporation on a new and very general approach to mathematical relations called Link Theory, with applications in both computation and physics. Etter has also served as President of the Alternative Natural Philosophy Association, an international scholarly organization, and as editor of its West Coast journal.

Thomas Etter died on April 1st, 2013, before he could provide his promised contribution to this Festschrift for his close friend Pierre Noyes, whose digital physics was inspirational and related to much of Tom's later work.

The paper of his which is presented here was selected from amongst the many in his archived material by his long-standing colleague Richard Shoup.

Speculation on
Consciousness as Relative Existence

Louis Gidney

Shiel Cottage, Strontian, Acharacle, Argyll PH36 4HZ, Scotland
E-mail: louis@gidney.com

A speculation which arose within the context of the nature of the Bit-string Combinatorial Heirarchy is offered for discussion. It seems to provide a strong clue to how systems must be constructed if they are to support consciousness. It depends on regarding consciousness as a case of 'relative existence'. So that:
'A is conscious of x' becomes: 'x exists for A'.

1. Constructing systems that are to support consciousness

I would like to put up the following speculation for further discussion. Its origins go back to conversations I had with the late Ted Bastin [1], a physicist, and the late Dorothy Emmet [2], a philosopher and subsequent conversations with Pierre Noyes and others.

'Relative existence' or 'existence–to' comes from re-thinking the basis of physics approximately along the process ideas of A. N. Whitehead [3]. I don't think much progress can be made on the 'hard problem' of how it is that there is experience at all, without re-interpreting physics.

This involves deleting 'absolute objective existence' from our vocabulary while satisfying our conviction that the world is real and connected, by envisaging it as a flux of pure activity which is non-substantial in its essence — more like energy than traditional matter.

We then have to imagine that spatial extension and material particles are produced by acts, within which they have only a relative (and transitory) 'existence–to' each other but no absolute existence. This is not consciousness, but it is the germ within 'matter' that makes it possible for 'mind' to arise — for systems to be conscious — for something to 'exist for' them.

The progression from this bare, undifferentiated, transitory 'relative existence' of a particle within the simplest quantum act, to all the richness of human consciousness, involves an entropic principle such that when the 'spontaneous' interactions between the parts of a system becomes constant (or constantly repetitive), the 'relative existence' of the parts to one another is neutralised or habituated, such that the formerly separate 'experiencing' of the parts merge into a unified experiencing activity directed outwards.

If this view is sound, it would follow that conscious entities have to be constructed in such a way as to take advantage of, or 'concentrate' relative existence in this way.

This would rule out the possibility that 'functionally equivalent' systems of silicon chips could support consciousness, because the processing that goes on in them does not have an intimate relation to underlying physical processes. However it does raise the intriguing question of what substances and/or states of matter might be suitable.

It seems to me that present understanding in this area is rather like the state of electromagnetism before the discovery of the electric current and the work of Faraday and Maxwell. The principle described here at least provides a 'handle' on how conscious devices should be made, which the purely behaviourist Turing test does not.

One can envisage devices manifesting species of 'relative existence' that are not much like consciousness. I doubt that the essence of consciousness has anything to do with intelligence — artificial or otherwise, though clearly some such appended means of communicating is necessary.

2. 'Existence-for-others' and how Dualism arises

Knowledge of physics (indeed all science) starts with experience and gives results which have meaning in further experiences. But it is exactly the having of experience that consciousness is about. Our concepts of what we mean by 'exist' and 'real' are also derived from experience *via* centuries of thought that may have gone astray and may now need revision. So the 'hard problem' of consciousness is intimately bound up with our ontological conceptions of the nature of the world and our place in it. (Schopenhauer's *'world knot'*).

Since we have been conditioned to a dualistic way of thinking by centuries of use, the way out is intricate and delicate. Because it is delicate, it requires sensitivity. I have coined the term *relative existence* to evoke a fresh return to observation of how we find the world to be in our experience.

Descartes' *'cogito'* strikes me as contrived — not least because he was adult. Babies do not pop into life, look around, conclude *'I think, therefore I am'*, then proceed to ruminate on whether some of the objects around them are conscious beings like themselves.

From evidence of maternal deprivation, 'wolf boys', *etc.*, it looks as if consciousness is not entirely an innate property of brains, but largely a product of interaction with others. Evidence from solitary confinement, *etc.*, suggests that interaction may even be needed to sustain consciousness.

If this is so, solipsism could not arise. For if human consciousness, unlike simple awareness, is a product of interaction between people, then the world must come to exist for us in infancy as a shared experience, and is public from the very start.

To regain an accurate ontological foundation for the sciences I think we have to look very carefully at the kind of existence things have for us in direct experience, including early experiences, and how we have proceeded from there to where our scientific conceptions are today.

My own suggestion is that things as experienced are not primarily felt to exist 'in themselves' nor 'for me'. This comes later. Rather the mode of existence things have as experienced, might suitably be called 'existence for others' or 'existence to others'. This does not mean that we have other people's experiences, but we do spend a lot of time imagining how other people see things. This gives those things what some might call an 'overlayer of meaning', and others might regard as essential.

Either way it is a type of existence that is neither objective nor subjective in its primary immediacy. It may be seen as subsuming both as special cases which have grown out of it :

First, 'existence-in-itself' may be seen as that special case of primary 'existence for others' where the 'other' in question is replaced initially by a supreme other, an all-knowing God, and later by His secular carbon-copy: the *C19* 'impartial objective observer'. Either of these bestows upon everyday things an 'existence feel' which is different from the purely personal one out of which awareness grew in the first place. Nothing else in experience (itself an activity) suggests the concept of absolute existence, or the need for it.

Second, subjective 'existence for me' arises from the 'for others' kind only after appropriate socialisation of the individual (not the same in all cultures).

If, as seems to me, a fresh and careful phenomenological analysis of experience more or less deconstructs the concept of 'absolute objective existence', then we have to find another way to satisfy our conviction that the world is real and enduring.

Are there serious conceptual difficulties in rethinking the physical world as essentially 'process' rather than 'existence' — along the lines of A. N. Whitehead, William James, *etc.*, with the benefit of more experimental data than they had? For if we can so rethink, then the conceptual scheme described would embrace consciousness in an uncontrived and natural way.

To what extent would we be guilty of anthropomorphism (and does it matter) if we project onto nature the idea that the relative existence of one thing to another can occur within fundamental physical processes?

After all we can't have it both ways:

> either we want to have a "property of 'matter' " that allows consciousness to arise from it, or not.

I suspect that *if we don't put it in, we don't get it out.* Evidently it is in there since we ourselves are conscious.

3. Summary

According to the conceptual scheme outlined above, inasmuch as a conscious subject is not an object (but in some sense its opposite) no amount of signal processing complexity in an artificial brain object can ever make it happen that there is something that exists 'for' that brain object, however it may behave or whatever it may report, unless the natural 'relative existence' inherent in physical processes is used and made to 'concentrate' in some sense.

These ideas provide a guideline to work to, where we now have none, except for a semi-magical belief that by imitating perceptual mechanisms somehow, complexity alone will one day convert objects into subjects.

References

[1] See, *e.g.* : Ted Bastin, 'Mathematics of a Hierarchy of Brouwerian Operations', Information Structures Unit, September 01, 1964, Cambridge Language Research Unit.
[Recreated with permission and edited by John Amson, 1998] in *Aspects I*, Proceedings ANPA19, pp.5–22 (1999).
[2] See, *e.g.* : Dorothy Emmet, *The Passage of Nature*, Macmillan, London (1992).
[3] A. N. Whitehead, *Process and Reality*, Macmillan USA; 2nd Revised edition (1979).

A Management View of ANPA (East) 1979 to 2012

Michael Horner

1 Chemin des Rannaux, Coppet, Vaud, Switzerland, CH–1296
E-mail: michael.horner@altimasa.net

Using the ANPA annual proceedings we analyze the authors and titles of ANPA papers and detect some long term trends. The author recalls some early conversations with Pierre Noyes and presents the results after following his advice for 27 years.

1. Introduction

The Alternative Natural Philosophy Association (ANPA) is an unusual international association with British and American roots. One view of ANPA is that it is a safety valve where academics can present and discuss alternative views which would be normally disallowed. A second view is that science has been promising a paradigm shift ever since Kuhn made the term part of common speech and that ANPA has made the shift. A third view is that it is a place where top quality minds get together once a year.

In this paper I introduce some foundational ANPA documents so that the reader can get a sense of the origins of ANPA, the founder members, and the content of the work. In another contribution [GET REF HERE] to this Festschrift the reader will find a 2008 report to the ANPA Advisory Board written by the late Clive Kilmister, one of the five Founder members of ANPA. For me that paper summarises in a few pages what ANPA had investigated from its beginning.

Next I use parts of the ANPA (East) annual proceedings to analyse ANPA since the beginning. From this I generate tables of interest about the people and their papers as well as the annual meetings.

Armed with these tables I classify all the papers and attempt to find trends in their content.

I respectfully submit this report to Pierre Noyes and thank him for his advice and support. As a result of being part of ANPA I have encountered many interesting ideas and people. Thank you Pierre.

2. Executive Summary

Since I conducted an analysis looking for trends I divided ANPA into 4 periods.

Period 1	ANPA 01	to	ANPA 09
Period 2	ANPA 10	to	ANPA 18
Period 3	ANPA 19	to	ANPA 24
Period 4	ANPA 25	to	ANPA 32

The Formative Years from ANPA 01 to ANPA 09

This period begins with the formation of ANPA in 1979.

Although formal meetings were held every year proceedings were only published for ANPA 07 and ANPA 09, both edited by Pierre Noyes. The proceedings for ANPA 09 set the proceedings style and in a way is a summary of work since the formation of ANPA in 1979. The preface to the proceedings was written by Pierre Noyes and suggests ANPA had reached a turning point by making the claim that "We have achieved a discrete reconciliation of quantum mechanics and relativity by going beyond the conceptual framework of Bohr and Einstein".

Of significance is the only paper by Frederick Parker-Rhodes in the data set used. "FPR" was the discoverer of the Combinatorial Hierarchy "CH" which is central to ANPA work.

The Growth Years from ANPA 10 to ANPA 18

The nine meetings and their proceedings in this period were solidly based on the founding concepts. The Combinatorial Hierarchy was examined and re-examined and Discrete Physics was a common topic. Membership judged by attendance at meetings increased and quality papers were delivered. Towards the end of the period the topics became wider but the founding ideas dominated.

The Philosophical Years from ANPA 19 to ANPA 24

During period 3 there were 11 proceedings published over 6 years.

This was due to 5 annual meetings having days dedicated to philosophical matters as opposed to scientific matters. Meetings were also organised in a different way so that fewer papers were presented on a given day and members therefore had more time to meet and discuss. This turned out to be a popular notion as was the notion of each day being the responsibility of people sharing organisational work with the president and the coordinator.

The Mature Years from ANPA 25 to ANPA 32

During periods 1, 2, and 3 the processes associated with arranging annual meetings and producing proceedings had been sorted out. Providing there was money we had people who knew what to do. The content of the meetings continued to evolve and younger people were beginning to be respected and have as much say as the founders. The separate philosophical proceedings were dropped but the content in this period was much less dominated by scientific topics than in the early years.

Summary of Contributions to ANPA Proceedings

Total number of Authors	109	First named authors only
Total number of Papers	444	Only proceedings of ANPA (East)
Authors with ≥ 11 Papers	10	
Total Papers by Founders	100	Including 1 paper by Parker-Rhodes
Number of Proceedings	30	
Number of Proceedings Editors	6	

Papers per Author

1	2	3	4	5	6	7	8	9	≥ 11	Total	Papers by all authors	Papers by Founders
54	15	8	10	2	3	4	1	2	10	109	Authors	
54	30	24	40	10	18	28	8	18	214	444	Papers	100
1											Parker Rhodes	1
									35		Kilmister	35
									32		Bastin	32
									25		Marcer	
									23		Kauffman	
									21		Noyes	21
									20		Rowlands	
									18		Manthey	
									18		Bowden	
									11		Amson	11
									11		Constable	

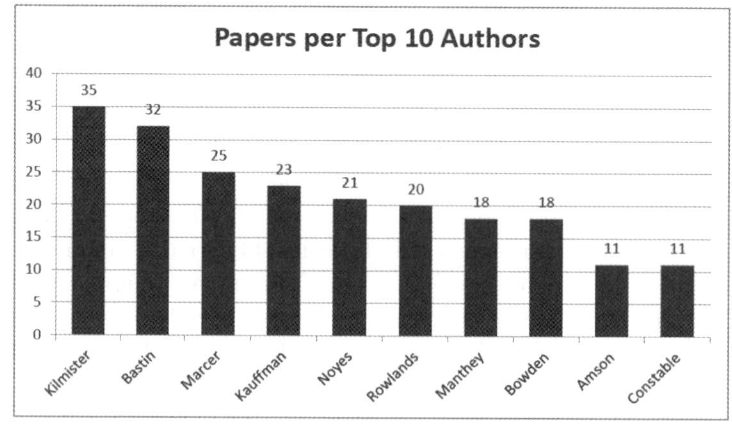

3. The Formation of ANPA

ANPA was founded by five people and according to the most often told anecdotes Pierre Noyes was responsible for crafting the ideas leading to the formal existence of ANPA. In this sense he was the founder but Pierre always claimed he merely brought together what was in the minds of all five. Clive Kilmister and Ted Bastin had worked together for some time in the 1950s and then also with Frederick Parker-Rhodes and John Amson in the 1960s. Pierre made up the quintet. Whilst Ted Bastin was visiting California in the early 1970s he gave a seminar in the Philosophy Department at Stanford on his work on self-organising systems, both theoretical and as simulated in electronic analogue computers devised with the help of Gordon Pask. It was at that seminar that he met Pierre Noyes and the ANPA seeds were laid. Ted and Clive had worked together for many years and had published their first paper in 1953. They had met when Clive was looking for a book in the University library which was signed out to a certain Dr Bastin. The book was by Eddington whose ideas influenced the original notions of ANPA.

Here is the story as told by Pierre Noyes :

Preface to Proceedings of ANPA 07 (1985)

Currently accepted ideas and results based on them in elementary particle physics and cosmology are exciting for some, frustrating for others, and incomprehensible to many who sense the excitement but don't know quite how to join in the game. The group represented by these proceedings share some mix of all three attitudes in varying proportions. Some of us think we are on the track of new physics, or even new philosophy, while others are sceptical. We are not trying to conceal these internal problems from the reader. Rather, we ask him to share our perplexity – along with our excitement.

The Alternative Natural Philosophy Association was founded in 1979 by Noyes, Kilmister and Bastin – who were immediately joined by Parker-Rhodes and Amson. The first four of us met at Kilmister's "Red Tiles Cottage" for our first international meeting (ANPA 1). ANPA 2–8 (1980-86) have been held (ANPA 8 only prospectively next month as of this writing) at King's College Cambridge each summer. The purposes of the organization have remained the same: "... to consider coherent models based on a minimal number of assumptions to bring together major areas of thought and experience within a scientific philosophy alternative to the prevailing scientific attitude. ... " A more complete statement closes this volume.

The original, and still the central, focus of our research stems from the discovery of the combinatorial hierarchy — *i.e.* the terminating sequence 3, 10, 137, $2^{127}-1+137 \sim 1.7 \times 10^{38}$ — which has numerical connection to the scale constants of physics. But as the number of ANPA members expands other ideas are also entering our discussions, as I hope this volume will make clear. The original organizational structure changes this year to an elective presidency, as indicated at the close of the volume. The consensus at ANPA 7 was that much convergence had occurred and that every effort should be made to get out a "Proceedings". Since no one else was forthcoming, I made the mistake of volunteering to take the responsibility of editing them. The result is presented here.

Combinatorial Hierarchy (CH)

Central to the content of the work of ANPA is the CH.

This had been discovered "two decades" before Pierre was writing the preface to the proceedings of ANPA 9 (1987) by Amson, Bastin, Kilmister and Frederick Parker-Rhodes.

Here is the abstract of the paper by FPR from the proceedings of ANPA 07.

AGNOSIA : A Philosophical Apologia for INDISTINGUISHABLES
by A. F. Parker-Rhodes,

As an introduction to my topic, I propose to offer a brief historical prelude. Somewhere around 1962 I hit upon a series of numbers of which Ted Bastin noticed that the last two (the generating procedure could not produce more than four) were close to two well-known physical constants, the reciprocals of the fine-structure constant and of the gravitation coupling constant.
He drew the attention of Clive Kilmister at King's College London, and John Amson at St Andrews, to this series, and these two began to work on the algebraic formulation of the series, whose self-terminating property intrigued the mathematicians as much as the contents of the series did the physicists. I also worked on the problem, albeit divergently, having noticed that it might be based also and perhaps more profoundly, on "indistinguishables". At a meeting where I expounded 'the germ' of this idea, there was a student who drew attention to the lack of mathematical rigour in my exposition. This set me to try and correct the deficiency. The work took some years, and led me much further from orthodox mathematics than I had expected. Nevertheless, it eventually reached a form in which I could hope to publish it. Pierre Noyes of Stanford University, who had meanwhile initiated the setting up of the Alternative Natural Philosophy Association to further the work, gave valuable assistance in the final stages, and was instrumental in getting it accepted by Jaako Ilintikka, general editor of the Synthese series published by Reidel of Dordrecht in 1981.

The approach through indistinguishables has been viewed with suspicion, on philosophical grounds, by physicists in ANPA; for these entities are not physical objects, and it has been customary to look only to physical things to explain physical phenomena. I challenged this position; but it must be done at length if it is to persuade the opposition.

Statement of Purpose

(The following is a summary version of the official Statement of Purpose)

1   The primary purpose of the Association is to consider coherent models based on a minimal number of assumptions, so as to bring together major areas of thought and experience within a Natural Philosophy alternative to the prevailing scientific attitude. The Combinatorial Hierarchy, as such a model, will form an initial focus of our discussions.

2   The Association will seek ways to use its knowledge and facilities for the benefit of humanity and will try to prevent such knowledge and facilities being used to the detriment of humanity.

Organization

1   An Executive Council is the governing body of the Association.

2   Membership is made up of the people who attend annual meetings. No one of the general public or a member should be excluded from attending meetings.

3   The President is the official representative of the Association in external affairs, and has the responsibility for calling meetings of the Executive Council, at least annually, for the determination of overall policy

4   The Treasurer is the responsible financial officer of the Association.

5   The President and the Co-ordinator may be paid an appropriate salary for their services, funds permitting.

6   The Executive Council has selected an Independent Advisory Board.

Note: The Statement of Purpose (and Organisation) was modified several times over the years. It appeared in the annual proceedings and in principle the current version is in the proceedings for ANPA 32 (2011).

4. Analysis Using ANPA Proceedings

Following the tradition set down by Pierre Noyes by publishing formal proceedings for ANPA 09, proceedings were published following all meetings after ANPA 09.

Typical proceedings are books of 300 pages.

Proceedings Covers

The content of the cover of ANPA 09 set the style for future years. A great deal of effort by the editors went into choosing cover designs and a theme for the annual proceedings.

DISCRETE AND COMBINATORIAL PHYSICS
PROCEEDINGS OF ANPA 9
H. Pierre Noyes, Editor
Proceedings of the 9th Annual International Meeting of the
ALTERNATIVE NATURAL PHILOSOPHY ASSOCIATION
Department of the History and Philosophy of Science Cambridge University, September
23-28, 1987
Published by ANPA WEST 25 Buena Vista Way, Mill Valley, CA

The name, the date and place of the annual meeting as well as the editor and the publisher are all recorded. The proceedings were prepared between annual meetings and handed out (sold) at the next annual meeting.

Tables of Content

The TOC in the proceedings list the author and the title of their paper.

In library terminology a document comes in several levels
  1   Title
  2   Title and Abstract
  3   Title and Abstract and full Text
  4   Title and Abstract and full text and Citations

In this paper I have only used the information on the covers and the information contained in the TOCs of the proceedings which is roughly level 2.
To fully digitalise all 30 proceedings means scanning about 10,000 pages!

Analysis of the proceedings in this limited way yields the following tables.
  • Proceedings and Meetings by Period
  • Meetings, Authors, papers by Period
  • Meetings, Proceedings and Editors
  • List of all Authors
  • Classified Papers

Method of Analysis

ANPA is an international organisation and the most obvious features of ANPA are meetings and annual proceedings. I have had to make some restrictions.

- I have based this analysis solely on the ANPA (East) annual proceedings : ANPA 07 (1985) until ANPA 32 (2011).
- There is no analysis of documents from ANPA (West).
  - This is a major shortcoming as ANPA (West) played a big part in ANPA.
- Also excluded at this stage are newsletters, correspondence and meeting reports.
  - Suggestions for documents to be included in future are welcome.
  - I am especially interested in copies of the timetables for each meeting as they actually happened since not all papers and therefore all authors appeared in the proceedings.

Furthermore although my basic unit of analysis is the combination of authors and the papers they wrote, I have made several simplifications :

- Names are only the names of the first authors when multiple authors are listed.
  - Some authors are therefore underrepresented in the author analysis.
- Only surnames are used.
  - I expected this to lead to some confusion but among the 100 or so authors the only ones indistinguishable were in the Parker-Rhodes family!
- Titles of papers have been shortened or modified.
  - This could lead to confusion when we get a complete set of all documents.

With the help of several ANPA members I have been able to assemble a nearly complete set of ANPA (East) proceedings spanning 1979 until 2011. There are 30 such documents.

- My analysis is based on information in the cover pages and the tables of content.
  - Information from the cover page for ANPA 09 is included as an example
- Tables of Content (TOC) were scanned in and are now machine editable.
  - Page numbers were left out and replaced by a classification of each entry.

Basic unit of Analysis

The fundamental item of data including "Classification" has the following form since each paper can only be in one class :

Author   Paper (Title)   Class

I have classified all papers and then analysed them overall and by period.

Classification

In order to classify the papers I decided to use the three basic ANPA concepts :
                Alternative (A),   Natural (N),   Philosophy (P)
This allows me to generate 7 categories.

1	2	3	4	5	6	7
A	**N**	**P**	**AN**	**AP**	**NP**	**ANP**
Alternative	Natural	Philosophy	Alternative Natural	Alternative Philosophy	Natural Philosophy	Alternative Natural Philosophy
NonStand.	Standard	Not Science	NonStand.	NonStand.	Classical science	Living systems
	Continuous Physical		Continuous Informational	Discrete Informational	Newton Einstein	Biology Consciousness
		Fundamentals	IT	Fundamentals CH	science QM, ST, GR	Fundamentals Philosophy

1. Alternative means alternative to the standard model(s)
2. Natural means the standard model which means physical and continuous
3. Philosophy means as opposed to Science and deals with fundamentals
4. Alternative Natural by dropping physical means informational
5. Alternative Philosophy by dropping continuous means discrete and informational
6. Natural Philosophy means classical physics from Newton onwards
7. Alternative Natural Philosophy means multi level systems and consciousness

I applied this classification system to all 444 papers and noted common trends.

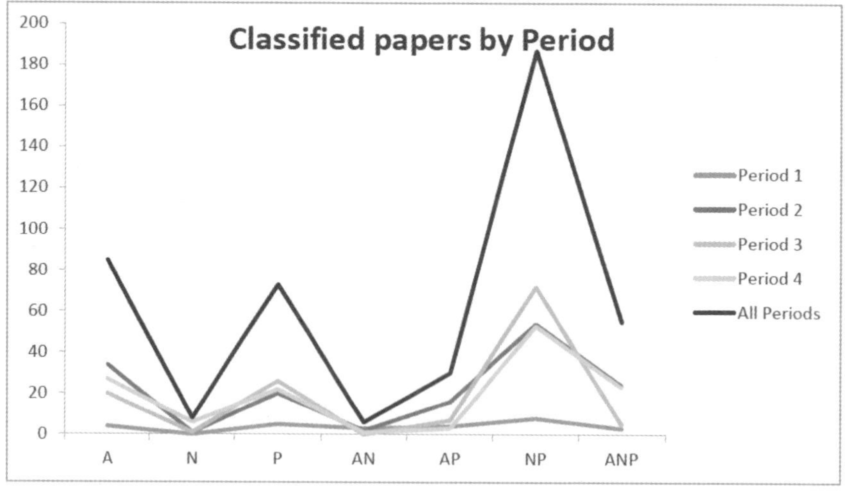

5. Some Results from the Analysis Using ANPA Proceedings

In my analysis of the Authors and their associated papers I noted that the founders had produced exactly 100 papers of the total of 444.

The ten most prolific authors produced 214 papers between them.

There were 109 authors and so 99 authors produced 230 papers.

The names of the ten most prolific authors are as follows.

1.	Kilmister
2.	Bastin
3.	Marcer
4.	Kauffman
5.	Noyes
6.	Rowlands
7.	Manthey
8.	Bowden
9.	Amson
10.	Constable

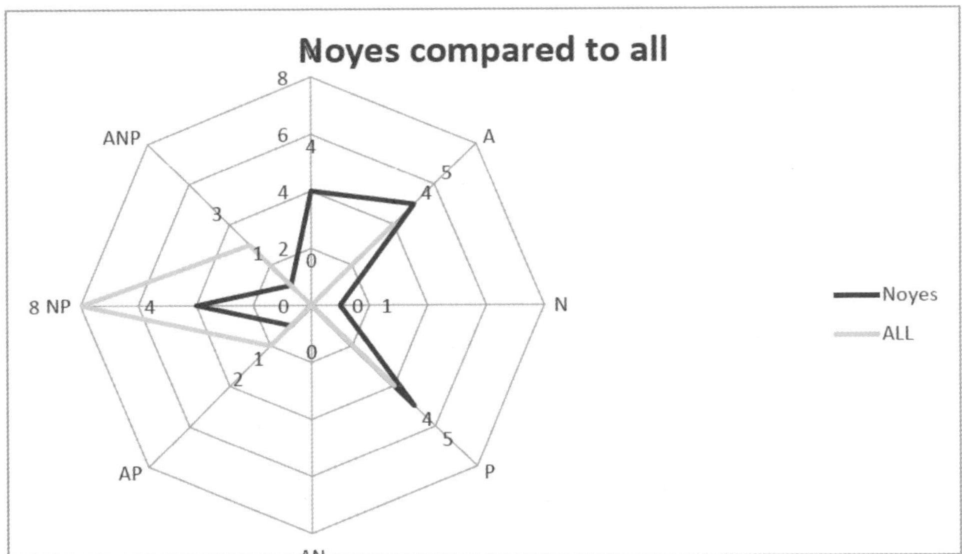

Proceedings and Meetings by Period

Year	Meeting	Count	No.	Title
1985	ANPA 07	1	7	Proceedings of the 7th annual conference of ANPA
1987	ANPA 09	1	9	Discrete and combinatorial Physics
	Period 1	2		
1988	ANPA 10	1	10	Discrete Physics and Beyond
1989	ANPA 11	1	11	Objects in Discrete Physics
1990	ANPA 12	1	12	Alternatives in Physics and Biology
1991	ANPA 13	1	13	Process and Combinatorial Hierarchy
1992	ANPA 14	1	14	Conservation and Invariance
1993	ANPA 15	1	15	Alternatives
1994	ANPA 16	1	16	Entelechies
1995	ANPA 17	1	17	Philosophies
1996	ANPA 18	1	18	Mereologies
	Period 2	9		
1997	ANPA 19	1	19	Aspects 1
1998	ANPA 20	1	20	Aspects 11
	ANPA 20 A	1		Paranormal Aspects
1999	ANPA 21	1	21	Participations
	ANPA 21 A	1		Participations Philosophical Aspects
2000	ANPA 22	1	22	Implications Scientific Aspects
	ANPA 22A	1		Implications Philosophical Aspects
2001	ANPA 23	1	23	Correlations
	ANPA 23A	1		Movements Philosophical Aspects
2002	ANPA 24	1	24	Boundaries
	ANPA 24A	1		Boundaries Philosophical Aspects
	Period 3	11		
2003	ANPA 25	1	25	Spin. Proceedings of ANPA 25
2004	ANPA 26	1	26	Against Bull
2005	ANPA 27	1	27	Conceptions
2006	ANPA 28	1	28	Foundations
2007	ANPA 29	1	29	Exactness
2008	ANPA 30	1	30	Reflexivity
2009	No meeting	0		
2010	ANPA 31	1	31	Contexts
2011	ANPA 32	1	32	Haecceity
2112	Not included	0	33	"33"
	Period 4	8		
	Total	30		

Meetings, Authors, Papers, by Period

YEAR	Meeting No	Authors	Papers	Authors	Total Papers
			444	109	444
		Period 1		26	27
1985	ANPA 07	10	10		
1987	ANPA 09	16	17		
		Period 2		139	151
1988	ANPA 10	11	11		
1989	ANPA 11	12	16		
1990	ANPA 12	15	16		
1991	ANPA 13	13	16		
1992	ANPA 14	15	15		
1993	ANPA 15	16	18		
1994	ANPA 16	21	22		
1995	ANPA 17	20	20		
1996	ANPA 18	16	17		
		Period 3		113	131
1997	ANPA 19	14	15		
1998	ANPA 20	16	19		
	ANPA 20 A	8	8		
1999	ANPA 21	12	14		
	ANPA 21 A	5	6		
2000	ANPA 22	14	17		
	ANPA 22A	6	8		
2001	ANPA 23	12	15		
	ANPA 23A	7	7		
2002	ANPA 24	11	12		
	ANPA 24A	8	10		
		Period 4		120	135
2003	ANPA 25	18	21		
2004	ANPA 26	17	21		
2005	ANPA 27	17	21		
2006	ANPA 28	19	21		
2007	ANPA 29	13	14		
2008	ANPA 30	12	12		
2010	ANPA 31	12	13		
2011	ANPA 32	12	12		

Meetings, Proceedings and Editors

	ANPA No	Editor	Title
1985	ANPA 07	Noyes	Proceedings of the 7th conference
1987	ANPA 09	Noyes	Discrete and combinatorial Physics
1988	ANPA 10	Kilmister	Discrete Physics & Beyond
1989	ANPA 11	Manthey	Objects in Discrete Physics
1990	ANPA 12	Manthey	Alternatives in Physics and Biology
1991	ANPA 13	Abdullah	Process and Combinatorial Hierarchy
1992	ANPA 14	Abdullah	Conservation & Invariance
1993	ANPA 15	Kilmister	Alternatives
1994	ANPA 16	Bowden	Entelechies
1995	ANPA 17	Bowden	Philosophies
1996	ANPA 18	Bowden	Mereologies
1997	ANPA 19	Arleta	Aspects1
1998	ANPA 20	Bowden	Aspects 11
1998	ANPA 20	Bowden	Paranormal
1999	ANPA 21	Bowden	Participations
1999	ANPA 21A	Bowden	Participations Philosophical Aspects
2000	ANPA 22	Bowden	Implications Scientific Aspects
2000	ANPA 22A	Arleta	Implications Philosophical Aspects
2001	ANPA 23	Bowden	Correlations
2001	ANPA 23A	Arleta	Movements Philosophical Aspects
2002	ANPA 24	Bowden	Boundaries
2002	ANPA 24A	Bowden	Boundaries Philosophical Aspects
2003	ANPA 25	Bowden	Spin
2004	ANPA 26	Bowden	Against Bull
2005	ANPA 27	Bowden	Conceptions
2006	ANPA 28	Bowden	Foundations
2007	ANPA 29	Bowden	Exactness
2008	ANPA 30	Arleta	Reflexivity
2010	ANPA 31	Arleta	Contexts
2011	ANPA 32	Arleta	Haecceity

30 proceedings

6 Editors

Noyes	Kilmister	Manthey	Abdullah	Arleta	Bowden
2	2	2	2	6	16

Arleta = Arleta Ford, Arleta Griffor.

List of all Authors

Order	Name	No of papers	Order	Name	no	Order	Name
1	Kilmister	35	38	Ord	3	75	Doughty
2	Bastin	32	39	Read	3	76	Drago
3	Marcer	25	40	Young	3	77	Dubois
4	Kauffman	23	41	Aspden	2	78	Durgun
5	Noyes	21	42	Buzzetti	2	79	Engstrom
6	Rowlands	20	43	Comfort	2	80	Fawn
7	Bowden	18	44	Gabrys	2	81	Gefert
8	Manthey	18	45	Glendza	2	82	Goodwin
9	Amson	11	46	Gregory	2	83	Goyal
10	Constable	11	47	Landsberg	2	84	Hadley
11	Deakin	9	48	Lindesay	2	85	Karmanov
12	Pope	9	49	Morikawa	2	86	Kessler
13	Arleta	8	50	Mountcastle	2	87	Lehmann
14	Croll	7	51	Oshins	2	88	Luscombe
15	Eisenhardt	7	52	A Parker-Rhodes	2	89	Maroney
16	McGoveran	7	53	Pratt	2	90	Maxwell
17	Zimmerman	7	54	Pstrudna	2	91	Mobius
18	Clement	6	55	Whitney	2	92	Nagra
19	Heather	6	56	Barrington	1	93	Osborne
20	Roscoe	6	57	Beck	1	94	F Parker-Rhodes
21	Etter	5	58	Bell	1	95	Raptis
22	Honey	5	59	Belt	1	96	Robson
23	Blizard	4	60	Benford	1	97	Rossiter
24	Grey	4	61	Bohm	1	98	Sanchez
25	Hiley	4	62	Booth	1	99	Selleri
26	Honig	4	63	Bortoft	1	100	Sorkin
27	Koberlain	4	64	Briddell	1	101	Starson
28	Kurth	4	65	Burgos	1	102	Stein
29	Matzke	4	66	Cambett	1	103	Thomas
30	Ransford	4	67	Carr	1	104	Watson
31	Shoup	4	68	Chatfield	1	105	Wegener
32	Wood	4	69	Chattergee	1	106	Wheeler
33	Abdullah	3	70	Chew	1	107	Wikman
34	Alvarez	3	71	Clarke	1	108	Wilks
35	Blake	3	72	Colvin	1	109	Wouthuysen
36	Gidney	3	73	Curtis	1		1 paper for
37	Hill	3	74	Deravi	1		Names above

Classification of Papers

YEAR	ANPA No	Papers	A	N	P	AN	AP	NP	ANP
		444	85	8	73	6	30	187	55
		Period 1							
1985	ANPA 07	10	2	0	2	1	3	2	0
1987	ANPA 09	17	2	0	3	2	1	6	3
	Totals 1	27	4	0	5	3	4	8	3
		Period 2							
1988	ANPA 10	11	1	0	2	0	1	4	3
1989	ANPA 11	16	4	0	5	0	2	4	1
1990	ANPA 12	16	5	0	5	0	0	5	1
1991	ANPA 13	16	6	1	1	0	2	5	1
1992	ANPA 14	15	3	0	1	1	2	3	5
1993	ANPA 15	18	2	0	1	1	3	7	4
1994	ANPA 16	22	6	0	2	0	1	10	3
1995	ANPA 17	20	4	0	2	0	2	9	3
1996	ANPA 18	17	3	0	1	0	3	7	3
	Totals 2	151	34	1	20	2	16	54	24
		Period 3							
1997	ANPA 19	15	0	0	2	0	2	9	2
1998	ANPA 20	27	4	0	8	0	0	14	1
1999	ANPA 21	20	1	0	3	0	2	14	0
2000	ANPA 22	25	4	0	1	0	1	19	0
2001	ANPA 23	22	7	0	5	0	2	7	1
2002	ANPA 24	22	4	1	7	0	0	9	1
	Totals 3	131	20	1	26	0	7	72	5
		Period 4							
2003	ANPA 25	21	5	1	6	0	2	4	3
2004	ANPA 26	21	5	0	3	0	1	8	4
2005	ANPA 27	21	3	0	3	0	0	12	3
2006	ANPA 28	21	6	2	3	1	0	4	5
2007	ANPA 29	14	1	1	2	0	0	7	3
2008	ANPA 30	12	2	0	2	0	0	6	2
2010	ANPA 31	13	3	1	2	0	0	6	1
2011	ANPA 32	12	2	1	1	0	0	6	2
	Totals 4	135	27	6	22	1	3	53	23

6. Distribution of Papers by the Ten Most Prolific Authors

	M	A	N	P	AN	AP	NP	ANP
Kilmister Papers	0	14	12	2	0	0	3	4
1 Foundational discussion				1				
2 Concluding Remarks			1					
3 What do the Bits in the CH Mean		1						
4 Discrete Physics and Lorentz Transformation			1					
5 Dirac								1
6 ANPA, The CH, PU and the UK			1					
7 What is the Combinatorial Hierarchy?			1					
8 Should Eddington be elected to ANPA posthumously?			1					
9 Calculating Combination			1					
10 What is the Meaning of the Combinatorial Hierarchy?			1					
11 Backwards Progress is Still Progress			1					
12 Space, Time, Discrete			1					
13 Everybody's doing it			1					
14 Labelling the Hierarchy			1					
15 Algebraic Structure in General Relativity								1
16 Theoretical Quantities: and Entropy								1
17 The Hierarchy Reviewed			1					
18 Discrimination with Aspect								1
19 False Conjectures about Cycles		1						
20 Comments on The Dirac Algebra							1	
21 Prospero and Caliban		1						
22 The Einstein Velocity Relation & the CH		1						
23 Thoughts on the continuum and CH		1						
24 Interpreting the Hierarchy: Level Change		1						
25 Time				1				
26 The Fine Structure Constant		1						
27 The Eddington Heritage - A Claim for ANPA		1						
28 Calculating the Fine Structure Constant		1						
29 My Persistence with ANPA		1						
30 Indistinguishables Reconsidered		1						
31 Tame Non-Associativity							1	
32 "New Foundations" for the Combinatorial Hierarchy		1						
33 Comments on the above							1	
34 The status of the Combinatorial Hierarchy		1						
35 Process, the Combinatorial Hierarchy and Similarity		1						

		M	A	N	P	AN	AP	NP	ANP
	Bastin Papers	9	7	5	2	1	5	2	1
1	Critique Process Hierarchy models					1			
2	Combinatorial Physics				1				
3	Towards a Synthesis		1						
4	Apologia		1						
5	Labels, Dimensions and the Total Ordering Operator		1						
6	A Philosophical Focus				1				
7	Spaces and Minds								1
8	Quantisation Exists implies a Maximum Velocity		1						
9	Heterogeneous Objects		1						
10	The Hierarchy and Physical Space		1						
11	Mathematics of a Hierarchy of Brouwerian Operations								1
12	Measurement and Spin							1	
13	The consequences of Aspects: broad physical picture						1		
14	The Participant Observer			1					
15	Paranormal theory of Etter and Shoup						1		
16	Why I find Rowlands and Cullerne exciting						1		
17	Participant Observer Philosophy						1		
18	Philosophy and Common sense at ANPA			1					
19	Some ANPA Pre-History	1							
20	Representing Measurable Numbers in the Hierarchy	1							
21	The Photon						1		
22	The disembodied Observer			1					
23	A Combinatorial Perspective on the Standard Model	1							
24	Process			1					
25	Early Philosophical Ideas	1							
26	The Origin of Discrete Particles	1							
27	Understanding Comes First			1					
28	The basic two-ity	1							
29	Entelechies		1						
30	Algebraic Approaches to the Particle Concept	1							
31	On the Physical Interpretation of the CH	1							
32	Self-organization and the notion of level.	1							

Marcer Papers	M	A	N	P	AN	AP	NP	ANP
	0	2	1	1	5	10	4	2
1 Physics of computation					1			
2 Thermodynamics of Computation					1			
3 ANPA has much to celebrate		1						
4 A Model for an Evolutionary Epistemology						1		
5 A Grand Unification for Wave field Physics							1	
6 Foundations of Computers & Conscious Machines						1		
7 Neuron, Computational model Utilising QM Devices?						1		
8 Mind, Hierarchy and Cosmos								1
9 Quantum holography and the fallacy of uncertainty						1		
10 The Ultimate Hypothesis						1		
11 How Quantum Mechanics Maps onto Reality							1	
12 From the Self Creation to the Creation of the Self								1
13 A Mathematical Definition of Intelligence						1		
14 Wider Perspectives: Nature, Cognition and Quantum Physics							1	
15 Quantum holography as a possible explanation						1		
16 Why Space has Three Dimensions:			1					
17 Self-reference, Scale of Quantum Mechanical Effects					1			
18 Origin of Matter & how 3+1 Space-Time came to be		1						
19 Fundamental Questions in Information Science - Part 1					1			
20 Key to a Theory of Everything							1	
21 How General is Nilpotency?						1		
22 Lie Computability					1			
23 Talking about Creation						1		
24 Evolutionary 'Anthropic' Semantic Principle						1		
25 Semantic Computational principle				1				

Constable Papers	M	A	N	P	AN	AP	NP	ANP
	0	0	0	0	5	6	0	0
1 Maximum and Minimum Values of Physical Variables						1		
2 The quantisation of gravitational potential						1		
3 The relationships between physical constants						1		
4 Theory of Quantised Variables:						1		
5 Investigation of the Theory of Quantised Variables						1		
6 The Theory of Quantised Variables						1		
7 First Steps into the Particle Zoo					1			
8 Probabilistic shadow					1			
9 A Digital Universe					1			
10 The Digital Universe: Gravity and other loose ends					1			
11 Ethereal thoughts					1			

		M	A	N	P	AN	AP	NP	ANP
	Kauffman Papers	0	0	1	0	0	19	3	0
1	Special Relativity & a Calculus of Distinctions						1		
2	Spin Networks, Topology & Discrete Physics							1	
3	Sign & Space							1	
4	Path Integrals and Discrete Physics							1	
5	*A* Non commutative Approach to Discrete Physics						1		
6	Comment on the paranormal						1		
7	Zero Numbers & the Discrete Ordered Calculus						1		
8	Is ordinary (mathematical} awareness paranormal						1		
9	Schrodinger's Cat and the Cheshire Cat						1		
10	Non-commutative Calculus and Discrete Physics						1		
11	*Non-Commutative Worlds*						1		
12	Pope's Pythagorean Interpretation of Special Relativity						1		
13	Non-Commutative Worlds: A Summary						1		
14	Paragraphs from Void						1		
15	Concepts & Conceptions			1					
16	Differential Geometry in Non-Commutative Worlds						1		
17	The Erdos Machine						1		
18	Relativity and the Bianchi Identity						1		
19	Fibonacci Form and the Quantized Mark						1		
20	Reflexivity, Eigenform and Foundations of Physics						1		
21	Higher Categories and Self-Reference						1		
22	Non-Commutative worlds and classical constraints						1		
23	Quantum & Topological Entanglement						1		

		M	A	N	P	AN	AP	NP	ANP
	Amson Papers	0	1	0	0	9	1	0	0
1	Bi-ourobourous. Recursive Hierarchy					1			
2	On Discrimination					1			
3	Discrimination Systems and Projective Geometries					1			
4	Discrimination System		1						
5	A Cyclically-Scaled Cosmology						1		
6	Trigonometry on the CH					1			
7	A Combinatoric Hierarchy with Indistinguishables					1			
8	Purely Combinatorial Hierarchy					1			
9	Another New Combinatorial Hierarchy					1			
10	Monster Holographic CH					1			
11	Combinatoric Bit-Hoop System					1			

		M	A	N	P	AN	AP	NP	ANP
	Noyes Papers	4	5	1	5	0	1	4	1
1	Relativistic Quantum Mechanics				1				
2	Discrete Physics				1				
3	Where are we			1					
4	Inevitable Universe: FPR Ontology & Physics				1				
5	Bit-String Scattering Theory				1				
6	From Bit Strings (part-way) to Quaternions		1						
7	On Measurement of x								1
8	Discrete Anti-Gravity		1						
9	Anti-gravity: key to 21st century physics							1	
10	Decoherence, determinism and chaos							1	
11	Measurement , Bit-Strings, Quaternions and RRQM			1					
12	Discrete Physics as an Ultimate Dynamical Theory			1					
13	A Short Introduction to Bit String Physics		1						
14	Some Thoughts on Commutation Relations							1	
15	Program Universe and recent Cosmological results		1						
16	Science and paranormal phenomena						1		
17	On Ed Jones microcosmology	1							
18	A Brief History of ANPA West	1							
19	Twenty Five Years of the Combinatorial Hierarchy		1						
20	On Biology as an Emergent Science							1	
21	Scientific Eschatology				1				

		M	A	N	P	AN	AP	NP	ANP
	Manthey Papers	0	2	4	1	2	2	6	1
1	Program Universe						1		
2	Bits in the Combinatorial Hierarchy		1						
3	Synchronisation: and Conservation Laws					1			
4	Are conservation laws absolute?					1			
5	*A* Vector Semantics for Actions								1
6	The Combinatorial Hierarchy recapitulated			1					
7	A Topsy Example			1					
8	Phase Web Paradigm							1	
9	Distributed Computation							1	
10	A Pure Process View of Anticipatory Systems							1	
11	A Combinatorial Big Bang Leading to Quaternions		1						
12	Toward a scientific account of the paranormal							1	
13	On the Efficacy of Mental Events							1	
14	Experiencing the Meditative State				1				
15	Towards the Homology of the Phase Web Paradigm						1		
16	Synchronisation: The Font of Structure			1					
17	Relations Yield Generalized Dirac Equation			1					
18	Riemann							1	

Rowlands Papers

	Rowlands Papers	M	A	N	P	AN	AP	NP	ANP
		0	2	0	2	0	0	16	0
1	Let's get down to Basics							1	
2	The Physical Significance of the Factor 2							1	
3	Representations of a Fundamental Theory				1				
4	The Origin and Meaning of 3-Dimensionality							1	
5	An Exploration of the Vacuum							1	
6	Removing redundancy in relativistic QM							1	
7	Nilpotent Operators and Particle States							1	
8	Minimalist Approach to QM and Particle Physics		1						
9	What is Vacuum? A Nilpotent Solution							1	
10	Quantum Mechanics and Physical Meaning							1	
11	Nonlocal ity							1	
12	Particle Structures and the Dirac Symmetries							1	
13	The Dirac Algebra and Charge Accommodation							1	
14	SU(5) and Grand Unification							1	
15	An Investigation of the Higgs Mechanism							1	
16	The Nilpotent Representation of the Dirac Algebra							1	
17	Symmetry Breaking and Dirac Algebra							1	
18	Comparison between two versions of Dirac Algebra							1	
19	Charge Accommodation & Combinatorial Hierarchy		1						
20	Fundamental Structures in Physics& Biology				1				

Bowden Papers

	Bowden Papers	M	A	N	P	AN	AP	NP	ANP
		5	0	1	0	1	7	3	1
1	On General Physical Systems Theories							1	
2	Hierarchical Tearing and the Discrete Holomovement							1	
3	Orthogonality and General Systems Theory						1		
4	The spatial transmission of information						1		
5	Constructive Post Modern Physics							1	
6	What is to be Done								1
7	Huygens' Principle Physics and Computers						1		
8	Classical Computation can be Counterfactual						1		
9	My take on the paranormal						1		
10	Huygens' Principle, Physics and Computers					1			
11	Some ANPA prehistory	1							
12	Proceedings of ANPA West: A Catalogue	1							
13	Goops						1		
14	On Goop						1		
15	Level Change in the Combinatorial Hierarchy	1							
16	Selected publications by ANPA members	1							
17	An extension to Eddington's Fundamental Theory			1					
18	Essential Frederick Construction	1							

Brief Biography

Michael Horner was educated in the UK education system and after National Service worked in Electronics and Computers. In the computer industry he became interested in information theory and management practice where he ran a Management Systems Research group and was involved with long term strategies. He took early retirement at age 55 to apply his ideas to small and medium enterprises.

Member of ANPA, 1985 onwards. Secretary to the ANPA Advisory Council. Treasurer and member of the ANPA Executive Council.

Critical Stability of Few-body Systems

V.A. Karmanov

Lebedev Physical Institute,
Moscow, Russia
E-mail: karmanov@sci.lebedev.ru

J. Carbonell

Institut de Physique Nucléaire
Université Paris-Sud, France
E-mail: carbonell@ipno.in2p3.fr

When a two-body system is bound by a zero-range interaction, the corresponding three-body system – considered in a non-relativistic framework – collapses, that is its binding energy is unbounded from below. In a paper by J.V. Lindesay and H.P. Noyes [1] it was shown that the relativistic effects result in an effective repulsion in such a way that three-body binding energy remains also finite, thus preventing the three-body system from collapse. Later, this property was confirmed in other works based on different versions of relativistic approaches. However, the three-body system exists only for a limited range of two-body binding energy values. For stronger two-body interaction, the relativistic three-body system still collapses.

A similar phenomenon was found in a two-body systems themselves: a two-fermion system with one-boson exchange interaction in a state with zero angular momentum $J = 0$ exists if the coupling constant does not exceed some critical value but it also collapses for larger coupling constant. For a $J = 1$ state, it collapses for any coupling constant value. These properties are called "critical stability". This contribution aims to be a brief review of this field pioneered by H.P. Noyes.

1. Introduction

The radius of nuclear forces – the interaction between protons and neutrons – is sensibly smaller than the size of nuclei themselves. Since the wave function at large distances r behaves as $\sim \exp(-|E_b|r)$, the latter is determined by the nuclear binding energy E_b. The binding energy, in its turn, is a cancellation of a large (negative) potential energy and large (positive) kinetic energy. Therefore E_b is much smaller than each of these energies and the nuclear radius $r \sim 1/|E_b|$ can be larger than the radius of the nuclear forces. To understand qualitatively some nuclear properties, one can consider the "zero-range interaction limit". To this aim, we approximate the nuclear interaction V by a potential well:

$$V(r) = \begin{cases} -U_0, & \text{if } r < a \\ 0, & \text{if } r > a \end{cases}$$

As it is well known from standard quantum mechanics textbooks (see e.g. [2] a bound state exists if some relation between the potential depth U_0 and its range a is fulfilled, that is if

$$U_0 > \frac{\pi^2 \hbar^2}{8ma^2}$$

$\hbar$ being the Planck constant and m the mass of the particle. If we let a tend to zero and U_0 to infinity, keeping constant the product $U_0 a^2$, we will get in this limit an infinitely deep zero range potential well, in which a two-body bound state exists.

The zero-range two-body interaction provides an important limiting case which qualitatively reflects characteristic properties of nuclear [3] and atomic [4] few-body systems. It turned out, however, that when using non-relativistic dynamics, it generates the Thomas collapse [5] of the three-body system. The latter means that the three-body binding energy tends to $-\infty$, when the interaction radius tends to zero keeping constant the product $U_0 a^2$ and consequently the two-body binding energy. As an illustration, we have solved numerically the three-body Faddeev equation in momentum space with the two-body amplitude for zero-range interaction as input. The corresponding three-body binding energy is kept finite by introducing a momentum cutoff L. The result for the three-body binding energy (in units of the particle mass m) is shown in Fig. 1. We see that when cutoff L is removed (L tends to infinity), the three-body binding energy $|E_3|$ increases monotonously without any limit. This is just the manifestation of the Thomas collapse. Several ways to regularize this interaction have been proposed in the literature [6,7]

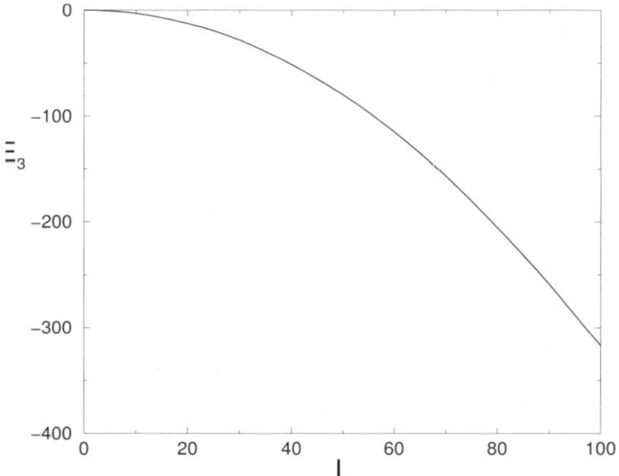

Fig. 1.   Three-body binding energy (in the units of mass m), for the zero-range two-body interaction and finite two-body binding energy, as a function of momentum cutoff L in the Faddeev equation.

It should be emphasized that the Thomas collapse was found in the non-relativistic framework, which should be applied only when the binding energy is much smaller than the particle mass. We see that the results displayed in Fig. 1 do not correspond to this situation: the module of the binding energy $|E_3|$ becomes much larger that particle mass. For example, for the cutoff $L \approx 50\,m$ the binding energy is $E_3 \approx -100\,m$. This is far beyond the domain where the non-relativistic treatment is valid. The answer to the question: "what happens with the three-body system in the limit of two-body zero-range interaction" should be obtained in a relativistic framework only.

This answer was first found in the paper by J.V. Lindesay and H.P. Noyes [1] in the so called "minimal relativistic model". It was shown that the relativistic effects result in an effective repulsion and can thus prevent the three-body system from collapse: the three-body binding energy remains also finite.

Later, this property was confirmed in other works based on different versions of relativistic approaches. In particular, two-body calculations showed that in the scalar case, relativistic effects were indeed strongly repulsive [8]. However, it was found [9] that this stabilization]had some restrictions: the three-body system exists only in a limited range of two-body binding energy. For stronger two-body interaction, the mass squared of three-body system M_3^2 though remaining finite, crosses zero and becomes negative. This means that the relativistic three-body system does not longer exists. Then a similar phenomenon was also found in the two-body systems: the two-fermion systems with one-boson exchange interaction also collapses if the coupling constant exceeds some critical value. These properties are called "critical stability" and they are forming now an interesting field of research. In what follows we will give a brief review of this developing field pioneered by H.P. Noyes.

2. Relativistic three-body system with zero-range interaction

In paper [1], relativistic three-body calculations with zero-range interaction have been performed in a minimal relativistic model. Later, a much more general and sophisticated approach to the relativistic few-body systems – Light-Front Dynamics – was developed (see for review [10,11]). In the framework of this relativistic approach the problem of three equal mass (m) bosons interacting via zero-range forces was reconsidered in the [9].

The relativistic three-body equation is derived in Section 2.1. In Section 2.2, their solutions are presented and some concluding remarks are given in Section 4.

2.1. *Equation*

Our starting point is the explicitly covariant formulation of the Light-Front Dynamics [10]. In non-relativistic approach the wave function $\psi(\vec{r}, t)$ is a probability amplitude defined at a given time t, say at $t = 0$. In four-dimensional Minkowski space one can define the wave function on any space-like plane to preserve the causality, or more generally on any space-like surface. The orientation of this plane is defined by a four-vector $\lambda = (\lambda_0, \vec{\lambda})$ orthogonal to this plane. We can change its orientation moving λ within the light cone in such a way that the plane where the wave function is defined remains space-like. Its limiting value is reached when λ lies on the light-cone surface. Then such a four-vector is denoted by $\omega = (\omega_0, \vec{\omega})$ and has the property $\omega^2 = \omega_0^2 - \vec{\omega}^2 = 0$.

The corresponding plane is given by the equation $\omega{\cdot}x = 0$. In the particular case $\omega = (1,0,0,-1)$ it turns into $t + z = 0$, setting hereafter $c = 1$. This equation coincides with the light-front equation and therefore the plane $t + z = 0$ is called the light-front plane. The dynamics determining the evolution of the wave function from one light-front plane to another one is the light-front dynamics. This approach was proposed by Dirac [12] and it has many advantages. Later, its explicitly covariant version was developed, when the light-front plane is defined by the covariant equation $\omega{\cdot}x = 0$ and no any particular axes like t or z is selected [10] We will just use the light-front dynamics as a relativistic approach.

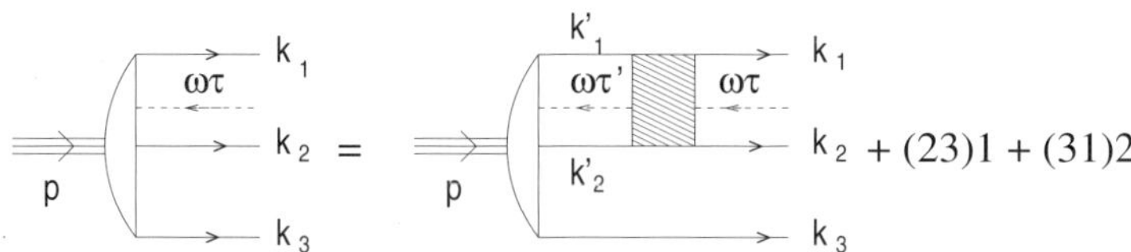

Fig. 2.   Three-body equation for the vertex function Γ.

The three-body equation is represented graphically in figure 2. It concerns the vertex function Γ, related to the wave function ψ in the standard way:

$$\psi(k_1, k_2, k_3, p, \omega\tau) = \frac{\Gamma(k_1, k_2, k_3, p, \omega\tau)}{\mathcal{M}^2 - M_3^2}, \quad \mathcal{M}^2 = (k_1 + k_2 + k_3)^2 = (p + \omega\tau)^2.$$

All four-momenta are on the corresponding mass shells ($k_i^2 = m^2$, $p^2 = M_3^2$, $(\omega\tau)^2 = 0$) and satisfy the conservation law $k_1 + k_2 + k_3 = p + \omega\tau$ involving $\omega\tau$. The four-momenta $\omega\tau$ and $\omega\tau'$ are drawn in figure 2 by dash lines. The off-energy shell character of the wave function is ensured by non-zero value of the scalar variable τ. In the standard approach [11], the minus-components of the momenta are not conserved and the only non-zero component of ω is $\omega_- = \omega_0 - \omega_z = 2$. Variable 2τ is just the non-zero difference of non-conserved components $2\tau = k_{1-} + k_{2-} + k_{3-} - p_-$.

Applying to figure 2 the covariant light-front graph techniques [10], we find the equation:

$$\Gamma(k_1, k_2, k_3, p, \omega\tau) = \frac{\lambda}{(2\pi)^3} \int \frac{d\tau'}{\tau'} \frac{d^3 k_1'}{2\varepsilon_{k_1'}} \frac{d^3 k_2'}{2\varepsilon_{k_2'}} \; \Gamma(k_1', k_2', k_3, p, \omega\tau')$$

$$\times \, \delta^{(4)}(k_1' + k_2' - \omega\tau' - k_1 - k_2 + \omega\tau) \, + \, (23)1 + (31)2, \quad (1)$$

where $\varepsilon_k = \sqrt{m^2 + \vec{k}^2}$. For the zero-range forces we are interested in, the interaction kernel appears as a constant λ. In (1) the contribution of interacting pair (12) is explicitly written while the contributions of the remaining pairs are simply denoted by $(23)1 + (31)2$.

Equation (1) can be rewritten in variables $\vec{R}_{i\perp}, x_i$, $(i = 1, 2, 3)$, where $\vec{R}_{i\perp}$ is the spatial component of the four-vector $R_i = k_i - x_i p$ orthogonal to $\vec{\omega}$ and $x_i = \frac{\omega \cdot k_i}{\omega \cdot p}$ [10]. For this aim we insert in r.h.-side of (1) the unity integral

$$1 = \int 2(\omega \cdot k_3')\delta^{(4)}(k_3' - k_3 - \omega\tau_3)d\tau_3 \, \frac{d^3 k_3'}{2\varepsilon_{k_3'}}$$

and recover the usual three-body space volume which, expressed in the variables $(\vec{R}_{i\perp}, x_i)$, reads

$$\int \delta^{(4)}\left(\sum_{i=1}^{3} k_i' - p - \omega\tau'\right) \prod_{i=1}^{3} \frac{d^3 k_i'}{2\varepsilon_{k_i'}} 2(\omega \cdot p)d\tau'$$
$$= \int \delta^{(2)}\left(\sum_{i=1}^{3} \vec{R}'_{\perp i}\right)\delta\left(\sum_{i=1}^{3} x_i' - 1\right)2\prod_{i=1}^{3} \frac{d^2 R'_{\perp i}dx_i'}{2x_i'}.$$

The Faddeev amplitudes Γ_{ij} are introduced in the standard way:

$$\Gamma(1,2,3) = \Gamma_{12}(1,2,3) + \Gamma_{23}(1,2,3) + \Gamma_{31}(1,2,3),$$

and equation (1) is equivalent to a system of three coupled equations for these components. With the symmetry relations $\Gamma_{23}(1,2,3) = \Gamma_{12}(2,3,1)$ and $\Gamma_{31}(1,2,3) = \Gamma_{12}(3,1,2)$, the system is reduced to a single equation for one of the amplitudes, say Γ_{12}.

In general, Γ_{12} depends on all variables $(\vec{R}_{i\perp}, x_i)$, constrained by the relations $\vec{R}_{1\perp} + \vec{R}_{2\perp} + \vec{R}_{3\perp} = 0$, $x_1 + x_2 + x_3 = 1$, but for a contact kernel it depends only on $(\vec{R}_{3\perp}, x_3)$ [13]. Equation (1) results into:

$$\Gamma_{12}(\vec{R}_\perp, x) =$$
$$\frac{\lambda}{(2\pi)^3} \int \left[\Gamma_{12}(\vec{R}_\perp, x) + 2\Gamma_{12}\left(\vec{R}'_\perp - x'\vec{R}_\perp, \, x'(1-x)\right)\right] \frac{1}{s_{12}' - M_{12}^2} \frac{d^2 R'_\perp dx'}{2x'(1-x')},$$
$$\tag{2}$$

in which

$$s_{12}' = (k_1' + k_2')^2 = \frac{R_\perp'^2 + m^2}{x'(1-x')}$$

is the effective on shell mass squared of the two-body subsystem, whereas $M_{12}^2 = (k_1' + k_2' - \omega\tau')^2 = (p - k_3)^2$ corresponds to its off-shell mass. It is expressed through $M_3^2, R_\perp^2, x$ as

$$M_{12}^2 = (1 - x)M_3^2 - \frac{R_\perp^2 + (1-x)m^2}{x}.$$
$$\tag{3}$$

These on- and off-shell masses s_{12}' and M_{12}^2 differ from each other, since $k_1' + k_2' + k_3 \neq p$. On the energy shell, at $\tau' = 0$, the value M_{12}^2 turns into s_{12}', what is never reached for a bound state problem.

Since the first term $\Gamma_{12}(\vec{R}_\perp, x)$ in the integrand does not depend on the integration variables $\vec{R}'_\perp, x'$, we can transform (2) as:

$$\Gamma_{12}(\vec{R}_\perp, x) =$$
$$\frac{1}{\lambda^{-1} - I(M_{12})} \frac{2}{(2\pi)^3} \int \Gamma_{12}\left(\vec{R}'_\perp - x'\vec{R}_\perp, x'(1-x)\right) \frac{1}{s'_{12} - M_{12}^2} \frac{d^2 R'_\perp \, dx'}{2x'(1-x')}, \quad (4)$$

where

$$I(M_{12}) = \frac{1}{(2\pi)^3} \int \frac{1}{s'_{12} - M_{12}^2} \frac{d^2 R'_\perp \, dx'}{2x'(1-x')}. \quad (5)$$

The integral (5) diverges logarithmically and we implicitly assume that a cutoff L is introduced.

The value of λ is found by solving the two-body problem with the same zero-range interaction under the condition that the two-body bound state mass has a fixed value M_2. From that we get $\lambda^{-1} = I(M_2)$ with I given by (5). It also diverges when the momentum space cutoff L tends to infinity (or, equivalently, the interaction range tends to zero). However, the difference $\lambda^{-1} - I(M_{12}) = I(M_2) - I(M_{12})$ which appears in (4) converges in the limit $L \to \infty$. The factor $F(M_{12}) = 1/[I(M_2) - I(M_{12})]$ gives the two-body off-shell scattering amplitude, depending on the off-shell two-body mass M_{12}, without any regularization. For $0 \leq M_{12}^2 < 4m^2$ explicit calculations gives:

$$F(M_{12}) = \frac{8\pi^2}{\dfrac{\arctan y_{M_{12}}}{y_{M_{12}}} - \dfrac{\arctan y_{M_2}}{y_{M_2}}},$$

where $y_{M_{12}} = \dfrac{M_{12}}{\sqrt{4m^2 - M_{12}^2}}$ and similarly for y_{M_2}. If $M_{12}^2 < 0$, the amplitude obtains the form:

$$F(M_{12}) = \frac{8\pi^2}{\dfrac{1}{2y'_{M_{12}}} \log \dfrac{1 + y'_{M_{12}}}{1 - y'_{M_{12}}} - \dfrac{\arctan y_{M_2}}{y_{M_2}}},$$

where $y'_{M_{12}} = \dfrac{\sqrt{-M_{12}^2}}{\sqrt{4m^2 - M_{12}^2}}$.

Finally, the equation for the Faddeev amplitude reads:

$$\Gamma_{12}(R_\perp, x) = F(M_{12}) \frac{1}{(2\pi)^3} \int_0^1 dx' \int_0^\infty \frac{\Gamma_{12}\left(R'_\perp, x'(1-x)\right) \, d^2 R'_\perp}{(\vec{R}'_\perp - x'\vec{R}_\perp)^2 + m^2 - x'(1-x')M_{12}^2}. \quad (6)$$

The three-body mass M_3 enters in this equation through the variable M_{12}^2, defined by (3).

By replacing $x'(1-x) \to x'$, equation (6) can be transformed into

$$\Gamma_{12}(R_\perp, x) = F(M_{12}) \frac{1}{(2\pi)^3} \int_0^{1-x} \frac{dx'}{x'(1-x-x')} \int_0^\infty \frac{d^2 R'_\perp}{\mathcal{M}'^2 - M_3^2} \Gamma_{12}\left(R'_\perp, x'\right), \quad (7)$$

with

$$\mathcal{M}'^2 = \frac{\vec{R}'^2_\perp + m^2}{x'} + \frac{\vec{R}^2_\perp + m^2}{x} + \frac{(\vec{R}'_\perp + \vec{R}_\perp)^2 + m^2}{1 - x - x'}$$

This equation is the same as the equation (11) from [13] except for the integration limits of $(\vec{R}'_\perp, x')$ variables. In [13] the integration limits follow from the condition $M_{12}^2 > 0$. They read

$$\int_{\frac{m^2}{M_3^2}}^{1-x} [\ldots] \, dx' \int_0^{k_\perp^{max}} [\ldots] \, d^2 R'_\perp \tag{8}$$

with $k_\perp^{max} = \sqrt{(1 - x')(M_3^2 x' - m^2)}$ and thus implicitly introduce a lower bound on the three-body mass $M_3 > \sqrt{2}m$. The same condition, though in a different relativistic approach, was used in [1]. The integration limits in (8) restrict the arguments of Γ_{12} to the domain

$$\frac{m^2}{M_3^2} \le x \le 1 - \frac{m^2}{M_3^2}, \quad 0 \le R_\perp \le k_\perp^{max}$$

and can be considered as a method of regularization. In this case, one no longer deals with the zero-range forces.

Being interested in studying the zero-range interaction, we do not cut off the variation domain of variables $R_\perp, x$:

$$0 \le x \le 1, \quad 0 \le R_\perp < \infty$$

The integration limits for these variables reflect the conservation law of the four-momenta in the three-body system and they are automatically fulfilled, as far as the $\delta^{(4)}$-function in (1) is taken into account. The off-shell variable M_{12}^2 may take negative values, when $R_\perp$ and x vary in their proper limits. Thus, if $M_3^2 > m^2$ one has $-\infty \le M_{12}^2 \le (M_3 - m)^2$ but if $M_3^2 < m^2$, M_{12}^2 is always negative $-\infty \le M_{12}^2 \le 0$.

We would like to notice that M_{12}^2 is not to be confused with the on-shell effective mass squared $s'_{12} = (k'_1 + k'_2)^2$ which is indeed always positive and even $s'_{12} \ge 4m^2$. As we will see in the next section, this point turns out to be crucial for the appearance of the relativistic collapse.

2.2. *Results*

The results of solving equation (6) are presented in what follows. Calculations were carried out with constituent mass $m = 1$ and correspond to the ground state. We represent in Fig. 3a the three-body bound state mass M_3 as a function of the two-body one M_2 (solid line) together with the dissociation limit $M_3 = M_2 + m$ (dotted line). The two-body zero binding limit $B_2 = 2m - M_2 \to 0$ is magnified in Fig. 3b. In this limit the three-boson system has a binding energy $B_3^{(c)} \approx 0.012$.

When M_2 decreases, the three-body mass M_3 decreases very quickly and vanishes at the two-body mass value $M_2 = M_2^{(c)} \approx 1.43$. Whereas the meaning of collapse as

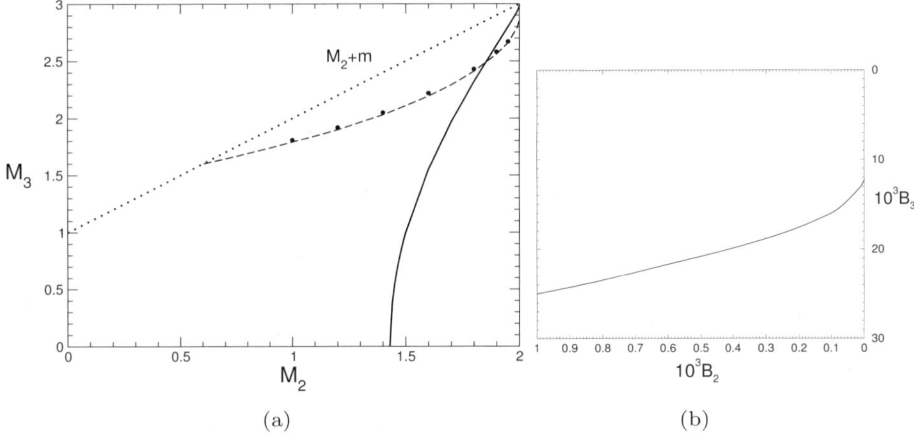

Fig. 3.   (a) Three-body bound state mass M_3 versus the two-body one M_2 (solid line). Dotted line represents the dissociation limit. Results obtained with integration limits (8) are in dash line. Bold dots are taken from [15]. (b) Zoom of the two-body zero binding limit region ($M_2 \to 2m$, $B_2 = 2m - M_2 \to 0$) corresponding to the solid line only.

used in the Thomas paper [5] implies unbounded nonrelativistic binding energies and cannot be used here, the zero bound state mass $M_3 = 0$ constitutes its relativistic counterpart. Indeed, for two-body masses below the critical value $M_2^{(c)}$, the three-body system no longer exists.

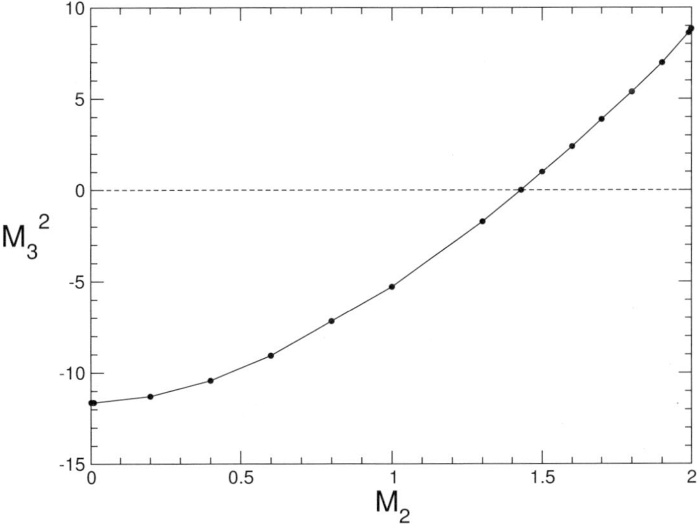

Fig. 4.   Three-body bound state mass squared M_3^2 versus M_2.

The results corresponding to integration limits (8) are included in Fig. 3a (dash line) for comparison. Values given in [13] were not fully converged. They have been corrected in [15] and are indicated by dots. In both cases the repulsive relativistic effects produce a natural cutoff in equation (6), leading to a finite spectrum and – in the Thomas sense – an absence of collapse, like it was already found in [1]. However, solid and dash curves strongly differ from each other, even in the zero binding limit.

We would like to remark that for $M_2 \leq M_2^{(c)}$, equation (6) posses square integrable solutions with negative values of M_3^2. They have no physical meaning but M_3^2 remains finite in all the two-body mass range $M_2 \in [0, 2]$. The results of M_3^2 are given in figure 4. When $M_2 \to 0$, M_3^2 tends to ≈ -11.6.

It is also worth noticing that the critical value of the two-body bound state mass $M_2^{(c)}$ as well as the three-body binding energy $B_3^{(c)}$ are universal quantities for bosonic systems. $M_2^{(c)} = 1.43\,m$ represents the maximal two-body binding energy $B_2 = 2m - M_2^{(c)} = 0.57\,m$ compatible with the existence of 3-boson bound states with mass $M_3 = 0$ ($B_3 = 3\,m$). $B_3^{(c)} = 0.012\,m$ represents the minimal binding energy that a three-boson system can have when two-body binding energy $B_2 = 0$ ($M_2 = 2\,m$).

3. Two-fermion system with Yukawa interaction

3.1. *States with $J = 0$*

So far we have considered the behavior of the three-boson relativistic bound system and its critical stability depending on the two-body binding energy. The conclusion are valid for the zero-range interactions, considered as input for the two-body sector, and we have supposed that the particles were spinles.

Now we will study a system of two fermions – spin $1/2$ particles – with more sophisticated interaction, resulting from spinless mesons exchange with mass μ. This model traces back to the very origin of the nuclear forces theory proposed by Yukawa. The interaction Lagrangian reads:

$$\mathcal{L}^{int} = g\,\bar{\psi}\psi\phi$$

Let us consider first the case of zero total angular momentum $J = 0$. We denote the fermion momenta as $\vec{k}_1$, $\vec{k}_2$. It is convenient to analyze the wave function in the reference frame where $\vec{k}_1 = \vec{k}_2 = 0$. Then the two-fermion wave function depends on the relative momentum $\vec{k} = \vec{k}_1 = -\vec{k}_2$ and on the spin projections of each fermion $\sigma_1, \sigma_2 = \pm 1/2$. Relative to the spin projections, it is a 2×2 matrix which has the following general form [16]:

$$\psi(\vec{k}, \vec{n}) = \frac{1}{\sqrt{2}}\left(f_1 + \frac{i\vec{\sigma}\cdot[\vec{k} \times \vec{n}]}{\sin\theta}f_2\right), \tag{1}$$

where $\vec{\sigma}$ are the Pauli matrices, $\vec{n} = \vec{\omega}/\omega_0$ and θ is the angle between $\vec{k}$ and $\vec{n}$. One cannot construct any other independent structures in addition to those appearing in (1). Therefore the 2×2 matrix $\psi(\vec{k}, \vec{n})$ contains only two independent matrix elements or, correspondingly, it is determined by the two coefficients f_1, f_2 of the independent structures. The normalization condition has the form:

$$\frac{m}{(2\pi)^3} \int (f_1^2 + f_2^2) \frac{d^3 k}{\varepsilon_k} = 1, \quad \varepsilon_k = \sqrt{m^2 + k^2}.$$

The equation for the wave function is reduced to a system of two coupled equations for $f_{1,2}$:

$$\left[4(k^2 + m^2) - M^2 \right] f_1(k, \theta)$$

$$= -\frac{m^2}{2\pi^3} \int \left[K_{11}(k, \theta; k', \theta') f_1(k', \theta') + K_{12}(k, \theta; k', \theta') f_2(k', \theta') \right] \frac{d^3 k'}{\varepsilon_{k'}},$$

$$\left[4(k^2 + m^2) - M^2 \right] f_2(k, \theta)$$

$$= -\frac{m^2}{2\pi^3} \int \left[K_{21}(k, \theta; k', \theta') f_1(k', \theta') + K_{22}(k, \theta; k', \theta') f_2(k', \theta') \right] \frac{d^3 k'}{\varepsilon_{k'}} \qquad (2)$$

with the kernels:

$$K_{ij} = \int_0^{2\pi} \frac{\kappa_{ij}}{(K^2 + \mu^2) m^2 \varepsilon_k \varepsilon_{k'}} \frac{d\phi'}{2\pi}, \qquad (3)$$

where

$$K^2 = k^2 + k'^2 - 2kk' \left(1 + \frac{(\varepsilon_k - \varepsilon_{k'})^2}{2\varepsilon_k \varepsilon_{k'}} \right) \cos\theta \cos\theta' - 2kk' \sin\theta \sin\theta' \cos\phi'$$

$$+ \left(\varepsilon_k^2 + \varepsilon_{k'}^2 - \frac{1}{2} M^2 \right) \left| \frac{k \cos\theta}{\varepsilon_k} - \frac{k' \cos\theta'}{\varepsilon_{k'}} \right| \qquad (4)$$

Here ϕ' is the azimuthal angle between $\vec{k}$ and $\vec{k}'$ in the plane orthogonal to $\vec{n}$ and

$$\cos\theta = \cos\vec{n}{\cdot}\vec{k}/k, \quad \cos\theta' = \cos\vec{n}{\cdot}\vec{k}'/k'.$$

The explicit expressions for κ_{ij} is given by [16,17]:

$$\kappa_{11} = -\alpha\pi \left[2k^2 k'^2 + 3k^2 m^2 + 3k'^2 m^2 + 4m^4 - 2kk' \varepsilon_k \varepsilon_{k'} \cos\theta \cos\theta' \right.$$

$$\left. -kk'(k^2 + k'^2 + 2m^2) \sin\theta \sin\theta' \cos\phi' \right],$$

$$\kappa_{12} = -\alpha\pi m(k^2 - k'^2)\left(k' \sin\theta' + k \sin\theta \cos\phi' \right),$$

$$\kappa_{21} = -\alpha\pi m(k'^2 - k^2)\left(k \sin\theta + k' \sin\theta' \cos\phi' \right),$$

$$\kappa_{22} = -\alpha\pi \left[\left(2k^2 k'^2 + 3k^2 m^2 + 3k'^2 m^2 + 4m^4 - 2kk' \varepsilon_k \varepsilon_{k'} \cos\theta \cos\theta' \right) \cos\phi' \right.$$

$$\left. -kk'(k^2 + k'^2 + 2m^2) \sin\theta \sin\theta' \right], \qquad (5)$$

where we denote $\alpha = g^2/(4\pi)$.

3.2. *Asymptotical behavior of the kernels*

The r.h.-sides of equations (2) contain the integrals over k' in infinite limits. The existence of a finite solution depends critically on the behavior of the kernels K_{ij} at large momenta. For the kernels (5) determining the $J = 0$ state, we get the following leading terms:

$$K_{11} \propto \begin{cases} \frac{1}{k}, & \text{if } k \to \infty, \; k' \text{ fixed} \\[2ex] \frac{1}{k'}, & \text{if } k' \to \infty, \; k \text{ fixed} \end{cases}$$

$$K_{12} \propto \begin{cases} \frac{1}{k}, & \text{if } k \to \infty, \; k' \text{ fixed} \\ c_{12}, & \text{if } k' \to \infty, \; k \text{ fixed} \end{cases}$$

$$K_{21} \propto \begin{cases} c_{21} = c_{12}, & \text{if } k \to \infty, \; k' \text{ fixed} \\ \frac{1}{k'}, & \text{if } k' \to \infty, \; k \text{ fixed} \end{cases}$$

$$K_{22} = \begin{cases} c_{22}, & \text{if } k \to \infty, \; k' \text{ fixed,} \\ c'_{22}, & \text{if } k' \to \infty, \; k \text{ fixed} \end{cases} \tag{6}$$

In the above equations the coefficients $c_{12} = c_{21}$, c_{22} and c'_{22} depend on θ, θ'. The coefficients c_{22}, c'_{22} are positive:

$$c_{22} = \frac{\alpha \pi \sin \theta \sin \theta'}{m(1 + \cos \theta)(\varepsilon_{k'} - k' \cos \theta')} > 0, \tag{7}$$

and c'_{22} is obtained from c_{22} by the replacement $k' \to k, \theta \leftrightarrow \theta'$.

Note that the second iteration of the kernel K_{11} converges at $k' \to \infty$:

$$\int^L K_{11} G_0 K_{11} \frac{d^3 k'}{\varepsilon_{k'}} \propto \int^L \frac{1}{k'} \frac{1}{k'^2} \frac{1}{k'} \frac{k'^2 dk'}{k'} = \int^L \frac{dk'}{k'^3} \propto const.$$

Here $G_0 \propto 1/k'^2$ is the intermediate propagator. The integrals

$$\int K_{21} G_0 K_{11} d^3 k' / \varepsilon_{k'} \quad , \quad \int K_{11} G_0 K_{12} d^3 k' / \varepsilon_{k'}$$

are also convergent, whereas the the second iteration of the kernel K_{22} diverges logarithmically:

$$\int^L K_{22} G_0 K_{22} \frac{d^3 k'}{\varepsilon_{k'}} \propto \int^L const \frac{1}{k'^2} const \frac{k'^2 dk'}{k'} = \int^L \frac{dk'}{k'} \propto \log(L).$$

The integrals

$$\int K_{12} G_0 K_{22} d^3 k' / \varepsilon_{k'} \quad , \quad \int K_{22} G_0 K_{21} d^3 k' / \varepsilon_{k'}$$

also diverge logarithmically. This is a manifestation of the logarithmical divergence of the box fermion diagram in LFD.

In the domain where both k, k' tend to infinity, but the ratio $k'/k = \gamma$ is fixed, we find for K_{11}:

$$K_{11} = -\frac{2\pi^2 \alpha'}{m} \begin{cases} \sqrt{\gamma} A_{11}(\theta, \theta', \gamma), & \text{if } \gamma \leq 1 \\ \dfrac{A_{11}(\theta, \theta', 1/\gamma)}{\sqrt{\gamma}}, & \text{if } \gamma \geq 1 \end{cases} \tag{8}$$

with the function $A_{11}(\theta, \theta', \gamma)$:

$$A_{11}(\theta, \theta', \gamma) =$$
$$\frac{1}{\sqrt{\gamma}} \int_0^{2\pi} \frac{d\phi}{2\pi} \frac{2\gamma(1 - \cos\theta \cos\theta') - (1 + \gamma^2)\sin\theta \sin\theta' \cos\phi}{(1 + \gamma^2)(1 + |\cos\theta - \cos\theta'| - \cos\theta \cos\theta') - 2\gamma \sin\theta \sin\theta' \cos\phi}, \quad (9)$$

where we set $\alpha' = \alpha/(2m\pi)$. In eq. (8) we extracted for convenience the factor $\sqrt{\gamma}$. In the limit $\gamma \to 0$ A_{11} has the behavior $A_{11}(\theta, \theta', \gamma) \propto \sqrt{\gamma}$.

In the same domain, the kernel K_{22} also has asymptotic (8) with the corresponding function A_{22} given by:

$$A_{22}(\theta, \theta', \gamma) =$$
$$-\frac{1}{\sqrt{\gamma}} \int_0^{2\pi} \frac{d\phi}{2\pi} \frac{(1 + \gamma^2)\sin\theta \sin\theta' - 2\gamma(1 - \cos\theta \cos\theta')\cos\phi}{(1 + \gamma^2)(1 + |\cos\theta - \cos\theta'| - \cos\theta \cos\theta') - 2\gamma \sin\theta \sin\theta' \cos\phi}. \quad (10)$$

In the limit $\gamma \to 0$ this function has the behavior $A_{22}(\theta, \theta', \gamma) \propto -1/\sqrt{\gamma}$.

Comparing the above formulas, we see that the dominating kernel is K_{22}. It does not decreases in any direction of the (k, k') plane, whereas in the domain $k \to \infty$, k' fixed, and vice versa, the kernels K_{11} decrease. In the domain $k'/k = \gamma$ fixed, $k \to \infty$, both kernels do not decrease, but K_{22} is proportional to the unbounded function A_{22}.

3.3. *The cutoff dependence of the binding energy*

We are now in position to investigate the stability of the bound states. To disentangle the two different sources of collapse, we will first consider the one channel problem for the component f_1 with the kernel K_{11}. We remove the second equation from (2) and deal with the single equation:

$$\left[4(\vec{k}^2 + m^2) - M^2 \right] f_1(k, z) = -\frac{m^2}{2\pi^3} \int K_{11}(k, z; k', z') f_1(k') \frac{d^3 k'}{\varepsilon_{k'}}. \quad (11)$$

Our further analysis is based on the collapse condition found by Smirnov [18]. It is obtained by analyzing the asymptotic of eq. (11) with the kernel represented by eq. (8). The solution is searched in the form

$$f_1(k, z) \propto \frac{f(z)}{k^{2+\beta}}. \quad (12)$$

In eq. (11) one should make the replacement of variables $k' = \gamma k$ and take the limit $k \to \infty$. Provided the kernel $K_{11}(k, k' = const)$ decreases like $1/\sqrt{k}$ or faster, one gets:

$$4k^2 \frac{f(z)}{k^{2+\beta}} = -\frac{m^2}{2\pi^3} \int K_{11}(k, z; k\gamma, z') \frac{f(z')}{(k\gamma)^{2+\beta}} \frac{k^3 \gamma^2 d\gamma 2\pi dz'}{k\gamma}.$$

Splitting the integral in two terms:

$$\int_0^\infty \ldots d\gamma = \int_0^1 \ldots d\gamma + \int_1^\infty \ldots d\gamma,$$

making in the second term the substitution $\gamma = 1/\gamma'$ and substituting here the kernel (8), we obtain the equation:

$$f(z) = 2m\alpha' \int_0^1 dz' f(z') \int_0^1 d\gamma \frac{A_{11}(\gamma, z, z')}{\sqrt{\gamma}} \cosh(\beta \log(\gamma)) \qquad (13)$$

Using the symmetry relative to $z \to 1 - z$, we replaced the integral $\int_{-1}^1 \ldots dz'$ by $2\int_0^1 \ldots dz'$.

In the above equations we neglected the binding energy, supposing that it is finite. For given α' the equation (13) gives the value of β, determining the wave function asymptotic (12) for the solution with finite energy. The function $\cosh(\beta \log(\gamma))$ in (13) has minimum at $\beta = 0$. When the factor α' in (13) increases, this is compensated by decrease of $\cosh(\beta \log(\gamma))$, so the value of β is approaching to 0. The maximal, critical value of α' is achieved when $\beta = 0$. So, if we solve the eigenvalue equation [18]:

$$\int_0^1 H(z, z') f(z') dz' = \lambda f(z), \quad \text{with} \quad H(z, z') = 2 \int_0^1 \frac{A_{11}(\gamma, z, z')}{\sqrt{\gamma}} d\gamma \qquad (14)$$

then the critical value of α' is related to λ as $\alpha'_c = \frac{1}{m\lambda}$, that gives for the coupling constant in the Yukawa model $\alpha = g^2/(4\pi) = 2\pi m\alpha'$ the following critical value:

$$\alpha_c = \frac{2\pi}{\lambda}. \qquad (15)$$

Note that if $A_{11}(\gamma, z, z') = A(\gamma)$ does not depend on z, z', one gets [18]:

$$\alpha'_c = \frac{1}{2m \int_0^1 \frac{A(\gamma)}{\sqrt{\gamma}} d\gamma}. \qquad (16)$$

For the potential $V(r) = -\alpha'/r^2$ one can find $A(\gamma) = 1$ and one gets the well known value $\alpha'_c = 1/(4m)$ [2]. In [19] we have estimated $\alpha_c = \pi$ by majorating the kernel A_{11} by $A_{11} = \sqrt{\gamma}$. Substitution of this function A_{11} into eq. (16) reproduces this result.

Solving eq. (14) numerically with the function $A_{11}(\gamma, z, z')$ given by eq. (8), we found the only eigenvalue:

$$\lambda = 1.748$$

that gives by eq. (15):

$$\alpha_c = 3.594 \quad \Longleftrightarrow \quad g_c = \sqrt{4\pi\alpha_c} = 6.720$$

in agreement with our numerical estimations [19].

In the two-channel problem, the kernel dominating in asymptotic is K_{22}. In the case $J = 0$ it is positive and corresponds to repulsion. Because of that, this channel does not lead to any collapse. This repulsion cannot prevent from the collapse in the first channel (for enough large α), since due to coupling between two channels the singular potential in the channel 1 "pumps out" the wave function from the channel

2 into the channel 1. So, in the coupled equations system (2) the situation with the cutoff dependence is the same as for one channel. A similar analysis of what we detailed in the one channel case, provided us the critical value of the coupling constant [19,17]

$$\alpha_c = 3.723 \quad \Longleftrightarrow \quad g_c = \sqrt{4\pi\alpha_c} = 6.840 \tag{17}$$

The critical stability of the Yukawa model has been also considered in the framework of the Bethe-Salpeter equation [20]. By using the methods developed in the previous section we have found [21,22,23] similar results of what we have obtained in the Light-Front dynamics. There very existence of a critical coupling constant for the J=0 state was confirmed, although with slightly different numerical value:

$$\alpha_c = \pi \quad \Longleftrightarrow \quad g_c = 2\pi \tag{18}$$

to be compared with (17).

3.4. *States with $J = 1$*

In general, the wave function of the $J = 1$ state is determined by six independent structures [24] It turns out that the following operator commutes with the kernel:

$$A^2 = (\vec{n}\cdot\vec{J})^2. \tag{19}$$

Since A^2 is a scalar, it commutes also with $\vec{J}$. Therefore, in addition to J, J_z, the solutions are labeled by a:

$$A^2\vec{\psi}^a(\vec{k},\vec{n}) = a^2\vec{\psi}^a(\vec{k},\vec{n}). \tag{20}$$

Though the wave function for $J = 1$ is determined by six components, the equation system is split in two subsystems with $a = 0$ and $a = 1$, containing 2 and 4 equations respectively [16]

The function $\vec{\psi}^0$ corresponding to $J = 1$, $a = 0$ has the following general decomposition:

$$\vec{\psi}^0(\vec{k},\vec{n}) = \sqrt{\frac{3}{2}}\left\{ g_1^{(0)}\vec{\sigma}\cdot\hat{\vec{k}} + g_2^{(0)}\frac{\vec{\sigma}\cdot(\hat{\vec{k}}\cos\theta - \vec{n})}{\sin\theta}\right\}\vec{n}, \tag{21}$$

where $\hat{\vec{k}}$ denotes the unit vector $\hat{\vec{k}} = \vec{k}/k$. Since it corresponds to $J^\pi = 1^+$, it is a pseudovector. Since $\vec{n}$ is a true vector ($J^\pi = 1^-$), it should be multiplied by a pseudoscalar. We can construct two pseudoscalars only: $\vec{\sigma}\cdot\hat{\vec{k}}$ and $\vec{\sigma}\cdot\hat{\vec{n}}$, what gives two terms. The particular structures in (21) are constructed in such a way to be orthogonal and normalized to 1.

The function $\vec{\psi}^1$ satisfies the orthogonality condition $\vec{\psi}^1\cdot\vec{n} = 0$. To satisfy this condition, it is convenient to introduce the vectors orthogonal to $\vec{n}$:

$$\hat{\vec{k}}_\perp = \frac{\hat{\vec{k}} - \cos\theta\vec{n}}{\sin\theta}, \quad \vec{\sigma}_\perp = \vec{\sigma} - (\vec{n}\cdot\vec{\sigma})\vec{n}.$$

Then the function $\vec{\psi}^{1}$ obtains the following general form:

$$\vec{\psi}^{1}(\vec{k},\vec{n}) =$$

$$g_{1}^{(1)}\frac{\sqrt{3}}{2}\vec{\sigma}_{\perp} + g_{2}^{(1)}\frac{\sqrt{3}}{2}\left(2\hat{\vec{k}}_{\perp}(\hat{\vec{k}}_{\perp}\cdot\vec{\sigma}_{\perp}) - \vec{\sigma}_{\perp}\right) + g_{3}^{(1)}\sqrt{\frac{3}{2}}\hat{\vec{k}}_{\perp}(\vec{\sigma}\cdot\vec{n}) + g_{4}^{(1)}\sqrt{\frac{3}{2}}i[\hat{\vec{k}}_{\perp}\times\vec{n}] \tag{22}$$

In summary, the system of six equations for the J=1 state is split in two subsystems: two equations for $a = 0$ and four for $a = 1$. The subsystem for $a = 0$ has the same structure than (2) with different kernels $K_{ij}^{(J=1)}$. The asymptotic of the kernel $K_{22}^{(J=1)}$ is the same than $-K_{22}^{(J=0)}$: it is negative and corresponds to attraction. The integral (14) for the kernel $H(z,z')$ with the function A_{22} given by (10) diverges logarithmically. Therefore it results in a collapse for any value of the coupling constant. This result coincides with conclusion of the paper [25].

3.5. *Numerical results*

The preceding analysis are confirmed by several numerical calculations. In all what follows, the constituent masses were taken equal to $m=1$ and the mass of the exchanged scalar $\mu=0.25$.

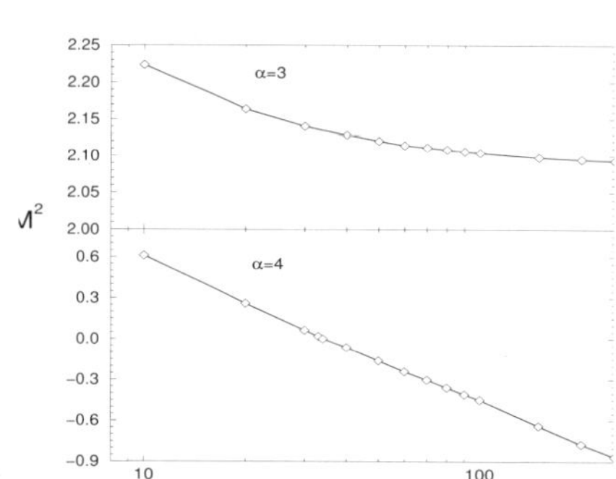

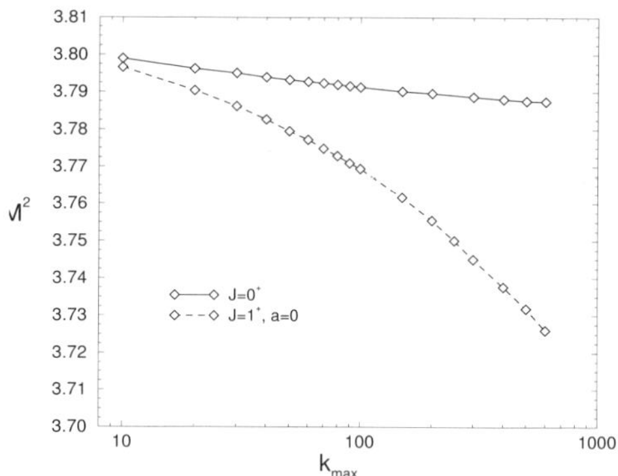

Fig. 5.   Cutoff dependence of the binding energy in the $J = 0$ state, in the one-channel problem (f_1), for two fixed values of the coupling constant below and above the critical value.

Fig. 6.   Cutoff dependence of the binding energy, for $J = 0$ and $J = 1, J_z = 0$ states, in full (two-channel) problem, for $\alpha = 1.184$.

Let us first present the results given by the one channel problem: a single equation for f_1 with kernel K_{11} in the $J = 0$ case. We have plotted in figure 5 the mass square M^2 of the two fermion system as a function of the cutoff k_{max} for two fixed values of the coupling constant below and above the critical value. In our calculations the cutoff appears directly as the maximum value k_{max} up to which the integrals in (2) are performed. One can see two dramatically different behaviors depending on the value of the coupling constant α. For $\alpha = 3$, i.e. $\alpha < \alpha_c = 3.594$, the result is convergent. For $\alpha = 4$, i.e. $\alpha > \alpha_c$, the result is clearly divergent. M^2 decreases logarithmically as a function of k_{max} and becomes even negative. This property is due only to the large k behavior of K_{11}. Though the negative values of M^2 which appear in Fig. 5 are physically meaningless, they are formally allowed by the equations (2). The first degree of M does not enter neither in the equation nor in the kernel, and M^2 crosses zero without any singularity. The value of the critical α does not depend on the exchange mass μ. For $\mu \ll m$, e.g. $\mu \approx 0.25$, its existence is not relevant in describing physical states since any solution with positive M^2, stable relative to cutoff, corresponds to $\alpha < \alpha_c$. For $\mu \sim m$ one can reach the critical α for positive, though small values of M^2.

We consider now the full Yukawa problem as given by the two coupled equations (2). In figure 6 are displayed the variations of M^2 for $J = 0$ and $J = 1, J_z = 0$ states as a function of the cutoff k_{max}. The value of the coupling constant for both J is $\alpha_c = 1.184$, the same that in Fig. 2 of [25], below the critical value. Our numerical values are in agreement with the results for the cutoff $\Lambda \leq 100$ presented in this figure [21], but our calculation at larger k_{max} leads to different conclusion for the $J = 0$ state. We first notice a qualitatively different behavior of the two states. In what concerns $J = 0$, the numerical results become more flat when k_{max} increases, – with less than a 0.5% variation in M^2 when changing k_{max} between $k_{max}=10$ and 300. This strongly suggests a convergence. We thus conclude to the stability of the state with $J = 0$, as expected from our analysis in Sect. 3.3.

On the contrary, for $J = 1, J_z = 0$ the value of $M^2(k_{max})$ continues to decrease faster than logarithmically and indicates, – as found in [25], – a collapse. As mentioned above, the asymptotic of the $K_{22}^{(J=1)}$ kernel is the same as the $K_{22}^{(J=0)}$ one but with an opposite sign, i.e. it is attractive, what leads to instability for any value of α. The same result was found when solving the $J = 0$ equations with the opposite sign of $K_{22}^{(J=0)}$.

3.6. *Positronium*

We applied our method to the positronium system in the $J = 0^-$, a bound state of electron and positron which exists in nature. We consider this important application in more detail.

The wave function is again determined by two components and has the form (1). The negative parity of the state comes from the intrinsic positron parity so that the corresponding kernels are those of the $J^\pi = 0^+$ two-fermion system. They were derived, for the Feynman gauge in [16] (eqs. (A8) in appendix A). They have the form (3) with the following values κ_{ij} instead of eqs. (5) for the scalar case:

$$\kappa_{11} = -2\pi\alpha(4k^2 k'^2 + 3m^2(k^2 + k'^2) + 2m^4) \tag{23}$$
$$\kappa_{12} = 2\pi\alpha mk'(k^2 - k'^2)\sin\theta'$$
$$\kappa_{21} = -2\pi\alpha mk(k^2 - k'^2)\sin\theta$$
$$\kappa_{22} = -2\pi\alpha[kk'(k^2 + k'^2 + 2m^2)\sin\theta\sin\theta' + 2\epsilon_k\epsilon_{k'}(\epsilon_k\epsilon_{k'} + kk'\cos\theta\cos\theta')\cos\phi']$$

Following Sect. 3.2, we substitute $k' = \gamma k$ and take the limit $k \to \infty$. The non-diagonal kernels tend to zero, whereas for K_{11} and K_{22} we reproduce (8) with the following kernels $A(\theta,\theta',\gamma)$:

$$A_{11}(\theta,\theta',\gamma) =$$
$$8\sqrt{\gamma}\int_0^{2\pi}\frac{d\phi}{2\pi}\frac{1}{(1+\gamma^2)(1+|\cos\theta - cos\theta'| - \cos\theta\cos\theta') - 2\gamma\sin\theta\sin\theta'\cos\phi}, \tag{24}$$

$$A_{22}(\theta,\theta',\gamma) =$$
$$\frac{2}{\sqrt{\gamma}}\int_0^{2\pi}\frac{d\phi}{2\pi}\frac{(1+\gamma^2)\sin\theta\sin\theta' + 2\gamma(1+\cos\theta\cos\theta')\cos\phi}{(1+\gamma^2)(1+|\cos\theta - cos\theta'| - \cos\theta\cos\theta') - 2\gamma\sin\theta\sin\theta'\cos\phi}, \tag{25}$$

where we denote $\alpha' = \alpha/(2m\pi)$.

At $\gamma \to 0$ A_{22} has the behavior: $A_{22}(\theta,\theta',\gamma) \propto +1/\sqrt{\gamma}$ (compare with $A_{22}(\theta,\theta',\gamma) \propto -1/\sqrt{\gamma}$ in eq. (10) for Yukawa model).

As discussed at the end of Sect. 3.3, the behavior $A_{22}(\theta,\theta',\gamma) \propto -1/\sqrt{\gamma}$ corresponds to repulsion, hence for positronium with $A_{22}(\theta,\theta',\gamma) \propto +1/\sqrt{\gamma}$ we have attraction. The integral (14) diverges and the spectrum is unbounded from below.

This conclusion is confirmed by numerical calculations. In Table 1 are presented the values of the coupling constant α as a function of the sharp cut-off k_{max} and for a fixed binding energy $B = 0.0225$. The dependence is very slow – 0.3% variation for $k_{max} \in [10, 300]$ – but it actually corresponds to a logarithmic divergence of $\alpha(k_{max})$ as it can be seen in Fig. 7. The origin of this instability is the coupling to the second component, whose kernel matrix element κ_{22} has an attractive, constant asymptotic limit. If one removes this component – which has a very small contribution in norm – calculations become stable and give for $\alpha_{NR} = 0.30$ the value $\alpha_{LFD} = 0.3975$.

We should emphasize that as one can see from Fig. 7, and from Table 1, the development of this instability vs. k_{max} is very slow. The value $k_{max} = 300$ (in units of electron masses) is very large. At this momentum the contributions having other origin (beyond QED), can make influence and change the behavior of the binding energy vs. k_{max}.

Table 1. Coupling constant α as a function of the sharp cut-off k_{max} for the $J = 0^-$ positronium state with binding energy $B = 0.0225$ a.u.

k_{max}	10	20	30	40	50	70	100	200	300
α	0.3945	0.3928	0.3918	0.3911	0.3905	0.3896	0.3887	0.3867	0.3854

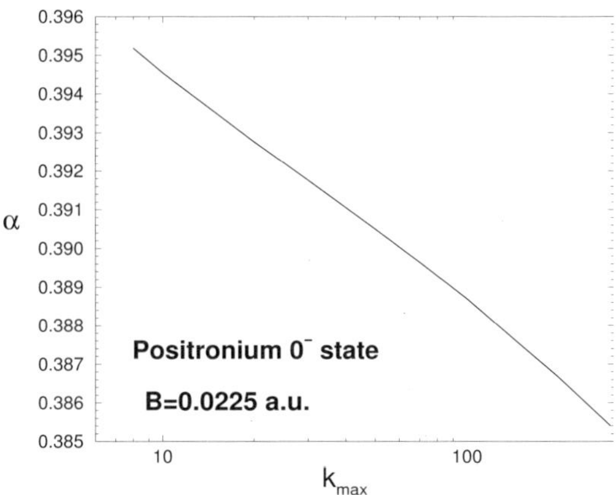

Fig. 7.   Coupling constant α as a function of the sharp cut-off k_{max} for the $J = 0^-$ positronium state with binding energy $B = 0.0225$ a.u.

Let us now consider also another gauge – the so called light-cone gauge [26] – which is often used in the light-front dynamics calculations. In the explicitly covariant version of LFD, the photon propagator in the light-cone gauge, is obtained from the Feynman one by the replacement (see eq. (2.65) from [10])

$$-g_{\mu\nu} \to -g_{\mu\nu} + \frac{\omega_\mu k_\nu + \omega_\nu k_\mu}{\omega \cdot k} \tag{26}$$

The behavior $1/\omega \cdot k \sim 1/x$ is singular and should be regularized [26]. There are two graphs corresponding to the photon exchange which differ from each other by the order of vertices in the light-front time, see e.g. Fig. 3 from [16]. The value of the momentum k transferred by photon is different in these LF graphs, see eq. (14) from [16], By performing the calculations, we have found that the second term in (26) gives additional contributions to eqs. (24) and (25) which turns into:

$$A_{11}(\theta,\theta',\gamma) = \int_0^{2\pi} \frac{d\phi}{2\pi D} \left[8\sqrt{\gamma} + \frac{4(1+\gamma^2)\sin\theta\sin\theta'\cos\phi}{\sqrt{\gamma}|\cos\theta - \cos\theta'|} \right] \tag{27}$$

$$A_{22}(\theta,\theta',\gamma) = \int_0^{2\pi} \frac{d\phi}{2\pi D} \left[\frac{2}{\sqrt{\gamma}}[(1+\gamma^2)\sin\theta\sin\theta' + 2\gamma(1+\cos\theta\cos\theta')\cos\phi] \right.$$
$$\left. + \frac{4(1+\gamma^2)\sin\theta\sin\theta'}{\sqrt{\gamma}|\cos\theta - \cos\theta'|} \right] \tag{28}$$

The singularity $\sim 1/|\cos\theta - \cos\theta'|$ appears from $1/x$ in (26) and, as mentioned, it should be regularized. In the limit $\gamma \to 0$, the extra contribution $\sim 1/\sqrt{\gamma}$ in $A_{11}(\theta,\theta',\gamma)$ is smoothen due to the integration over ϕ, whereas it does not change the behavior of $A_{22}(\theta,\theta',\gamma)$ which remains of the form $A_{22}(\theta,\theta',\gamma) \propto +1/\sqrt{\gamma}$. As explained above, this corresponds to a spectrum unbounded from below. This is manifested by an unbounded increasing of the binding energy B as a function of the cutoff k_{max} or by a decreasing – down to zero – of the coupling constant α for a fixed value of the binding energy, as it is shown in Fig. 7 and in the Table 1. We would like to again that this dependence on k_{max}, fatal for the very existence of stable bound states, is very weak and so not at all easy to find its evidence in numerical calculations, specially when using non-uniform mappings.

Due to this very slow k_{max}-dependence we have fixed the cut-off to an arbitrary value $k_{max} = 10$ and considered the case $\alpha = 3$. The non relativistic binding energy is $B = 0.0225$ and we found, for the ladder LFD in the Feynman gauge [16], a value $B_{LFD} = 0.0132$, that is a strong repulsive effect. This repulsion, observed in most of the kernels examined both for bosons and fermions, however contradicts the leading order QED corrections [27]

$$B_{QED} = \frac{\alpha^2}{4}\left[1 + \frac{21}{16}\alpha^2 + o(\alpha^4)\right] \approx 0.02516,$$

which are attractive. This indicates that the ladder light-front kernel, in the Feynman gauge, is unable to predict even the sign for the relativistic corrections of such a genuine system. It remains to see if this failure is a consequence of the relative simplicity of the ladder sum or it has other reason.

4. Conclusions

In the relativistic framework of Light-Front Dynamics, we have studied the critical stability of three equal-mass bosons, interacting via zero-range forces and the two-fermion system interacting *via* ladder scalar, pseudoscalar and vector exchanges.

The three equal-mass bosons interact via zero-range forces constrained to provide finite two-body mass M_2. We have found that the three-body bound state exists for two-body mass values in the range $M_2^{(c)} = 1.43\,m \leq M_2 \leq 2\,m$. At the zero two-body binding limit, the three-body binding energy is $B_3^{(c)} \approx 0.012\,m$ and represent the minimal binding energy for a three bosons system with contact interactions. The Thomas collapse is avoided in the sense that three-body mass M_3 is finite, in agreement with [1,13].

However, another kind of catastrophe happens. Although removing infinite binding energies, the relativistic dynamics generates zero three-body mass M_3 at a critical value $M_2 = M_2^{(c)}$. For stronger interaction, i.e. when $0 \leq M_2 < M_2^{(c)}$, there are no physical solutions of the Light-Front equations with real value of M_3. In this domain, M_3^2 becomes negative.

If in the non-relativistic dynamics the system collapses when its binding energy tends to $-\infty$, in the relativistic approach the system does not exist when its mass squared is negative. This fact can be interpreted as the relativistic counterpart of the non-relativistic Thomas collapse.

We extended this study to two-fermion system interacting by exchange of scalar pseudo scalar and vector particles. In [16] we have separately examined the different types of these couplings and found very different behaviors concerning the stability of the solutions themselves and their relation with the corresponding non relativistic reductions.

In particular, the scalar coupling (Yukawa model) is found to be stable without any kernel regularization for the $J^\pi = 0^+$ state and coupling constants below some critical value $\alpha < \alpha_c = 3.72$. For values above α_c the system collapses. For $J^\pi = 1^+$ state the solution is unstable. The comparison with the non relativistic solutions shows always repulsive effects.

Electromagnetic coupling presents the stronger anomalies. It has been applied to positronium 0^+ state. It is found to be unstable and, once regularized by means of sharp cut-off, the ladder approximation in the Feynman gauge gives relativistic corrections of opposite sign compared to QED perturbative results. This failure shows, probably, the poorness of the ladder approximation in one of the rare cases in which it can be confronted to experimental results.

As a final remarks, we would like to emphasize again that, as it was first pointed out in [1], the relativistic dynamics allows to exist, in principle, systems which would not exist in the non-relativistic framework. Their existence is determined by the properties and strength of relativistic interaction and we dentoe this fact by "critical stability". The pioneering work of Pierre Noyes [1] opened thus a fruitful and interesting field in the theory of few-body systems.

. . . .

References

[1] James V. Lindesay and H. Pierre Noyes, *Zero range scattering theory II. Minimal relativistic three-particle equations and the Efimov effect*, Preprint SLAC-PUB-2932(rev.), 1986.

[2] L.D. Landau, E.M. Lifshits, *Quantum mechanics*, Pergamon press, 1965.

[3] G.E. Brown, A.D. Jackson, The nucleon-nucleon interaction, North-Holland, Amsterdam, 1976.

[4] Y.N. Demkov, V.N. Ostrovskii, Zero-range potentials and their applications in atomic physics, Plenum Press, New-York 1988.

[5] L.H. Thomas, Phys. Rev. **47** (1935) 903.

[6] S. K. Adhikari, T. Frederico, I.D. Goldman, Phys. Rev. Lett. **74** (1995) 487; T. Frederico, L. Tomio, A. Delfino, A.E.A Amorin, Phys. Rev. **A60** (1999) R9.

[7] D.V. Fedorov, A.S. Jensen, Phys. Rev. **A63** (2001) 063608; Nucl. Phys. A697 (2002) 783.

[8] M. Mangin-Brinet, J. Carbonell, Phys. Lett. **B474**, (2000) 237

[9] J. Carbonell, V.A. Karmanov, Phys.Rev. **C67** (2003) 037001.

[10] J. Carbonell, B. Desplanques, V.A. Karmanov, J.-F. Mathiot, Phys. Reports, **300** (1998) 215.

[11] S.J. Brodsky, H.-C. Pauli, S.S. Pinsky, Phys. Reports, **301** (1998) 299.

[12] P.A.M. Dirac, Rev. Mod. Phys. **21** (1949) 392.

[13] T. Frederico, Phys. Lett. **B282** (1992) 409.

[14] B.L.G. Bakker, L.A. Kondratyuk, M.V. Terentyev, Nucl. Phys, **B158** (1979) 497.

[15] W.R.B. de Araujo, J.P.B.C. de Melo, T. Frederico, Phys. Rev. **C52** (1995) 2733.

[16] M. Mangin-Brinet, J. Carbonell, V.A. Karmanov, Phys. Rev. **C68** (2003) 055203.

[17] M. Mangin-Brinet, J. Carbonell, V.A. Karmanov, Phys.Rev. **D64** (2001) 125005.

[18] A.V. Smirnov, privite communication of Feb. 20, 2001.

[19] M. Mangin-Brinet, J. Carbonell, V.A. Karmanov, Phys.Rev. **D64** (2001) 027701.

[20] E. Salpeter, H. Bethe, Phys.Rev. **84** (1951) 1232–1242.

[21] J. Carbonell, V. Karmanov, Eur.Phys.J. **A46** (2010) 387–397.

[22] J. Carbonell, V. Karmanov, Few Body Syst. **49** (2011) 205–222.

[23] J. Carbonell, V.A. Karmanov, F. de Soto, Few-Body Systems (2013); arXiv:1211.5474

[24] J. Carbonell and V.A. Karmanov, Nucl. Phys. **A581** (1995) 625.

[25] St. Glazek, A. Harindranath, S. Pinsky, J. Shigemutsu and K. Wilson, Phys. Rev. **D47** (1993) 1599.

[26] G.P. Lepage, S.J. Brodsky, Phys. Rev. **D22** (1980) 2157.

[27] H.A. Bethe and E.E. Salpeter, Quantum Mechanics of one- and two-electron atoms, A Plenum/Roseta Ed., (1977).

Non-Commutative Worlds
and Classical Constraints

Louis H. Kauffman

Department of Mathematics, Statistics and Computer Science
University of Illinois at Chicago
851 South Morgan Street, Chicago, Illinois, 60607-7045 USA,
E-mail: kauffman@uic.edu

To Pierre Noyes for his 90-th birthday. This paper generalizes our joint work on discrete electromagetism and the Feynman-Dyson derivation of electromagnetism from commutator calculus.

1. Introduction

Aspects of gauge theory, Hamiltonian mechanics, relativity and quantum mechanics arise naturally in the mathematics of a non-commutative framework for calculus and differential geometry. In this paper, we first give a review of our previous results (that started in work with Pierre Noyes [21]) about discrete physics and non-commutative worlds. The simplest discrete system corresponds directly to the square root of minus one, seen as an oscillation between one and minus one. This way thinking about i as an *iterant* is explained below. By starting with a discrete time series of positions, one has immediately a non-commutativity of observations since the measurement of velocity involves the tick of the clock and the measurement of position does not demand the tick of the clock. Commutators that arise from discrete observation suggest a non-commutative calculus, and this calculus leads to a generalization of standard advanced calculus in terms of a non-commutative world. In a non-commutative world, all derivatives are represented by commutators. We then give our version of a gauge-theoretic generalization of the Feynman-Dyson derivation of the formalism of electromagnetic gauge theory. This generalization is based on demanding a transfer to the non-commutative world of a standard formula about time derivatives in advanced calculus. As we shall see, this one demand is enough to make the formalism of electromagnetism appear in a world where the derivatives are all represented by commutators. The rest of the paper investigates algebraic constraints that bind the commutative and non-commutative worlds and contains a description of the connection between our considerations and the work of Tony Deakin [3, 4].

Section 2 is a self-contained miniature version of the whole story in this paper, starting with the square root of minus one seen as a discrete oscillation, a clock. We proceed from there and analyze the position of the square root of minus one in relation to discrete systems and quantum mechanics. We end this section by fitting together these observations into the structure of the Heisenberg commutator

$$[p, q] = i\hbar.$$

Section 3 is a review of the context of non-commutative worlds. This section generalizes the concepts in Section 2 and places them in the wider context of non-commutative worlds. The key to this generalization is our method of embedding discrete calculus into non-commutative calculus. Section 4 gives a complete treatment of our generalization of the Feynman-Dyson derivation of Maxwell's equations in a non-commutative framework. This section is the first foray into the consequences of constraints. This version of the Feynman-Dyson derivation depends entirely on the postulation of a full time derivative in the non-commutative world that matches the corresponding formula in ordinary commutative advanced calculus. The result of the derivation is a generalization of the Maxwell Equations that includes gauge theory. Section 5 discusses constraints on non-commutative worlds that are imposed by asking for correspondences between forms of classical differentiation and the derivatives represented by commutators in a correspondent non-commutative world. This discussion of constraints parallels work of Tony Deakin [3, 4] and will be continued in joint work of the author and Deakin. At the level of the second constraint we encounter issues related to general relativity. Section 6 continues the constraints discussion in Section 5, showing how to generalize to higher-order constraints. We obtain a commutator formula for the third order constraint.

2. Quantum Mechanics - The Square Root of Minus One is a Clock

The purpose of this section is to place i, the square root of minus one, and its algebra in a context of discrete recursive systems. We begin by starting with a simple periodic process that is associated directly with the classical attempt to solve for i as a solution to a quadratic equation. We take the point of view that solving $x^2 = ax + b$ is the same (when $x \neq 0$) as solving

$$x = a + b/x,$$

and hence is a matter of finding a fixed point. In the case of i we have

$$x^2 = -1$$

and so desire a fixed point

$$x = -1/x.$$

There are no real numbers that are fixed points for this operator and so we consider the oscillatory process generated by

$$R(x) = -1/x.$$

The fixed point would satisfy

$$i = -1/i$$

and multiplying, we get that

$$ii = -1.$$

The widest mathematical generalization of an eigenvalue is an *eigenform* [9, 10, 13, 15–18, 20], a fixed point in some domain. Thus we have conceptualized the square root of minus one as an eigenform for the operator $R(x) = -1/x$. The square root of minus one is a perfect example of an eigenform that occurs in a new and wider domain than the original context in which its recursive process arose. The process has no fixed point in the original domain.

On the other hand the iteration of R yields

$$1, R(1) = -1, R(R(1)) = +1, R(R(R(1))) = -1, +1, -1, +1, -1, \cdots .$$

Looking at the oscillation between $+1$ and -1, we see that there are naturally two phase-shifted viewpoints. We denote these two views of the oscillation by $[+1, -1]$ and $[-1, +1]$. These viewpoints correspond to whether one regards the oscillation at time zero as starting with $+1$ or with -1. See Figure 1.

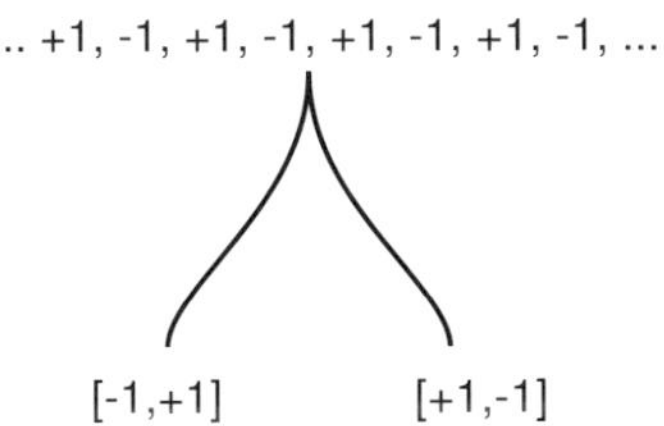

Fig. 1.   **A Basic Oscillation**

We shall let $I\{+1, -1\}$ stand for an undisclosed alternation or ambiguity between $+1$ and -1 and call $I\{+1, -1\}$ an *iterant*. There are two iterant views: $[+1, -1]$ and $[-1, +1]$. For simplicity, we refer to each of the iterant views as *iterants*. Conceptually, one should understand that the ordering of an iterant is a choice made by an observer.

Given an iterant $[a, b]$, we can think of $[b, a]$ as the same process with a shift of one time step. These two iterant views, seen as points of view of an alternating process, will become the square roots of negative unity, i and $-i$.

We introduce a temporal shift operator η such that

$$[a, b]\eta = \eta[b, a]$$

and

$$\eta\eta = 1$$

for any iterant $[a, b]$, so that concatenated observations can include a time step of one-half period of the process

$$\cdots abababab \cdots .$$

We combine iterant views term-by-term as in

$$[a, b][c, d] = [ac, bd].$$

We now define i by the equation

$$i = [1, -1]\eta.$$

This makes i both a value and an operator that takes into account a step in time.

We calculate

$$ii = [1, -1]\eta[1, -1]\eta = [1, -1][-1, 1]\eta\eta = [-1, -1] = -1.$$

Thus we have constructed the square root of minus one by using an iterant viewpoint. In this view i represents a discrete oscillating temporal process and it is an eigenform for $R(x) = -1/x$, participating in the algebraic structure of the complex numbers. In fact the corresponding algebra structure of linear combinations $[a, b] + [c, d]\eta$ is isomorphic with 2×2 matrix algebra and iterants can be used to construct $n \times n$ matrix algebra. We treat this generalization elsewhere [9, 11, 17, 18, 20, 25–28]. For now, we point out the translation by the formulas below.

$$[a, d] + [b, c]\eta = \begin{pmatrix} a & b \\ c & d \end{pmatrix}$$

where

$$[x, y] = \begin{pmatrix} x & 0 \\ 0 & y \end{pmatrix}$$

and

$$\eta = \begin{pmatrix} 0 & 1 \\ 1 & 0 \end{pmatrix}.$$

The reader can do some exercise to see that the details of matrix multiplication dovetail with the iterant algebra.

The Temporal Nexus. *We take as a matter of principle that the usual real variable t for time is better represented as it so that time is seen to be a process, an observation and a magnitude all at once.* This principle of "imaginary time" is justified by the eigenform approach to the structure of time and the structure of the square root of minus one.

As an example of the use of the Temporal Nexus, consider the expression $x^2 + y^2 + z^2 + t^2$, the square of the Euclidean distance of a point (x, y, z, t) from the origin in Euclidean four-dimensional space. Now replace t by it, and find

$$x^2 + y^2 + z^2 + (it)^2 = x^2 + y^2 + z^2 - t^2,$$

the squared distance in hyperbolic metric for special relativity. By replacing t by its process operator value it we make the transition to the physical mathematics of special relativity.

2.1. *The Wave Function in Quantum Mechanics and The Square Root of Minus One*

In quantum modeling, the state of a physical system is represented by a vector in a Hilbert space. As time goes on the vector undergoes a unitary evolution in the Hilbert space. Observable quantities correspond to Hermitian operators H and vectors v that have the property that the application of H to v results in a new vector that is a multiple of v by a real factor λ. Thus

$$Hv = \lambda v.$$

As we have seen, eigenforms generalize eigenvalues and in fact, since $\frac{d}{dx}(e^x) = e^x$ we see that the function e^x is an eigenform for the operator $\frac{d}{dx}$. In this way, one sees that the mathematics of quantum mechanics is an interplay of eigenforms and eigenvalues.

One can regard a wave function such as $\psi(x,t) = exp(i(kx - wt))$ as containing a micro-oscillatory system with the special synchronizations of the iterant view $i = [+1, -1]\eta$. It is these synchronizations that make the big eigenform of the exponential work correctly with respect to differentiation, allowing it to create the appearance of rotational behaviour, wave behaviour and the semblance of the continuum. In other words, that one can take a temporal view of the well-known equation of Euler:

$$e^{i\theta} = cos(\theta) + isin(\theta)$$

by regarding the i in this equation as an iterant, as discrete oscillation between -1 and $+1$. One can blend the classical geometrical view of the complex numbers with the iterant view by thinking of a point that orbits the origin of the complex plane, intersecting the real axis periodically and producing, in the real axis, a periodic oscillation in relation to its orbital movement in the two dimensional space. The special synchronization is the algebra of the time shift embodied in

$$\eta\eta = 1$$

and

$$[a, b]\eta = \eta[b, a]$$

that makes the algebra of $i = [1, -1]\eta$ imply that $i^2 = -1$. This interpretation does not change the formalism of these complex-valued functions, but it does change one's point of view and we now show how the properties of i as a discrete dynamical system are found in any such system.

2.2. *Time Series and Discrete Physics*

We have just reformulated the complex numbers and expanded the context of matrix algebra to an interpretation of i as an oscillatory process and matrix elements as combined spatial and temporal oscillatory processes (in the sense that $[a, b]$ is not affected in its order by a time step, while $[a, b]\eta$ includes the time dynamic in its interactive capability, and 2×2 matrix algebra is the algebra of iterant views $[a, b] + [c, d]\eta$).

We now consider elementary discrete physics in one dimension. Consider a time series of positions

$$x(t) : t = 0, \Delta t, 2\Delta t, 3\Delta t, \cdots .$$

We can define the velocity $v(t)$ by the formula

$$v(t) = (x(t + \Delta t) - x(t))/\Delta t = Dx(t)$$

where D denotes this discrete derivative. In order to obtain $v(t)$ we need at least one tick Δt of the discrete clock. Just as in the iterant algebra, we need a time-shift operator to handle the fact that once we have observed $v(t)$, the time has moved up by one tick.

We adjust the discrete derivative. We shall add an operator J that in this context accomplishes the time shift:

$$x(t)J = Jx(t + \Delta t).$$

We then redefine the derivative to include this shift:

$$Dx(t) = J(x(t + \Delta t) - x(t))/\Delta t.$$

This readjustment of the derivative rewrites it so that the temporal properties of successive observations are handled automatically.

Discrete observations do not commute. Let A and B denote quantities that we wish to observe in the discrete system. Let AB denote the result of first observing B and then observing A. The result of this definition is that a successive observation of the form $x(Dx)$ is distinct from an observation of the form $(Dx)x$. In the first case, we first observe the velocity at time t, and then x is measured at $t + \Delta t$. In the second case, we measure x at t and then measure the velocity.

We measure the difference between these two results by taking a commutator

$$[A, B] = AB - BA$$

and we get the following computations where we write $\Delta x = x(t + \Delta t) - x(t)$.

$$x(Dx) = x(t)J(x(t + \Delta t) - x(t)) = Jx(t + \Delta t)(x(t + \Delta t) - x(t)).$$

$$(Dx)x = J(x(t + \Delta t) - x(t))x(t).$$

$$[x, Dx] = x(Dx) - (Dx)x = (J/\Delta t)(x(t + \Delta t) - x(t))^2 = J(\Delta x)^2/\Delta t.$$

This final result is worth recording:

$$[x, Dx] = J(\Delta x)^2/\Delta t.$$

From this result we see that the commutator of x and Dx will be constant if $(\Delta x)^2/\Delta t = K$ is a constant. For a given time-step, this means that

$$(\Delta x)^2 = K\Delta t$$

so that

$$\Delta x = \pm\sqrt{(K\Delta t)}.$$

This is a Brownian process with diffusion constant equal to K.

Thus we arrive at the result that any discrete process viewed in this framework of discrete observation has the basic commutator

$$[x, Dx] = J(\Delta x)^2/\Delta t,$$

generalizing a Brownian process and containing the factor $(\Delta x)^2/\Delta t$ that corresponds to the classical diffusion constant. It is worth noting that the adjustment that we have made to the discrete derivative makes it into a commutator as follows:

$$Dx(t) = J(x(t + \Delta t) - x(t))/\Delta t = (x(t)J - Jx(t))\Delta t = [x(t), J]/\Delta t.$$

By replacing discrete derivatives by commutators we can express discrete physics in many variables in a context of non-commutative algebra. We enter this generalization in the next section of the paper.

We now use the temporal nexus (the square root of minus one as a clock) and rewrite these commutators to match quantum mechanics.

2.3. *Simplicity and the Heisenberg Commutator*

Finally, we arrive at the simplest place. Time and the square root of minus one are inseparable in the temporal nexus. The square root of minus one is a symbol and algebraic operator for the simplest oscillatory process. As a symbolic form, i is an eigenform satisfying the equation

$$i = -1/i.$$

One does not have an increment of time all alone as in classical t. One has it, a combination of an interval and the elemental dynamic that is time. With this understanding, we can return to the commutator for a discrete process and use it for the temporal increment.

We found that discrete observation led to the commutator equation

$$[x, Dx] = J(\Delta x)^2/\Delta t$$

which we will simplify to

$$[q, p/m] = (\Delta x)^2/\Delta t$$

taking q for the position x and p/m for velocity, the time derivative of position and ignoring the time shifting operator on the right hand side of the equation.

Understanding that Δt should be replaced by $i\Delta t$, and that, by comparison with the physics of a process at the Planck scale one can take

$$(\Delta x)^2/\Delta t = \hbar/m,$$

we have

$$[q, p/m] = (\Delta x)^2/i\Delta t = -i\hbar/m,$$

whence

$$[p, q] = i\hbar,$$

and we have arrived at Heisenberg's fundamental relationship between position and momentum. This mode of arrival is predicated on the recognition that only *it* represents a true interval of time. In the notion of time there is an inherent clock or an inherent shift of phase that is making a synchrony in our ability to observe, a precise dynamic beneath the apparent dynamic of the observed process. Once this substitution is made, once the correct imaginary value is placed in the temporal circuit, the patterns of quantum mechanics appear. In this way, quantum mechanics can be seen to emerge from the discrete.

The problem that we have examined in this section is the problem to understand the nature of quantum mechanics. In fact, we hope that the problem is seen to disappear the more we enter into the present viewpoint. A viewpoint is only on the periphery. The iterant from which the viewpoint emerges is in a superposition of indistinguishables, and can only be approached by varying the viewpoint until one is released from the particularities that a point of view contains.

3. Review of Non-Commutative Worlds

Now we begin the introduction to non-commutative worlds and a general discrete calculus. Our approach begins in an algebraic framework that naturally contains the formalism of the calculus, but not its notions of limits or constructions of spaces with specific locations, points and trajectories. Many patterns of physical law fit well into such an abstract framework. In this viewpoint one dispenses with continuum spacetime and replaces it by algebraic structure. Behind that structure, space stands ready to be constructed, by discrete derivatives and patterns of steps, or by starting with a discrete pattern in the form of a diagram, a network, a lattice, a knot, or a simplicial complex, and elaborating that structure until the specificity of spatio-temporal locations appear.

Poisson brackets allow one to connect classical notions of location with the non-commutative algebra used herein. Below the level of the Poisson brackets is a treatment of processes and operators as though they were variables in the same context as the variables in the classical calculus. In different degrees one lets go of the notion of classical variables and yet retains their form, as one makes a descent into the discrete. The discrete world of non-commutative operators is a world linked to our familiar world of continuous and commutative variables. This linkage is traditionally exploited in quantum mechanics to make the transition from the classical to the quantum. One can make the journey in the other direction, from the discrete and non-commutative to the "classical" and commutative, but that journey requires powers of invention and ingenuity that are the subject of this exploration. It is our conviction that the world is basically simple. To find simplicity in the complex requires special attention and care.

In starting from a discrete point of view one thinks of a sequence of states of the world $S, S', S'', S''', \cdots$ where S' denotes the state succeeding S in discrete time. It is natural to suppose that there is some measure of difference $DS^{(n)} = S^{(n+1)} - S^{(n)}$, and some way that states S and T might be combined to form a new state ST. We can thus think of world-states as operators in a non-commutative algebra with a temporal derivative $DS = S' - S$. At this bare level of the formalism the derivative does not satisfy the Leibniz rule. In fact it is easy to verify that $D(ST) = D(S)T + S'D(T)$. Remarkably, the Leibniz rule, and hence the formalisms of Newtonian calculus can be restored with the addition of one more operator J. In this instance J is a temporal shift operator with the property that $SJ = JS'$ for any state S. We then see that if $\nabla S = JD(S) = J(S' - S)$. then $\nabla(ST) = \nabla(S)T + S\nabla(T)$ for any states S and T. In fact $\nabla(S) = JS' - JS = SJ - JS = [S, J]$, so that this adjusted derivative is a commutator in the general calculus of states. This, in a nutshell, is our approach to non-commutative worlds. We begin with a very general framework that is a non-numerical calculus of states and operators. It is then fascinating and a topic of research to see how physics and mathematics fit into the frameworks so constructed.

A simplest and fundamental instance of these ideas is seen in the structure of $i = \sqrt{-1}$. We view i as an *iterant* , a discrete elementary dynamical system repeating in time the values $\{\cdots - 1, +1, -1, +1, \cdots \}$. One can think of this system as resulting from the attempt to solve $i^2 = -1$ in the form $i = -1/i$. Then one iterates the transformation $x \longrightarrow -1/x$ and finds the oscillation from a starting value of $+1$ or -1. In this sense i is identical in concept to a *primordial time.* Furthermore the algebraic structure of the complex numbers emerges from two conjugate views of this discrete series as $[-1, +1]$ and $[+1, -1]$. We introduce a temporal shift operator η such that $\eta[-1, +1] = [+1, -1]\eta$ and $\eta^2 = 1$ (sufficient to this purpose). Then we can define $i = [-1, +1]\eta$, endowing it with one view of the discrete oscillation and the sensitivity to shift the clock when interacting with itself or with another operator. See Section 2 for the details of this reconstruction of the complex numbers. The point of the reconstruction for our purposes is that i becomes inextricably identified with elemental time, and so the physical substitution of it for t (Wick rotation) becomes, in this epistemology, an act of recognition of the nature of time.

Constructions are performed in a Lie algebra $\mathcal{A}$. One may take $\mathcal{A}$ to be a specific matrix Lie algebra, or abstract Lie algebra. If $\mathcal{A}$ is taken to be an abstract Lie algebra, then it is convenient to use the universal enveloping algebra so that the Lie product can be expressed as a commutator. In making general constructions of operators satisfying certain relations, it is understood that one can always begin with a free algebra and make a quotient algebra where the relations are satisfied.

On $\mathcal{A}$, a variant of calculus is built by defining derivations as commutators (or more generally as Lie products). For a fixed N in $\mathcal{A}$ one defines

$$\nabla_N : \mathcal{A} \longrightarrow \mathcal{A}$$

by the formula

$$\nabla_N F = [F, N] = FN - NF.$$

∇_N is a derivation satisfying the Leibniz rule.

$$\nabla_N(FG) = \nabla_N(F)G + F\nabla_N(G).$$

Discrete Derivatives are Replaced by Commutators. There are many motivations for replacing derivatives by commutators. If $f(x)$ denotes (say) a function of a real variable x, and $\tilde{f}(x) = f(x + h)$ for a fixed increment h, define the *discrete derivative* Df by the formula $Df = (\tilde{f} - f)/h$, and find that the Leibniz rule is not satisfied. One has the basic formula for the discrete derivative of a product:

$$D(fg) = D(f)g + \tilde{f}D(g).$$

Correct this deviation from the Leibniz rule by introducing a new non-commutative operator J with the property that

$$fJ = J\tilde{f}.$$

Define a new discrete derivative in an extended non-commutative algebra by the formula

$$\nabla(f) = JD(f).$$

It follows at once that

$$\nabla(fg) = JD(f)g + J\tilde{f}D(g) = JD(f)g + fJD(g) = \nabla(f)g + f\nabla(g).$$

Note that

$$\nabla(f) = (J\tilde{f} - Jf)/h = (fJ - Jf)/h = [f, J/h].$$

In the extended algebra, discrete derivatives are represented by commutators, and satisfy the Leibniz rule. One can regard discrete calculus as a subset of non-commutative calculus based on commutators.

Advanced Calculus and Hamiltonian Mechanics or Quantum Mechanics in a Non-Commutative World. In $\mathcal{A}$ there are as many derivations as there are elements of the algebra, and these derivations behave quite wildly with respect to one another. If one takes the concept of *curvature* as the non-commutation of derivations, then $\mathcal{A}$ is a highly curved world indeed. Within $\mathcal{A}$ one can build a tame world of derivations that mimics the behaviour of flat coordinates in Euclidean space. The description of the structure of $\mathcal{A}$ with respect to these flat coordinates contains many of the equations and patterns of mathematical physics.

The flat coordinates Q^i satisfy the equations below with the P_j chosen to represent differentiation with respect to Q^j:

$$[Q^i, Q^j] = 0$$

$$[P_i, P_j] = 0$$

$$[Q^i, P_j] = \delta_{ij}.$$

Here δ_{ij} is the Kronecker delta, equal to 1 when $i = j$ and equal to 0 otherwise. Derivatives are represented by commutators.

$$\partial_i F = \partial F / \partial Q^i = [F, P_i],$$

$$\hat{\partial}_i F = \partial F / \partial P_i = [Q^i, F].$$

Our choice of commutators guarantees that the derivative of a variable with respect to itself is one and that the derivative of a variable with respect to a distinct variable is zero. Furthermore, the commuting of the variables with one another guarantees that mixed partial derivatives are independent of the order of differentiation. This is a flat non-commutative world.

Temporal derivative is represented by commutation with a special (Hamiltonian) element H of the algebra:

$$dF/dt = [F, H].$$

(For quantum mechanics, take $i\hbar dA/dt = [A, H]$.) These non-commutative coordinates are the simplest flat set of coordinates for description of temporal phenomena in a non-commutative world.

180   *Louis H. Kauffman*

Hamilton's Equations are Part of the Mathematical Structure of Non-Commutative Advanced Calculus.

$$dP_i/dt = [P_i, H] = -[H, P_i] = -\partial H/\partial Q^i$$

$$dQ^i/dt = [Q^i, H] = \partial H/\partial P_i.$$

These are exactly Hamilton's equations of motion. The pattern of Hamilton's equations is built into the system.

Schroedinger's Equation is Discrete. Here is how the Heisenberg form of Schroedinger's equation fits in this context. Let $J = (1 + i\hbar H \Delta t)$. Then $\nabla \psi = [\psi, J/\Delta t]$, and we calculate

$$\nabla \psi = \psi[(1 + i\hbar H \Delta t)/\Delta t] - [(1 + i\hbar H \Delta t)/\Delta t]\psi = i\hbar[\psi, H].$$

This is exactly the form of the Heisenberg equation.

Dynamical Equations Generalize Gauge Theory and Curvature. One can take the general dynamical equation in the form

$$dQ^i/dt = \mathcal{G}_i$$

where $\{\mathcal{G}_1, \cdots, \mathcal{G}_d\}$ is a collection of elements of $\mathcal{A}$. Write $\mathcal{G}_i$ relative to the flat coordinates via $\mathcal{G}_i = P_i - A_i$. This is a definition of A_i and $\partial F/\partial Q^i = [F, P_i]$. The formalism of gauge theory appears naturally. In particular, if

$$\nabla_i(F) = [F, \mathcal{G}_i],$$

then one has the curvature

$$[\nabla_i, \nabla_j]F = [R_{ij}, F]$$

and

$$R_{ij} = \partial_i A_j - \partial_j A_i + [A_i, A_j].$$

This is the well-known formula for the curvature of a gauge connection. Aspects of geometry arise naturally in this context, including the Levi-Civita connection (which is seen as a consequence of the Jacobi identity in an appropriate non-commutative world).

One can consider the consequences of the commutator $[Q^i, \dot{Q}^j] = g_{ij}$, deriving that

$$\ddot{Q}^r = G_r + F_{rs}\dot{Q}^s + \Gamma_{rst}\dot{Q}^s\dot{Q}^t,$$

where G_r is the analogue of a scalar field, F_{rs} is the analogue of a gauge field and Γ_{rst} is the Levi-Civita connection associated with g_{ij}. This decompositon of the acceleration is uniquely determined by the given framework.

Remark. While there is a large literature on non-commutative geometry, emanating from the idea of replacing a space by its ring of functions, work discussed herein is not written in that tradition. Non-commutative geometry does occur here, in the sense of geometry occuring in the context of non-commutative algebra. Derivations are represented by commutators. There are relationships between the present work and the traditional non-commutative geometry, but that is a subject for further exploration. In no way is this paper intended to be an introduction to that subject. The present summary is based on [19, 21–29] and the references cited therein.

The following references in relation to non-commutative calculus are useful in comparing with the present approach [2, 5, 7, 33]. Much of the present work is the fruit of a long series of discussions with Picrre Noyes. paper [31] also works with minimal coupling for the Feynman-Dyson derivation. The first remark about the minimal coupling occurs in the original paper by Dyson [1], in the context of Poisson brackets. The paper [8] is worth reading as a companion to Dyson. It is the purpose of this summary to indicate how non-commutative calculus can be used in foundations.

4. Non-commutative Electromagnetism and Gauge Theory

One can use this context to revisit the Feynman-Dyson derivation of electromagnetism from commutator equations, showing that most of the derivation is independent of any choice of commutators, but highly dependent upon the choice of definitions of the derivatives involved. Without any assumptions about initial commutator equations, but taking the right (in some sense simplest) definitions of the derivatives one obtains a significant generalization of the result of Feynman-Dyson. We give this derivation in the next subsection of the present paper, using diagrammatic algebra to clarify the structure. In this derivation we use X to denote the position vector rather than Q, as above.

Electromagnetic Theorem With the appropriate [see below] definitions of the operators, and taking

$$\nabla^2 = \partial_1^2 + \partial_2^2 + \partial_3^2, \ \ H = \dot{X} \times \dot{X} \ \text{ and } \ E = \partial_t \dot{X}, \ \text{ one has}$$

(1) $\ddot{X} = E + \dot{X} \times H$

(2) $\nabla \bullet H = 0$

(3) $\partial_t H + \nabla \times E = H \times H$

(4) $\partial_t E - \nabla \times H = (\partial_t^2 - \nabla^2)\dot{X}$

The key to the proof of this Theorem is the definition of the time derivative. This definition is as follows

$$\partial_t F = \dot{F} - \Sigma_i \dot{X}_i \partial_i(F) = \dot{F} - \Sigma_i \dot{X}_i [F, \dot{X}_i]$$

for all elements or vectors of elements F. The definition creates a distinction between space and time in the non-commutative world. A calculation reveals that

$$\ddot{X} = \partial_t \dot{X} + \dot{X} \times (\dot{X} \times \dot{X}).$$

This suggests taking $E = \partial_t \dot{X}$ as the electric field, and $B = \dot{X} \times \dot{X}$ as the magnetic field so that the Lorentz force law

$$\ddot{X} = E + \dot{X} \times B$$

is satisfied.

This result is applied to produce many discrete models of the Theorem. These models show that, just as the commutator $[X, \dot{X}] = Jk$ describes Brownian motion in one dimension, a generalization of electromagnetism describes the interaction of triples of time series in three dimensions.

Taking $\partial_t F = \dot{F} - \Sigma_i \dot{X}_i \partial_i(F) = \dot{F} - \Sigma_i \dot{X}_i[F, \dot{X}_i]$ as a definition of the partial derivative with respect to time is a natural move in this context because there is *no time variable t* in this non-commutative world. A formal move of this kind, matching a pattern from the commutative world to the mathematics of the non-commutative world is the theme of the Section 5 of this paper. In that section we consider the well known way to associate an operator to a product of commutative variables by taking a sum over all permutations of products of the operators corresponding to the individual variables. This provides a way to associate operator expressions with expressions in the commuative algebra, and hence to let a classical world correspond or map to a non-commutative world. To bind these worlds more closely, we can ask that the formulas for taking derivatives in the commuative world should have symmetrized operator product correspondences in the non-commutative world. In Section 5 we show how the resulting constraints are related to having a quadratic Hamiltonian (first order constraint) and to having a version of general relativity [3, 4] (second order constraint). Such constraints can be carried to all orders of derivatives, but the algebra of such constraints is, at the present time, in a very primitive state.

In the subsection to follow, we carry out the details of the generalized Feynman-Dyson derivation. The reader will find a derivation of a set of equations that have the form of the Maxwell Equations. This theory can be identified as a non-commutative version of electromagnetism if we make the same commutator assumptions as Feynman and Dyson. Otherwise it is a gauge theoretic generalization of electromagnetism put in non-commutative form. This formulation requires further investigation and will be the subject of papers subsequent to the present work.

4.1. *Generalized Feynman Dyson Derivation*

We assume that specific time-varying coordinate elements X_1, X_2, X_3 of the algebra $\mathcal{A}$ are given. *We do not assume any commutation relations about X_1, X_2, X_3.*

In this section we no longer avail ourselves of the commutation relations that are in back of the original Feynman-Dyson derivation. We do take the definitions of the derivations from that previous context. Surprisingly, the result is very similar to the one of Feynman and Dyson, as we shall see.

Here $A \times B$ is the non-commutative vector cross product:

$$(A \times B)_k = \Sigma^3_{i,j-1} \epsilon_{ijk} A_i B_j.$$

(We will drop this summation sign for vector cross products from now on.) Then, with $B = \dot{X} \times \dot{X}$, we have

$$B_k = \epsilon_{ijk} \dot{X}_i \dot{X}_j = (1/2)\epsilon_{ijk}[\dot{X}_i, \dot{X}_j].$$

The epsilon tensor ϵ_{ijk} is defined for the indices $\{i, j, k\}$ ranging from 1 to 3, and is equal to 0 if there is a repeated index and is otherwise equal to the sign of the permutation of 123 given by ijk. We represent dot products and cross products in diagrammatic tensor notation as indicated in Figure 2 and Figure 3. In Figure 2 we indicate the epsilon tensor by a trivalent vertex. The indices of the tensor correspond to labels for the three edges that impinge on the vertex. The diagram is drawn in the plane, and is well-defined since the epsilon tensor is invariant under cyclic permutation of its indices.

We will define the fields E and B by the equations

$$B = \dot{X} \times \dot{X} \ \text{ and } \ E = \partial_t \dot{X}.$$

We will see that E and B obey a generalization of the Maxwell Equations, and that this generalization describes specific discrete models. The reader should note that this means that a significant part of the *form* of electromagnetism is the consequence of choosing three coordinates of space, and the definitions of spatial and temporal derivatives with respect to them. The background process that is being described is otherwise arbitrary, and yet appears to obey physical laws once these choices are made.

In this section we will use diagrammatic matrix methods to carry out the mathematics. In general, in a diagram for matrix or tensor composition, we sum over all indices labelling any edge in the diagram that has no free ends. Thus matrix multiplication corresponds to the connecting of edges between diagrams, and to the summation over common indices. With this interpretation of compositions, view the first identity in Figure 2. This is a fundamental identity about the epsilon, and corresponds to the following lemma.

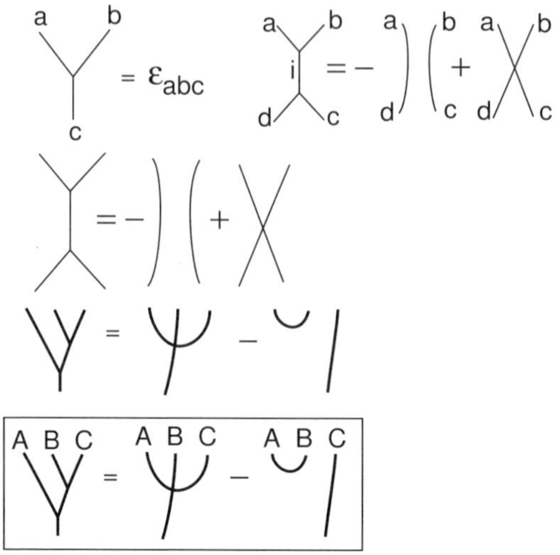

Fig. 2.   **Epsilon Identity**

Lemma. (View Figure 2) Let ϵ_{ijk} be the epsilon tensor taking values 0, 1 and -1 as follows: When ijk is a permutation of 123, then ϵ_{ijk} is equal to the sign of the permutation. When ijk contains a repetition from $\{1, 2, 3\}$, then the value of epsilon is zero. Then ϵ satisfies the labeled identity in Figure 2 in terms of the Kronecker delta.

$$\Sigma_i\, \epsilon_{abi}\epsilon_{cdi} = -\delta_{ad}\delta_{bc} + \delta_{ac}\delta_{bd}.$$

The proof of this identity is left to the reader. The identity itself will be referred to as the *epsilon identity*. The epsilon identity is a key structure in the work of this section, and indeed in all formulas involving the vector cross product.

The reader should compare the formula in this Lemma with the diagrams in Figure 2. The first two diagram are two versions of the Lemma. In the third diagram the labels are capitalized and refer to vectors A, B and C. We then see that the epsilon identity becomes the formula

$$A \times (B \times C) = (A \bullet C)B - (A \bullet B)C$$

for vectors in three-dimensional space (with commuting coordinates, and a generalization of this identity to our non-commutative context. Refer to Figure 3 for the diagrammatic definitions of dot and cross product of vectors. We take these definitions (with implicit order of multiplication) in the non-commutative context.

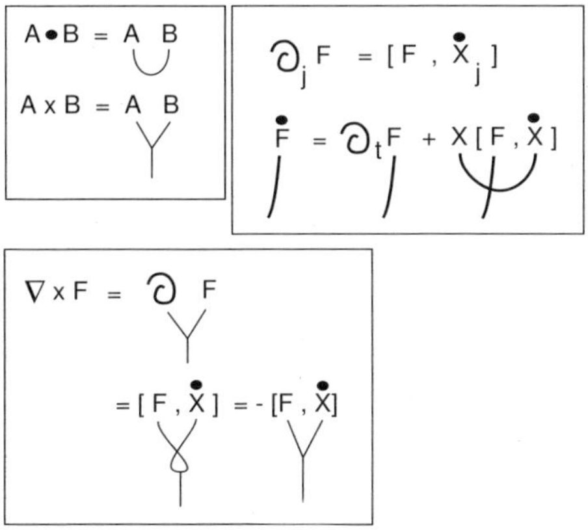

Fig. 3.   **Defining Derivatives**

Remarks on the Derivatives.

(1) Since we do not assume that $[X_i, \dot{X}_j] = \delta_{ij}$, nor do we assume $[X_i, X_j] = 0$, it will not follow that E and B commute with the X_i.

(2) We define

$$\partial_i(F) = [F, \dot{X}_i],$$

and the reader should note that, these spatial derivations are no longer flat in the sense of section 3 (nor were they in the original Feynman-Dyson derivation). See Figure 3 for the diagrammatic version of this definition.

(3) We define $\partial_t = \partial/\partial t$ by the equation

$$\partial_t F = \dot{F} - \Sigma_i \dot{X}_i \partial_i(F) = \dot{F} - \Sigma_i \dot{X}_i [F, \dot{X}_i]$$

for all elements or vectors of elements F. We take this equation as the global definition of the temporal partial derivative, even for elements that are not commuting with the X_i. This notion of temporal partial derivative ∂_t is a least relation that we can write to describe the temporal relationship of an arbitrary non-commutative vector F and the non-commutative coordinate vector X. See Figure 3 for the diagrammatic version of this definition.

(4) In defining

$$\partial_t F = \dot{F} - \Sigma_i \dot{X}_i \partial_i(F),$$

we are using the definition itself to obtain a notion of the variation of F with respect to time. The definition itself creates a distinction between space and time in the non-commutative world.

(5) The reader will have no difficulty verifying the following formula:

$$\partial_t(FG) = \partial_t(F)G + F\partial_t(G) + \Sigma_i \partial_i(F)\partial_i(G).$$

This formula shows that ∂_t does not satisfy the Leibniz rule in our non-commutative context. This is true for the original Feynman-Dyson context, and for our generalization of it. All derivations in this theory that are defined directly as commutators do satisfy the Leibniz rule. Thus ∂_t is an operator in our theory that does not have a representation as a commutator.

(6) We define divergence and curl by the equations

$$\nabla \bullet B = \Sigma_{i=1}^{3} \partial_i(B_i)$$

and

$$(\nabla \times E)_k = \epsilon_{ijk}\partial_i(E_j).$$

See Figure 3 and Figure 5 for the diagrammatic versions of curl and divergence.

Now view Figure 4. We see from this Figure that it follows directly from the definition of the time derivatives (as discussed above) that

$$\ddot{X} = \partial_t \dot{X} + \dot{X} \times (\dot{X} \times \dot{X}).$$

This is our motivation for defining

$$E = \partial_t \dot{X}$$

and

$$B = \dot{X} \times \dot{X}.$$

With these definitions in place, we have

$$\ddot{X} = E + \dot{X} \times B,$$

giving an analog of the Lorentz force law for this theory.

Just for the record, look at the following algebraic calculation for this derivative:

$$\dot{F} = \partial_t F + \Sigma_i \dot{X}_i[F, \dot{X}_i]$$

$$= \partial_t F + \Sigma_i(\dot{X}_i F \dot{X}_i - \dot{X}_i \dot{X}_i F)$$

$$= \partial_t F + \Sigma_i(\dot{X}_i F \dot{X}_i - \dot{X}_i F_i \dot{X}) + \dot{X}_i F_i \dot{X} - \dot{X}_i \dot{X}_i F$$

Hence

$$\dot{F} = \partial_t F + \dot{X} \times F + (\dot{X} \bullet F)\dot{X} - (\dot{X} \bullet \dot{X})F$$

(using the epsilon identity). Thus we have

$$\ddot{X} = \partial_t \dot{X} + \dot{X} \times (\dot{X} \times \dot{X}) + (\dot{X} \bullet \dot{X})\dot{X} - (\dot{X} \bullet \dot{X})\dot{X},$$

whence

$$\ddot{X} = \partial_t \dot{X} + \dot{X} \times (\dot{X} \times \dot{X}).$$

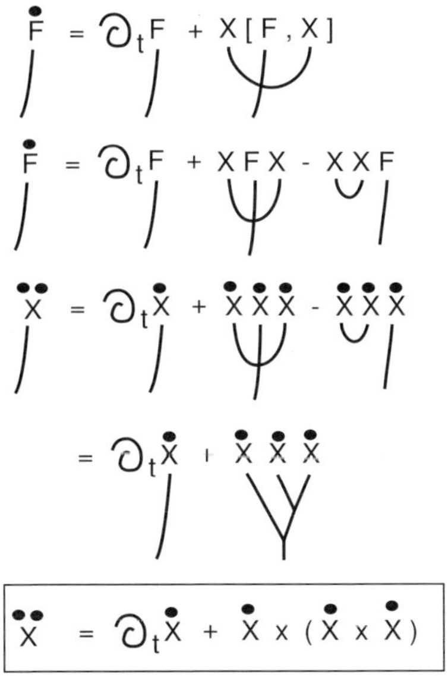

$$\dot{F} = \partial_t F + X[F, X]$$

$$\dot{F} = \partial_t F + XFX - XXF$$

$$\ddot{X} = \partial_t \dot{X} + \dot{X}\dot{X}\dot{X} - \dot{X}\dot{X}\dot{X}$$

$$= \partial_t \dot{X} + \dot{X}\,\dot{X}\,\dot{X}$$

$$\boxed{\ddot{X} = \partial_t \dot{X} + \dot{X} \times (\dot{X} \times \dot{X})}$$

Fig. 4. **The Formula for Acceleration**

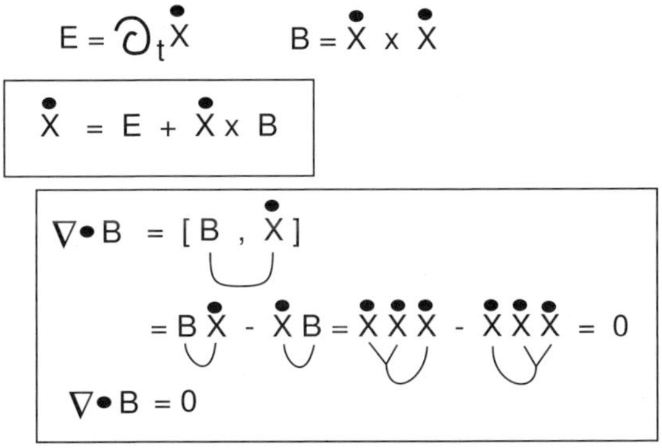

$$E = \partial_t \dot{X} \qquad B = \dot{X} \times \dot{X}$$

$$\boxed{\ddot{X} = E + \dot{X} \times B}$$

$$\nabla \cdot B = [B, \dot{X}]$$

$$= B\dot{X} - \dot{X}B = \dot{X}\dot{X}\dot{X} - \dot{X}\dot{X}\dot{X} = 0$$

$$\nabla \cdot B = 0$$

Fig. 5. **Divergence of B**

In Figure 5, we give the derivation that B has zero divergence.

Figures 6 and 7 compute derivatives of B and the Curl of E, culminating in the formula

$$\partial_t B + \nabla \times E = B \times B.$$

In classical electromagnetism, there is no term $B \times B$. This term is an artifact of our non-commutative context. In discrete models, as we shall see at the end of this section, there is no escaping the effects of this term.

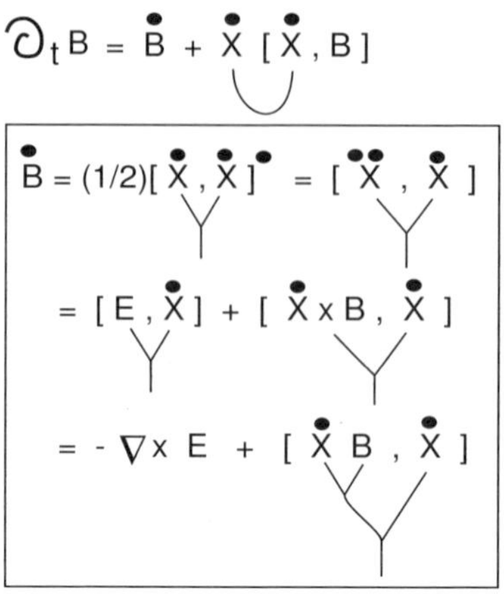

$$\partial_t B = \dot{B} + \dot{X}\,[\dot{X}, B]$$

$$\dot{B} = (1/2)[\dot{X}, \dot{X}]\,\dot{} = [\ddot{X}, \dot{X}]$$
$$= [E, \dot{X}] + [\dot{X} \times B, \dot{X}]$$
$$= -\nabla \times E + [\dot{X} B, \dot{X}]$$

Fig. 6.   **Computing $\dot{B}$**

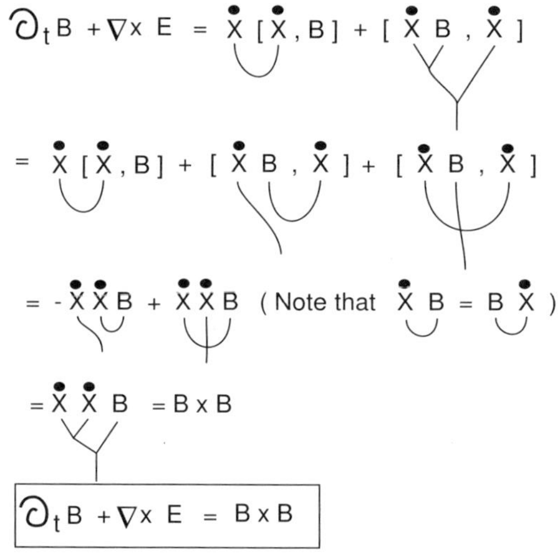

$$\partial_t B + \nabla \times E = \dot{X}\,[\dot{X}, B] + [\dot{X} B, \dot{X}]$$
$$= \dot{X}\,[\dot{X}, B] + [\dot{X} B, \dot{X}] + [\dot{X} B, \dot{X}]$$
$$= -\dot{X}\dot{X} B + \dot{X}\dot{X} B \quad (\text{ Note that } \dot{X} B = B \dot{X})$$
$$= \dot{X}\,\dot{X}\,B = B \times B$$

$$\boxed{\partial_t B + \nabla \times E = B \times B}$$

Fig. 7.   **Curl of E**

Finally, Figure 8 gives the diagrammatic proof that

$$\partial_t E - \nabla \times B = (\partial_t^2 - \nabla^2)\dot{X}.$$

This completes the proof of the Electromagnetic Theorem, which we restate below for reference.

Electromagnetic Theorem With the above definitions of the operators, and taking

$$\nabla^2 = \partial_1^2 + \partial_2^2 + \partial_3^2, \ \ B = \dot{X} \times \dot{X} \ \text{ and } \ E = \partial_t \dot{X} \ \text{ we have}$$

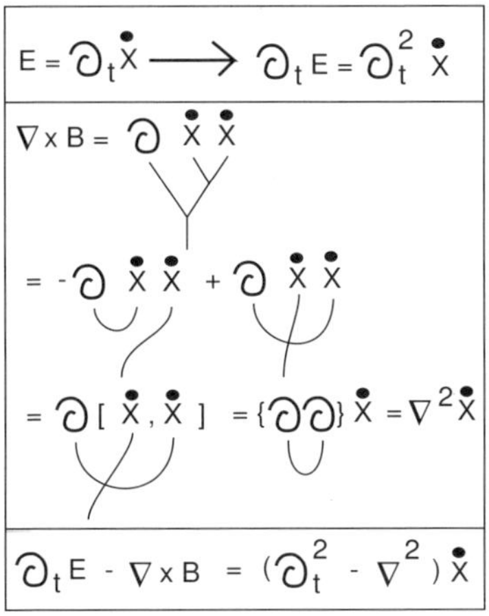

Fig. 8.   **Curl of B**

(1) $\ddot{X} = E + \dot{X} \times B$

(2) $\nabla \bullet B = 0$

(3) $\partial_t B + \nabla \times E = B \times B$

(4) $\partial_t E - \nabla \times B = (\partial_t^2 - \nabla^2)\dot{X}$

Remark. Note that this Theorem is a non-trivial generalization of the Feynman-Dyson derivation of electromagnetic equations. In the Feynman-Dyson case, one assumes that the commutation relations

$$[X_i, X_j] = 0$$

and

$$[X_i, \dot{X}_j] = \delta_{ij}$$

are given, *and* that the principle of commutativity is assumed, so that if A and B commute with the X_i then A and B commute with each other. One then can interpret ∂_i as a standard derivative with $\partial_i(X_j) = \delta_{ij}$. Furthermore, one can verify that E_j and B_j both commute with the X_i. From this it follows that $\partial_t(E)$ and $\partial_t(B)$ have standard intepretations and that $B \times B = 0$. The above formulation of the Theorem adds the description of E as $\partial_t(\dot{X})$, a non-standard use of ∂_t in the original context of Feyman-Dyson, where ∂_t would only be defined for those A that commute with X_i. In the same vein, the last formula $\partial_t E - \nabla \times B = (\partial_t^2 - \nabla^2)\dot{X}$ gives a way to express the remaining Maxwell Equation in the Feynman-Dyson context.

Remark. Note the role played by the epsilon tensor ϵ_{ijk} throughout the construction of generalized electromagnetism in this section. The epsilon tensor is the structure constant for the Lie algebra of the rotation group $SO(3)$. If we replace the epsilon tensor by a structure constant f_{ijk} for a Lie algebra $\mathcal{G}$ of dimension d such that the tensor is invariant under cyclic permutation ($f_{ijk} = f_{kij}$), then most of the work in this section will go over to that context. We would then have d operator/variables $X_1, \cdots X_d$ and a generalized cross product defined on vectors of length d by the equation

$$(A \times B)_k = f_{ijk} A_i B_j.$$

The Jacobi identity for the Lie algebra $\mathcal{G}$ implies that this cross product will satisfy

$$A \times (B \times C) = (A \times B) \times C + [B \times (A] \times C)$$

where

$$([B \times (A] \times C)_r = f_{klr} f_{ijk} A_i B_k C_j.$$

This extension of the Jacobi identity holds as well for the case of non-commutative cross product defined by the epsilon tensor. It is therefore of interest to explore the structure of generalized non-commutative electromagnetism over other Lie algebras (in the above sense). This will be the subject of another paper.

4.2. *Discrete Thoughts*

In the hypotheses of the Electromagnetic Theorem, we are free to take any non-commutative world, and the Electromagnetic Theorem will satisfied in that world. For example, we can take each X_i to be an arbitrary time series of real or complex numbers, or bitstrings of zeroes and ones. The global time derivative is defined by

$$\dot{F} = J(F' - F) = [F, J],$$

where $FJ = JF'$. This is the non-commutative discrete context discussed in sections 2, 3 and 4. We will write

$$\dot{F} = J\Delta(F)$$

where $\Delta(F)$ denotes the classical discrete derivative

$$\Delta(F) = F' - F.$$

With this interpretation X is a vector with three real or complex coordinates at each time, and

$$B = \dot{X} \times \dot{X} = J^2 \Delta(X') \times \Delta(X)$$

while

$$E = \ddot{X} - \dot{X} \times (\dot{X} \times \dot{X}) = J^2 \Delta^2(X) - J^3 \Delta(X'') \times (\Delta(X') \times \Delta(X)).$$

Note how the non-commutative vector cross products are composed through time shifts in this context of temporal sequences of scalars. The advantage of the generalization now becomes apparent. We can create very simple models of generalized electromagnetism with only the simplest of discrete materials. In the case of the model in terms of triples of time series, the generalized electromagnetic theory is a theory of measurements of the time series whose key quantities are

$$\Delta(X') \times \Delta(X)$$

and

$$\Delta(X'') \times (\Delta(X') \times \Delta(X)).$$

It is worth noting the forms of the basic derivations in this model. We have, assuming that F is a commuting scalar (or vector of scalars) and taking $\Delta_i = X_i' - X_i$,

$$\partial_i(F) = [F, \dot{X}_i] - [F, J\Delta_i] = FJ\Delta_i - J\Delta_i F = J(F'\Delta_i - \Delta_i F) = \dot{F}\Delta_i$$

and for the temporal derivative we have

$$\partial_t F = J[1 - J\Delta' \bullet \Delta]\Delta(F)$$

where $\Delta = (\Delta_1, \Delta_2, \Delta_3)$.

5. Constraints - Classical Physics and General Relativity

The program here is to investigate restrictions in a non-commutative world that are imposed by asking for a specific correspondence between classical variables acting in the usual context of continuum calculus, and non-commutative operators corresponding to these classical variables. By asking for the simplest constraints we find the need for a quadratic Hamiltonian and a remarkable relationship with Einstein's equations for general relativity [3, 4]. There is a hierarchy of constraints of which we only analyze the first two levels. An appendix to this paper indicates a direction for exploring the algebra of the higher constraints.

If, for example, we let x and y be classical variables and X and Y the corresponding non-commutative operators, then we ask that x^n correspond to X^n and that y^n correspond to Y^n for positive integers n. We further ask that linear combinations of classical variables correspond to linear combinations of the corresponding operators. These restrictions tell us what happens to products. For example, we have classically that $(x + y)^2 = x^2 + 2xy + y^2$. This, in turn must correspond to $(X + Y)^2 = X^2 + XY + YX + Y^2$. From this it follows that $2xy$ corresponds to $XY + YX$. Hence xy corresponds to

$$\{XY\} = (XY + YX)/2.$$

By a similar calculation, if $x_1, x_2, \cdots, x_n$ are classical variables, then the product $x_1 x_2 \cdots x_n$ corresponds to

$$\{X_1 X_2 \cdots X_n\} = (1/n!)\Sigma_{\sigma \in S_n} X_{\sigma_1} X_{\sigma_2} \cdots X_{\sigma_n}.$$

where S_n denotes all permutations of $1, 2, \cdots, n$. Note that we use curly brackets for these symmetrizers and square brackets for commutators as in $[A, B] = AB - BA$.

We can formulate constraints in the non-commutative world by asking for a correspondence between familiar differentiation formulas in continuum calculus and the corresponding formulas in the non-commutative calculus, where all derivatives are expressed via commutators. We will detail how this constraint algebra works in the first few cases. Exploration of these constraints has been pioneered by Tony Deakin [3, 4]. The author of this paper and Tony Deakin are planning a comprehensive paper on the consequences of these constraints in the interface between classical and quantum mechanics.

Recall that the temporal derivative in a non-commutative world is represented by commutator with an operator H that can be intrepreted as the Hamiltonian operator in certain contexts.

$$\dot{\Theta} = [\Theta, H].$$

For this discussion, we shall take a collection $Q^1, Q^2, \cdots, Q^n$ of operators to represent spatial coordinates $q^1, q^2, \cdots, q^n$. The Q^i commute with one another, and the derivatives with respect to Q^i are represented by operators P^i so that

$$\partial\Theta/\partial Q^i = \Theta_i = [\Theta, P^i].$$

We also write

$$\partial\Theta/\partial P^i = \Theta^i = [Q^i, \Theta].$$

To this purpose, we assume that $[Q^i, P^j] = \delta^{ij}$ and that the P^j commute with one another (so that mixed partial derivatives with respect to the Q^i are independent of order of differentiation).

Note that

$$\dot{Q}^i = [Q^i, H] = H^i.$$

It will be convenient for us to write H^i in place of $\dot{Q}^i$ in the calculations to follow.

The First Constraint. The *first constraint* is the equation

$$\dot{\Theta} = \{\dot{Q}^i\Theta_i\} = \{H^i\Theta_i\}.$$

This equation expresses the symmetrized version of the usual calculus formula $\dot{\theta} = \dot{q}^i\theta_i$. It is worth noting that the first constraint is satisfied by the quadratic Hamiltonian

$$H = \frac{1}{4}(g_{ij}P^iP^j + P^iP^jg_{ij})$$

where $g_{ij} = g_{ji}$ and the g_{ij} commute with the Q^k. We leave the verification of this point to the reader, and note that the fact that the quadratic Hamiltonian does satisfy the first constraint shows how the constraints bind properties of classical physics (in this case Hamiltonian mechanics) to the non-commutative world.

The Second Constraint. The *second constraint* is the symmetrized analog of the second temporal derivative:

$$\ddot{\Theta} = \{\dot{H}^i \Theta_i\} + \{H^i H^j \Theta_{ij}\}.$$

However, by differentiating the first constraint we have

$$\ddot{\Theta} = \{\dot{H}^i \Theta_i\} + \{H^i \{H^j \Theta_{ij}\}\}$$

Thus the second constraint is equivalent to the equation

$$\{H^i \{H^j \Theta_{ij}\}\} = \{H^i H^j \Theta_{ij}\}.$$

We now reformulate this version of the constraint in the following theorem.

Theorem. The second constraint in the form $\{H^i \{H^j \Theta_{ij}\}\} = \{H^i H^j \Theta_{ij}\}$ is equivalent to the equation

$$[[\Theta_{ij}, H^j], H^i] = 0.$$

Proof. We can shortcut the calculations involved in proving this Theorem by looking at the properties of symbols A, B, C such that $AB = BA$, $ACB = BCA$. Formally these mimic the behaviour of $A = H^i, B = H^j, C = \Theta_{ij}$ in the expressions $H^i H^j \Theta_{ij}$ and $H^i \Theta_{ij} H^j$ since $\Theta_{ij} = \Theta_{ji}$, and the Einstein summation convention is in place. Then

$$\{A\{BC\}\} = \frac{1}{4}(A(BC + CB) + (BC + CB)A)$$

$$= \frac{1}{4}(ABC + ACB + BCA + CBA),$$

$$\{ABC\} = \frac{1}{6}(ABC + ACB + BAC + BCA + CAB + CBA).$$

So

$$\{ABC\} - \{A\{BC\}\} = \frac{1}{12}(-ABC - ACB + 2BAC - BCA + 2CAB - CBA)$$

$$= \frac{1}{12}(ABC - 2ACB + CAB)$$

$$= \frac{1}{12}(ABC - 2BCA + CBA)$$

$$= \frac{1}{12}(A(BC - CB) + (CB - BC)A)$$

$$= \frac{1}{12}(A[B,C] - [B,C]A)$$

$$= \frac{1}{12}[A, [B,C]].$$

Thus the second constraint is equivalent to the equation

$$[H^i, [H^j, \Theta_{ij}]] = 0.$$

This in turn is equivalent to the equation

$$[[\Theta_{ij}, H^j], H^i] = 0,$$

completing the proof of the Theorem.

Remark. If we define

$$\nabla^i(\Theta) = [\Theta, H^i] = [\Theta, \dot{Q}^i]$$

then this is the natural covariant derivative that was described in the introduction to this paper. Thus the second order constraint is

$$\nabla^i(\nabla^j(\Theta_{ij})) = 0.$$

Note that

$$\nabla^i(\nabla^j(\Theta_{ij})) = [[\Theta_{ij}, H^j], H^i]$$

$$= -[[H^i, \Theta_{ij}], H^j] - [[H^j, H^i], \Theta_{ij}]$$

$$= \nabla^j(\nabla^i(\Theta_{ij})) + [[H^i, H^j], \Theta_{ij}]$$

$$= \nabla^i(\nabla^j(\Theta_{ij})) + [[H^i, H^j], \Theta_{ij}].$$

Hence the second order constraint is equivalent to the equation

$$[[H^i, H^j], \Theta_{ij}] = 0.$$

This equation weaves the curvature of ∇ with the flat derivatives of Θ.

A Relationship with General Relativity. Again, if we define

$$\nabla^i(\Theta) = [\Theta, H^i] = [\Theta, \dot{Q}^i]$$

then this is the natural covariant derivative that was described in the introduction to this paper. Thus the second order constraint is

$$\nabla^i(\nabla^j(\Theta_{ij})) = 0.$$

If we use the quadratic Hamiltonian $H = \frac{1}{4}(g_{ij}P^iP^j + P^iP^jg_{ij})$ as above, then with $\Theta = g^{lm}$ the second constraint becomes the equation

$$g^{uv}(g^{jk}g^{lm}_{,jku})_v = 0.$$

Deakin and Kilmister [4] observe that this last equation specializes to a fourth order version of Einstein's field equation for vacuum general relativity:

$$K_{ab} = g^{ef}(R_{ab;ef} + \frac{2}{3}R_{ae}R_{fb}) = 0$$

where $a, b, e, f = 1, 2, \cdots n$ and R is the curvature tensor corresponding to the metric g_{ab}. This equation has been studied by Deakin in [4]. It remains to be seen what the consequences for general relativity are in relation to this formulation, and it remains to be seen what the further consequences of higher order constraints will be.

The algebra of the higher order constraints is under investigation at this time.

6. On the Algebra of Constraints

We have the usual advanced calculus formula $\dot{\theta} = \dot{q}^i\theta_i$. We shall define $h^j = \dot{q}^i$ so that we can write $\dot{\theta} = h^i\theta_i$. We can then calculate successive derivatives with $\theta^{(n)}$ denoting the n-th temporal derivative of θ.

$$\theta^{(1)} = h^i\theta_i$$

$$\theta^{(2)} = h^{i(1)}\theta_i + h^ih^j\theta_{ij}$$

$$\theta^{(3)} = h^{i(2)}\theta_i + 3h^{i(1)}h^j\theta_{ij} + h^ih^jh^k\theta_{ijk}$$

The equality of mixed partial derivatives in these calculations makes it evident that one can use a formalism that hides all the superscripts and subscripts $(i, j, k, \cdots)$. In that simplified formalism, we can write

$$\theta^{(1)} = h\theta$$

$$\theta^{(2)} = h^{(1)}\theta + h^2\theta$$

$$\theta^{(3)} = h^{(2)}\theta + 3h^{(1)}h\theta + h^3\theta$$

$$\theta^{(4)} = h^4\theta + 6h^2\theta h^{(1)} + 3\theta h^{(1)2} + 4h\theta h^{(2)} + \theta h^{(3)}$$

Each successive row is obained from the previous row by applying the identity $\theta^{(1)} = h\theta$ in conjunction with the product rule for the derivative.

This procedure can be automated so that one can obtain the formulas for higher order derivatives as far as one desires. These can then be converted into the non-commutative constraint algebra and the consequences examined. Further analysis of this kind will be done in a sequel to this paper.

The interested reader may enjoy seeing how this formalism can be carried out. Below we illustrate a calculation using $Mathematica^{TM}$, where the program already knows how to formally differentiate using the product rule and so only needs to be told that $\theta^{(1)} = h\theta$. This is said in the equation $T'[x] = H[x]T[x]$ where $T[x]$ stands of θ and $H[x]$ stands for h with x a dummy variable for the differentiation. Here $D[T[x], x]$ denotes the derivative of $T[x]$ with respect to x, as does $T'[x]$,

In the calculation below we have indicated five levels of derivative. The structure of the coefficients in this recursion is interesting and complex territory. For example, the coefficients of $H[x]^n T[x]H'[x] = h^n\theta h'$ are the triangular numbers $\{1, 3, 6, 10, 15, 21, \cdots\}$ but the next series are the coefficients of $H[x]^n T[x]H'[x]^2 = h^n\theta h'^2$, and these form the series

$$\{1, 3, 15, 45, 105, 210, 378, 630, 990, 1485, 2145, \cdots\}.$$

This series is eventually constant after four discrete differentiations. This is the next simplest series that occurs in this structure after the triangular numbers. To penetrate the full algebra of constraints we need to understand the structure of these derivatives and their corresponding non-commutative symmetrizations.

$$T'[x]\!:=\!H[x]T[x]$$

$$D[T[x], x]$$
$$D[D[T[x], x], x]$$
$$D[D[D[T[x], x], x], x]$$
$$D[D[D[D[T[x], x], x], x], x]$$
$$D[D[D[D[D[T[x], x], x], x], x], x]$$
$$H[x]T[x]$$

$$H[x]^2 T[x] + T[x]H'[x]$$

$$H[x]^3 T[x] + 3H[x]T[x]H'[x] + T[x]H''[x]$$

$$H[x]^4 T[x] + 6H[x]^2 T[x]H'[x] + 3T[x]H'[x]^2 + 4H[x]T[x]H''[x] + T[x]H^{(3)}[x]$$

$$H[x]^5 T[x] + 10H[x]^3 T[x]H'[x] + 15H[x]T[x]H'[x]^2 + 10H[x]^2 T[x]H''[x] + 10T[x]H'[x]H''[x] + 5H[x]T[x]H^{(3)}[x] + T[x]H^{(4)}[x]$$

6.1. *Algebra of Constraints*

In this section we work with the hidden index conventions described before in the paper. In this form, the classical versions of the first two constraint equations are

(1) $\dot{\theta} = \theta h$
(2) $\ddot{\theta} = \theta h^2 + \theta \dot{h}$

In order to obtain the non-commutative versions of these equations, we replace h by H and θ by Θ where the capitalized versions are non-commuting operators. The first and second constraints then become

(1) $\{\dot{\Theta}\} = \{\Theta H\} = \frac{1}{2}(\Theta H + H \Theta)$
(2) $\{\ddot{\Theta}\} = \{\Theta H^2\} + \{\Theta \dot{H}\} = \frac{1}{3}(\Theta H^2 + H \Theta H + H^2 \Theta) + \frac{1}{2}(\Theta \dot{H} + \dot{H} \Theta)$

Proposition. The Second Constraint is equivalent to the commutator equation

$$[[\Theta, H], H] = 0.$$

Proof. We identify

$$\{\dot{\Theta}\}^{\bullet} = \{\ddot{\Theta}\}$$

and

$$\{\dot{\Theta}\}^{\bullet} = \{\{\Theta H\}H\} + \{\Theta \dot{H}\}.$$

So we need

$$\{\Theta H^2\} = \{\{\Theta H\}H\}.$$

The explicit formula for $\{\{\Theta H\}H\}$ is

$$\{\{\Theta H\}H\} = \frac{1}{2}(\{\Theta H\}H + H\{\Theta H\}) = \frac{1}{4}(\theta H H + H \Theta H + H \Theta H + H H \Theta).$$

Thus we require that

$$\frac{1}{3}(\Theta H^2 + H \Theta H + H^2 \Theta) = \frac{1}{4}(\theta H H + H \Theta H + H \Theta H + H H \Theta)$$

which is equivalent to

$$\Theta H^2 + H^2 \Theta - 2 H \Theta H = 0.$$

We then note that

$$[[\Theta, H], H] = (\Theta H - H \Theta)H - H(\Theta H - H \Theta) = \Theta H^2 + H^2 \Theta - 2 H \Theta H.$$

Thus the final form of the second constraint is the equation

$$[[\Theta, H], H] = 0.//$$

The Third Constraint. We now go on to an analysis of the third constraint. The third constraint consists in the two equations

(1) $\{\dddot{\Theta}\} = \{\Theta H^3\} + 3\{\Theta H \dot{H}\} + \{\Theta \ddot{H}\}$
(2) $\{\dddot{\Theta}\} = \{\ddot{\Theta}\}^\bullet$ where

$$\{\ddot{\Theta}\}^\bullet = \{\{\Theta H\}H^2\} + 2\{\Theta H \dot{H}\} + \{\{\Theta H\}\dot{H}\} + \{\Theta \ddot{H}\}.$$

Proposition. The Third Constraint is equivalent to the commutator equation

$$[H^2, [H, \Theta]] = [\dot{H}, [H, \Theta]] - 2[H, [\dot{H}, \Theta]].$$

Proof. We demand that $\{\dddot{\Theta}\} = \{\ddot{\Theta}\}^\bullet$ and this becomes the longer equation

$$\{\Theta H^3\} + 3\{\Theta H \dot{H}\} + \{\Theta \ddot{H}\} = \{\{\Theta H\}H^2\} + 2\{\Theta H \dot{H}\} + \{\{\Theta H\}\dot{H}\} + \{\Theta \ddot{H}\}.$$

This is equivalent to the equation

$$\{\Theta H^3\} + \{\Theta H \dot{H}\} = \{\{\Theta H\}H^2\} + \{\{\Theta H\}\dot{H}\}.$$

This, in turn is equivalent to

$$\{\Theta H^3\} - \{\{\Theta H\}H^2\} = \{\{\Theta H\}\dot{H}\} - \{\Theta H \dot{H}\}.$$

This is equivalent to

$$(1/4)(H^3\Theta + H^2\Theta H + H\Theta H^2 + \Theta H^3)$$
$$-(1/6)(H^2(H\Theta + \Theta H) + H(H\Theta + \Theta H)H + (H\Theta + \Theta H)H^2)$$
$$= (1/2)(\dot{H}(1/2)(H\Theta + \Theta H) + (1/2)(H\Theta + \Theta H)\dot{H})$$
$$-(1/6)(\dot{H}H\Theta + \dot{H}\Theta H + H\dot{H}\Theta + H\Theta\dot{H} + \Theta H\dot{H} + \Theta\dot{H}H).$$

This is equivalent to

$$3(H^3\Theta + H^2\Theta H + H\Theta H^2 + \Theta H^3)$$
$$-2(H^3\Theta + 2H^2\Theta H + 2H\Theta H^2 + \Theta H^3)$$
$$= 3(\dot{H}H\Theta + \dot{H}\Theta H + H\Theta\dot{H} + \Theta H\dot{H})$$
$$-2(\dot{H}H\Theta + \dot{H}\Theta H + H\dot{H}\Theta + H\Theta\dot{H} + \Theta H\dot{H} + \Theta\dot{H}H).$$

This is equivalent to

$$H^3\Theta - H^2\Theta H - H\Theta H^2 + \Theta H^3$$

$$= (\dot{H}H\Theta + \dot{H}\Theta H + H\Theta\dot{H} + \Theta H\dot{H}) - 2(H\dot{H}\Theta + \Theta\dot{H}H).$$

The reader can now easily verify that

$$[H^2, [H, \Theta]] = H^3\Theta - H^2\Theta H - H\Theta H^2 + \Theta H^3$$

and that

$$[\dot{H}, [H, \Theta]] - 2[H, [\dot{H}, \Theta]] = (\dot{H}H\Theta + \dot{H}\Theta H + H\Theta\dot{H} + \Theta H\dot{H}) - 2(H\dot{H}\Theta + \Theta\dot{H}H).$$

Thus we have proved that the third constraint equations are equivalent to the commutator equation

$$[H^2, [H, \Theta]] = [\dot{H}, [H, \Theta]] - 2[H, [\dot{H}, \Theta]].$$

This completes the proof of the Proposition. //

Discussion. Each successive constraint involves the explicit formula for the higher derivatives of Θ coupled with the extra constraint that

$$\{\Theta^{(n)}\}^\bullet = \{\Theta^{(n+1)}\}.$$

We conjecture that each constraint can be expressed as a commutator equation in terms of Θ, H, and the derivatives of H, in analogy to the formulas that we have found for the first three constraints. This project will continue with a deeper algebraic study of the constraints and their physical meanings.

. . . .

References

[1] Dyson, F. J. [1990], Feynman's proof of the Maxwell Equations, *Am. J. Phys.* 58 (3), March 1990, 209-211.

[2] Connes,Alain [1990], *Non-commutative Geometry* Academic Press.

[3] Deakin, A.M. [1999] Where does Schroedinger's equation really come from?, in "Aspects II - Proceedings of ANPA 20", edited by K.G. Bowden, published by ANPA.

[4] Deakin, A. M., Progress in constraints theory, in "Contexts - Proceedings of ANPA 31", edited by Arleta D. Ford, published by ANPA, pp. 164 - 201.

[5] Dimakis, A. and Müller-Hoissen [1992], F., Quantum mechanics on a lattice and q-deformations, *Phys. Lett.* 295B, p.242.

[6] Foster, J. and Nightingale, J.D. [1995] "A Short Course in General Relativity", Springer-Verlag.

[7] Forgy,Eric A. [2002] Differential geometry in computational electromagnetics, PhD Thesis, UIUC.

[8] Hughes, R. J. [1992], On Feynman's proof of the Maxwell Equations, *Am. J. Phys.* 60, (4), April 1992, 301-306.

[9] Kauffman, L. [1985], Sign and Space, In Religious Experience and Scientific Paradigms. Proceedings of the 1982 IASWR Conference, Stony Brook, New York: Institute of Advanced Study of World Religions, (1985), 118-164.

[10] Kauffman, L. [1987], Self-reference and recursive forms, Journal of Social and Biological Structures (1987), 53-72.

[11] Kauffman, L. [1987], Special relativity and a calculus of distinctions. Proceedings of the 9th Annual Intl. Meeting of ANPA, Cambridge, England (1987). Pub. by ANPA West, pp. 290-311.

[12] Kauffman, L. [1987], Imaginary values in mathematical logic. Proceedings of the Seventeenth International Conference on Multiple Valued Logic, May 26-28 (1987), Boston MA, IEEE Computer Society Press, 282-289.

[13] Kauffman, L. H., Knot Logic, In *Knots and Applications* ed. by L. Kauffman, World Scientific Pub. Co., (1994), pp. 1-110.

[14] Kauffman, Louis H. [2002], Biologic. *AMS Contemporary Mathematics Series*, Vol. 304, (2002), pp. 313 - 340.

[15] Kauffman,Louis H., Eigenform, *Kybernetes - The Intl J. of Systems and Cybernetics* 34, No. 1/2 (2005), Emerald Group Publishing Ltd, p. 129-150.

[16] Louis H. Kauffman, Reflexivity and Eigenform – The Shape of Process. - Kybernetes, Vol 4. No. 3, July 2009.

[17] Kauffman, Louis H., Reflexivity, Eigenform and Foundations of Physics. In Reflexivity, Proceedings of ANPA 30, Arleta D. Ford, Editor , Published by ANPA, June 2010, pp. 158-222.

[18] Kauffman, Louis H., Reflexivity and Foundations of Physics, In *Search for Fundamental Theory - The VIIth International Symposium Honoring French Mathematical Physicist Jean-Pierre Vigier, Imperial College, London, UK, 12-14 July 2010* , edited by R. Amaroso, P. Rowlands and S. Jeffers, AIP - American Institute of Physics Pub., Melville, N.Y., pp.48-89.

[19] Kauffman,Louis H.[1991,1994], *Knots and Physics*, World Scientific Pub.

[20] Kauffman, Louis H. [2002], Time imaginary value, paradox sign and space, in *Computing Anticipatory Systems, CASYS - Fifth International Conference*, Liege, Belgium (2001) ed. by Daniel Dubois, AIP Conference Proceedings Volume 627 (2002).

[21] Kauffman,Louis H. and Noyes,H. Pierre [1996], Discrete Physics and the Derivation of Electromagnetism from the formalism of Quantum Mechanics, *Proc. of the Royal Soc. Lond. A*, **452**, pp. 81-95.

[22] Kauffman,Louis H. and Noyes,H. Pierre [1996], Discrete Physics and the Dirac Equation, *Physics Letters A*, 218 ,pp. 139-146.

[23] Kauffman,Louis H. and Noyes,H.Pierre (In preparation)

[24] Kauffman, Louis H.[1996], Quantum electrodynamic birdtracks, *Twistor Newsletter Number 41*

[25] Kauffman, Louis H. [1998], Noncommutativity and discrete physics, *Physica D* 120 (1998), 125-138.

[26] Kauffman, Louis H. [1998], Space and time in discrete physics, *Intl. J. Gen. Syst.* Vol. 27, Nos. 1-3, 241-273.

[27] Kauffman, Louis H. [1999], A non-commutative approach to discrete physics, in *Aspects II - Proceedings of ANPA 20*, 215-238.

[28] Kauffman, Louis H. [2003], Non-commutative calculus and discrete physics, in *Boundaries- Scientific Aspects of ANPA 24*, 73-128.

[29] Kauffman, Louis H. [2004], Non-commutative worlds, *New Journal of Physics 6*, 2-46.

[30] Louis H. Kauffman, Differential geometry in non-commutative worlds, in "Quantum Gravity - Mathematical Models and Experimental Bounds", edited by B. Fauser, J. Tolksdorf and E. Zeidler, Birkhauser (2007), pp. 61 - 75.

[31] Montesinos, M. and Perez-Lorenzana, A., [1999], Minimal coupling and Feynman's proof, arXiv:quant-phy/9810088 v2 17 Sep 1999.

[32] Spencer-Brown,G., *Laws of Form,* Julian Press, New York (1969).

[33] Müller-Hoissen,Folkert [1998], Introduction to non-commutative geometry of commutative algebras and applications in physics, in *Proceedings of the 2nd Mexican School on Gravitation and Mathematical Physics*, Kostanz (1998) ¡http://kaluza.physik.uni-konstanz.de/2MS/mh/mh.html¿.

[34] Tanimura,Shogo [1992], Relativistic generalization and extension to the non-Abelian gauge theory of Feynman's proof of the Maxwell equations, *Annals of Physics, vol. 220*, pp. 229-247.

Brief Biography

Louis H. Kauffman received a BS from the Massschusetts Institute of Technology (MIT) in 1966 and a PhD from Princeton University in 1972. He has been teaching at the University of Illinois at Chicago since 1971. His research is in knot theory, foundations of mathematical form, cybernetics and discrete physics. His main discoveries are state models for knot invariants (including models for the Alexander polynomial and the Jones polynomial), a two variable polynomial invariant of knots and links (the Dubrovnik (Kauffman) polynomial), a formulation of knot invariants in terms of Hopf algebras, a wide generalization of knot theory called Virtual Knot Theory, certain invariants of virtual knots that generalize or shed light on the Jones polynomial, formulations of quantum computing in terms of knot theory and the Jones polynomial, formulations of discrete physics that utilize iterant techniques developed by him in relation to Spencer-Brown's Laws of Form. Kauffman is the author of a number of books on knot theory including Formal Knot Theory (Princeton University Press 1983 and Dover 2005), On Knots (Princton University Press 1987), Temperley Lieb Recoupling Theory and Invariants of Three-Manifolds (Princeton University Press 1991), Knots and Physics (World Scientific 1991 first edition, 2012 fourth edition). Kauffman is the recipient of the Warren McCulloch Award (1993) of the American Society for Cybernetics and the Alternative Natural Philosophy Award (1996) of ANPA — the Alternative Natural Philosophy Association. He is editor–in–chief and founding editor of the Journal on Knot Theory and its Ramifications, and the editor of the book series on Knots and Everything of World Scientific Publishing Company.

Report on ANPA
to the ANPA Advisory Board
2008

Clive W. Kilmister (1924–2010)

Note from Mike Horner

Clive, as a founder of ANPA, was at its heart from the formation in 1979 until just before his death. He served in several important capacities including President, Proceedings Editor, Member of the Executive Council and member of the Advisory Board. In addition he was the most prolific ANPA author. The latter was surprising since Clive did not use personal computers and after writing out his papers in longhand he would make corrections and then use a mechanical typewriter to produce his final text. As secretary of the Advisory Board I worked with Clive on the 2008 report to the Advisory Board and it is included in this festschrift as it summarizes a large part of ANPA achievements from the very beginning up until 2008, a period of more than 50 years.

Advisory Board (2008) :   Mike Horner (Chairman), Profs G.F.Chew (Berkeley), C.Isham (Imperial College), M.Redhead (Cambridge and LSE), N. Cartwright (LSE), C.W.Kilmister (London), H.Pierre Noyes (Stanford), Rev. Dr. Philip Luscombe (Wesley House, Cambridge).

1. Introduction

It is over 28 years since ANPA was set up and for some considerable part of that time you [the present members] have served on the Advisory Board. It seems time to take stock and to give you the opportunity, if you so wish, to comment on what has been achieved and, particularly, what should be tackled next. The original statement of aims for ANPA has been, over the years, slightly modified to read:

> *The primary purpose of the Association is to consider coherent models based on a minimal number of assumptions, so as to bring together major areas of thought and experience within a natural philosophy alternative to the prevailing scientific attitude. The Combinatorial Hierarchy (CH), as such a model, will form an initial focus of our discussions.*

This has been taken quite widely so that there have been a variety of contributions, though in many cases with a strong algebraic approach. This report is not concerned with these varied developments, valuable though some have been, as they have often been for only a few years. Rather, it will deal with the developments related directly to the combinatorial hierarchy since these have been a continuing theme. In section 2 the original Parker-Rhodes construction is described again, both as a reminder and to comment on it in a way he would not have done. In sections 3 and 4 the two basic problems in understanding the construction as a piece of physics are spelled out. Then sections 5 and 6 describe how and to what extent these problems have been tackled. Finally sections 7,8 and 9 list the most recent developments.

2. The Parker-Rhodes Construction

Parker-Rhodes began with the idea of two entities and the act of distinguishing one from the other. He expressed this in terms of what were later called bit-strings *i.e.* vectors over the field $\mathbf{Z}_2$ with two elements. He also used the properties of the field as he felt the need. Thus he takes the two entities as (1,0), (0,1) and the act of distinguishing them, which he called discrimination, as addition over $\mathbf{Z}_2$ is then (1,1). The corresponding act on two identical entities then produced the zero string which was therefore marked out as a signal of this and would not come up in other ways. It is now clear that bit-strings were only one representation; for example one can use 1,2 as the two entities and 12 as their discrimination. The three entities are an example of what later was called a *discriminately-closed subset* (dcs) that is, a set closed under the discrimination of any two different elements. So far the construction has generated three dcss: [1], [2], and [1,2,12] in a conventional algebraic exercise. Parker-Rhodes next step was level change. For each dcs he was able to find a unique non-singular linear transformation (in his representation a square matrix over $\mathbf{Z}_2$) which left invariant the elements of the dcs and only those. (Observe that this is not what would be more usual in algebra, where operators might be chosen to leave the whole dcs unchanged.) These transformations, which he represented as bit-strings of length 4, form the three elements of the next level. The discrimination at this level then gave rise to 7 more dcss making 10 in all.

In his original representation these dcss look different from the original three so he derived $10 = 7 + 3$ by addition. This was once seen as a problem but the difference is essentially one caused by the formalism and vanishes in a more abstract formulation. Now in going up a further level change the seven operators characterizing these dcss are no longer uniquely determined. In fact there are 74,088 different sets of such operators. About 90% of them lead to $2^7-1 = 127$ dcss bringing the total to 137; the remainder lead to fewer. Parker-Rhodes enjoined the construction to choose one of the 90%. A further level change then leads to $2^{127}-1 = 10^{38}$ dcss. There will not be enough operators to characterize these so the construction halts. These four levels constitute the Combinatorial Hierarchy (CH). It seemed in the 1960s — to those that went on to found ANPA — that the nearness of 137 and 10^{38} to $1/\alpha = \hbar c/e^2$ and the corresponding gravitational inverse coupling constant together with the stop at four marked this out as better than numerology. (The 3 and the 10 might, it was thought, have an interpretation for nuclear forces.)

3. The Identification Problem

There are evidently two enormous problems about making this at all respectable. One of them, essentially algebraic, is to ascertain how much of Parker-Rhodes' odd construction is necessary, how much is correct and how it can be connected with physics. This is deferred to the next section. The other problem considered in this section is: if it be granted that the algebra is physically based, how could it be that 137 dcss are related to $1/\alpha$? The corresponding gravitational problem may be left on one side for the present because the experimental accuracy is much less and the precise definition of the constant is not certain.

Although a physical justification of the algebra would suffice to remove the taint of numerology there is a further hazard. This results from the heritage of Eddington's failed attempts to find $1/\alpha$, also by counting elements of an algebra. His algebraic approach seems simply wrong, but this is not the reason that his attempt to justify his numerical results by solving a well known problem by two methods and comparing parameters always failed. The ANPA belief is that this is bound to happen because physical theories have different levels of complexity. The fault in Eddington's identification arguments is that they are trying to settle at a complex (logically advanced) level matters which can be settled only at a simple (logically primitive) one.

It has long been recognized that certain numerical constants arise at a simple level without their being empirically determined. For example, Kant draws attention to the dimensionality of space, 3, not as the result of an empirical observation, since to specify such an observation will first need a description in three-dimensional space.

Instead of Kant's *a priori* cognition it seems better to argue like this: the discovery of Planck's constant, as he pointed out himself, makes possible an absolute scale for physical measurements. There are numerical magnitudes in modern physics that seemed ideal stepping stones into this scale. These are scale constants whose values determine the theory rather than vice versa. One such is $1/\alpha$. If at a more complex stage it comes to be written as $\hbar c/e^2$ then $\hbar$, c, e are derived numbers got from $1/\alpha$ and other scale constants.

The present ANPA view about the combinatorial hierarchy is that the algebra is a precondition for imagining the physical world and stands separate from the formulae in which that imagination is expressed — ordinary physics. The question is then whether the physically detailed picture can fit into the frame. One aspect of such fitting in will then be whether the numerical values of scale constants are correct.

There remains one difficulty about this way of looking at things: if the expression $\hbar c/e^2$ is to be assigned only at a more complex level, how can one tell that a calculation of 137 or something near it is this scale constant rather than some other?

The answer is in two steps. Firstly, the value 137 serves to identify the constant in question. Using that means that this stage of the calculation deserves no credit. The second step is to make a more careful examination of the algebra to determine whether 137 is indeed an accurate value or not. The credit for the calculation hinges on the agreement of the more accurate value with the experimental.

4. Physics and Algebra

The first step in understanding the CH as part of physics is to recognize its progressive nature. It begins with two elements and then grows. But this is not analogous to the construction of a group from a set of generators and defining relations. In that case there is simply a whole group waiting to be constructed piecemeal. The difference is emphasized in the second level change which can be performed in any of 74,088 ways which do not lead to the same CH in the end. More so at the next level change. The notion of a static picture being filled in is ruled out and instead the idea of a dynamic process is introduced. The word process is often used and yet it is extraordinarily difficult to nail down its precise meaning. Whitehead is an obvious source and he traces it back to Plato. But fortunately the CH can be studied using only a very abstract form of the idea, as a universe that is constructed progressively. Physical principles are expected to emerge from the formulation of the process. These will be valid for all experimental procedures and will guide them. The process is one of elements constructed in a sequence; say, of "elements coming into play". They have no meaning to start with but will acquire meaning as a consequence of the construction process. The formalism develops with the construction. This approach is reminiscent of Brouwer but the theory is not intuitionistic. It is possible to abstract Parker-Rhodes combinatorial hierarchy in the same way; all that is needed is that it starts with two elements.

5. Process Generation of a New CH

The preceding sections have detailed the original problems about the CH and proposed ways in which they can be dealt with. It turns out that from the idea of "coming into play" there follows a stripped down version of the mathematics. To be a sequence implies that the process determines whether a new element is the same as one already in play or not. The determination of whether b is the same as a (call it ab) is a further element in play (say c). Discrimination is the binary operation ab = c. If this were a group operation then the group would be one of those considered by Parker-Rhodes with the identity element as the signal that a, b were the same. It is not assumed that a group will necessarily result. The nature of the binary operation follows from the elements in play idea. If a, b are in fact the same element, then no new element can arise in the discrimination so either aa = a or aa is not an element. The first possibility will not do because the next step in the process would be to check whether the element on the right-hand side were indeed the same as on the left, that is, exactly the same problem again. So aa is a non element which must be unique so that when it appears there is no need for further discrimination. Call this signal "z". Then if a, b are different it is necessary that ab $\neq$ z. If ab = c, then in general ba $\neq$ c since ba is the new element generated by determining whether a new a is in fact the same as b, but for obvious physical reasons ba and ab are closely related so ba is written as c* and c, c* are called dual. It is "duality" that allows the construction of this more abstract version of the mathematics.

Two elements may now be (i) the same, (ii) duals, (iii) different. (To avoid confusion (iii) could be called "essentially different".) To determine whether two elements are duals or not will require a signal, just as equality did. y is used for this signal, so that uu* = y for any u in play. These considerations are enough to determine the new bottom level of the hierarchy which turns out to be the non-associative and non-commutative structure often called Q^* which can be derived from quaternions by interpreting $-$a as a* (which turns out to be consistent with the definition of duality) and by replacing the diagonal elements in the Cayley table by the identity z. The reasons for level change and the way in which it is constructed are exactly the same as in the original CH. It is important to relate this new structure to the original CH. This is not part of the process; it is a deliberate simplification of the structure which yet has enough detail to give useful guidance, rather as a homology groups do in algebraic topology. This new structure called the "skeleton" of Q^*, is the result of defining a new binary operation directly between dcss by the general rule

$$[\text{u, v, w, } \ldots][\text{u}', \text{v}', \text{w}', \ldots] = [\text{uu}', \text{uv}', \text{vu}', \text{vv}', \ldots].$$

If ab = c in Q^* then $[\text{a}, ta^*][\text{b}, \text{b}^*] = [\text{c}, \text{c}^*]$ and these three generate the quadratic group with $[\text{z, y}]$ as the identity element. Thus the original CH is the skeleton of the new one. Exhibiting it as a structure embedded in the new hierarchy explains why it serves as a fairly accurate approximation to the correct results.

6. The Fine Structure Constant

In the original CH the constant $1/137$ appears as a probability, *viz.* if the construction were constrained to remain at the first three levels then $1/137$ would be the probability that some random pair of dcss were the same. Such a constraint is impossible in a process construction but there is a definite probability that the construction does in fact remain at the first three levels. This is calculated by assuming a "generalized ergodic hypothesis" — that when there are several possibilities in the process each will occur with equal probability. At the first level a dcs may be one of the three at that level or none of them (because it is at a higher level). Each has a probability of $1/4$. Similarly at the next two levels there are probabilities $1/8$ and $1/128$. (This last figure so long as Parker-Rhodes' choice of a suitable set of matrices is made.) The probability of being at none of these three levels is therefore ϵ where $1/\epsilon = 4.8.128 = 4096$. Thus the probability that the constraint condition happens to be fulfilled is $1 - \epsilon$.

Then $(1 - \epsilon)/137 = 1/137.0334\ldots$. This simply corrects Parker-Rhodes' identification. The process idea also prohibits the "choice" tactic and requires attaching probabilities to the different choices out of the 74,088 possible sets. This is carried out in the skeleton first and yields a value of $1/\alpha = 137.0351\ldots$ The full structure then uses this calculation but with a further use of the generalized ergodic principle to get $137.036011393\ldots$ which agrees to one part in 107 with the recent accurate determination by G. Gabrielse from the Landé g-factor for the electron, giving $1/\alpha$ as $137.035999710\ldots$ This is a striking instance of one numerical feature of the physical picture fitting into the frame.

7. The Construction of Space and Time

Space and time have not been assumed in the discrete picture. An investigation is needed to determine the extent to which the physical description of events in a space-time continuum fits into the frame. This is done by constructing space and time, to the extent to which this is possible, to establish the relation between the discrete system and the usually assumed continuum. When two elements have come into play and produced a third the three generate a dcs. If a new element comes into play it is possible for the system to go up a level so that it becomes possible to ascertain whether the new element belongs to the dcs or not. Exploiting this in section 6 led to the calculation of the fine-structure constant. If the system does not go up a level the tactic of discriminating the new element with each of the three existing ones can lead to an infinite regress and so is ruled out. The only remaining possibility is to leave the existing dcs as it is and start again. In due course the new elements will generate a second dcs. The new structure will then consist of pairs of elements related by the way in which they have been constructed. This is called the transverse construction in contrast to the earlier upward construction. The transverse construction turns out to be needed in the construction of space.

The sequential construction introduces the notion of time ordering but can develop into a true notion of time only by introducing the notion of duration. The time ordering may be represented by the ordinal integers. This representation is relative to a particular dcs. Another dcs will then assign ordinals with the same ordering but not necessarily the same ordinals. This is regraduation by a monotonic increasing function. Regraduations must be two-way and for functions from and into the integers such inverse functions will not exist in general. A dense ordering is needed, so the rational ordinals are needed. This is not the introduction of the rational field, merely a (potentially infinite) set of markers. Now duration is the lapse of time from a to b, a number of some kind, not merely an ordinal. If the lapse of time from order-parameter a to b is `f(a,b)`, a member of some number field, then the notion of duration requires that `f (a,b) + f(b,c) = f(a,c)`, so that `f(a,b) = f(a,c)-f(b,c)`.

Taking c as some fixed order parameter, it follows that $f(a,b) = g(a)-g(b)$ where $g(a) = f(a,c)$. If now `a'` is a later order than a the notion of lapse requires `f(a',b)`

< f(a,b), so that g(a') < g(a). The number field must be dense and the rational field is chosen (although the usual choice is the reals, Herman Weyl has cogently argued against the identification of the intuitive passage of time with the reals in *Das Kontinuum*, 1917.) Then it follows that there is a regraduation of order-parameters so that, with the new parameters, f(a,b) = a-b.

Duration leads to spatial distance by way of special relativity, so reversing the usual treatment in which three-dimensional space is assumed initially. Two separate dcss are related by two functional relations, one between the times and the other between the individual elements, so a permutation. The time aspect leads easily into a rational version of the so-called k-calculus for one-dimensional special relativity, popularized by Bondi but going back *via* Milne to Einstein.

One step needs mentioning; the argument rests on the solution of a functional equation by Whitrow which he carries out by means of the calculus. In such a situation the approximation to the reals by the rationals is suspect. A way out is chosen: adjoin virtual elements to the rational field rather as geometers define points at infinity. Over this virtual extension define functions as "reasonable" if their definitions automatically go over into the virtual elements. Then Whitrow's result holds for reasonable functions. The k-calculus algebra holds for a regraduated time only, not for duration. This is relevant to the problem of preferred inertial frames. These require the new regraduated time yet they refer to duration. The empirical fact of the existence of preferred inertial frames is logically equivalent to the equality of the new time with duration. This is satisfactory in correctly locating the existence at a logically primitive stage instead of later on when Newton's and Maxwell's equations have been derived.

The remaining functional relation between two dcss involves permutation of the elements at each time instant. The denseness of the time requires a correspond-ing structure for the permutations and this forces a probabilistic treatment. This is found to land up with SU2 as a basic group and so, finally, to the (rational) Lorentz group acting on a (rational) 3+1 dimensional space-time. This obviates the need for the rather weak argument in Milne's k-calculus treatment to extend the results to three dimensions. The distinction between real and rational fields can sometimes be physically important but seems irrelevant here so evidently the space-time descrip-tion fits well into the frame determined by the combinatorial hierarchy augmented by duration.

8. Relation to Non-relativistic Quantum Mechanics

Quantum mechanics presents a different kind of problem from that posed by the construction of space. The Old Quantum Theory was irredeemably inconsistent in inserting discreteness (*via h*) into a continuum theory. A careful analysis shows the inconsistency reappearing at each later stage of the New Quantum Theory, albeit in slightly different forms (collapse of the wave function etc.). The ANPA view is simply that discreteness is present from the start in the process approach and so the inconsistency is absent. That the discreteness is that of quantum mechanics is evidenced by the accurate calculation of $\hbar c/e^2$.

But there is a widespread view that quantum theory is such a highly efficient calculating tool that it has to be accepted even if it has confusions. There are different ANPA views on this. One view is that the major part of the great numerical success comes in spectroscopy so they were already there before the elaboration of the theory. Another is that quantum theory can no more be accepted than were the epicycles of celestial mechanics because it lacks explanatory power. Finally some believe that a substitute for quantum mechanics which will fit into the frame is worthwhile and comes from a more detailed study of commutators.

9. Relation to High Energy Particle Theory

It is usually assumed that in some vague way non-relativistic quantum theory is the foundation on which high energy particle theory is built. Some basic ideas are taken over but essentially particle theory is combinatorial. This fact is recognized in the group structures usually used to classify particles. Whether or not a particular groups fit the frame can serve as their justification or otherwise. This work is currently active but already it is clear that the transverse construction leads to 36 pairs of elements forming an exact copy of the structure of six quarks and six colours (i.e. 3 colours and their anti-colours). So the quark model fits the frame.

10. Conclusion

The other varied developments fostered by ANPA are numerous. There has been work on Clifford algebras in themselves and on their applicability in quantum mechanics and in questions of synchronizations in concurrent computing (which has been thought to have connections to physics ideas). Indeed the computer and information technology in general has been seen as a field which yields valuable insights into (quantum) physics. Attempts have been made to relate Parker-Rhodes' hierarchy and its developments to Category theory. Also his later work on Indistinguishables has been followed up. The relation of physics (as seen from an ANPA position) to problems of general philosophy (and theology) has not been neglected. It is probably true to say that in each of these cases more work needs doing.

Accounts of them can be found in the various Annual Proceedings.

They could form the basis of a second report if so desired.

CWK

September 2008

Brief Biography

Clive W. Kilmister (1924 – May 2, 2010) was a British Mathematician who specialised in the mathematical foundations of Physics, especially Quantum Mechanics and Relativity and published widely in these fields.   He was one of the discoverers of the Combinatorial Hierarchy, along with A. F. Parker-Rhodes, E. W. Bastin, and J. C. Amson. He was strongly influenced by astrophysicist Arthur Eddington and was well known for his elaboration and elucidation of Eddington's fundamental theory.

Kilmister attended Queen Mary College London for both his under- and post-graduate degrees. His PhD was supervised by cosmologist George McVittie (himself a student of Eddington), and his dissertation was entitled *The Use of Quaternions in Wave-Tensor Calculus* which related to Eddington's work. Kilmister received his doctoral degree in 1950. His own students included Brian Tupper (1959, King's College London, now professor emeritus of general relativity and cosmology at University of New Brunswick Fredericton, Samuel Edgar (1977, University of London), and Tony Crilly (reader in mathematical sciences at Middlesex University and author of *The Big Questions: Mathematics* (1981).

Kilmister was elected as a member of the London Mathematical Society during his doctoral studies (March 17, 1949). Upon graduation, he began his career as an Assistant Lecturer in the Mathematics Department of King's College in 1950.

The entirety of his academic career was spent at King's. In 1954, Kilmister founded the King's Gravitational Theory Group, in concert with Hermann Bondi and Felix Pirani, which focused on Einstein's theory of general relativity. At retirement, Kilmister was both a Professor of Mathematics and Head of the King's College Mathematics Department.

Honours, positions, and titles:

Member, London Mathematical Society, 1949-2010
President, British Society for the History of Mathematics, 1974-6
President, British Society for the Philosophy of Science, 1981-2
President, Mathematical Association, 1979-80
Gresham Professor of Geometry, 1972-88
Committee Member, International Society on General Relativity and
    Gravitation, 1971-4
Founding Member, Alternative Natural Philosophy Association, 1979

Reflections on
Fundamentals and Foundations of Physics

James Lindesay

Department of Physics
Howard University, Washington, DC 20059
E-mail: jlindesay@Howard.edu

Perspectives on three and a half decades of joint research with Pierre Noyes on fundamentals of physics are presented. At the core of the discussion is the no/yes nature of quantum detections, consistent with phase coherence and quantum interference. The incorporation of special relativity into non-perturbative scattering theory provides our initial insights into the subtleties connecting quantum physics to classical observations. We furthermore continue to explore quantum effects in general relativity, and tackle a perspective on dark energy as imparting a quantum imprint upon the fluctuations in the energy density of the universe that ultimately result in galaxy formation and life. We will end by exploring some of the late time implications that dark energy has on whether intelligent life can be perpetual, a favorite topic of Pierre's inspired by an article by his colleague Freeman Dyson.

1. Introduction

When I was a graduate student with H. Pierre Noyes back in the mid-late 70s, one of the first things he drilled into my academic psyche was that physics is an experimental science with particular subtleties. Modern physics depends crucially upon the detection of events, and any given detection is a no/yes event at some level of accuracy. Thus, appropriate descriptions of discrete detections is at the core of understanding any physical model at its most fundamental level.

Pierre has the creative spirit of a true founder of the Alternative Natural Philosophy Association (ANPA). On several occasions in the past, once we succeeded in solving a complicated problem that had defied a simple explanation, and that explanation had become accepted by the established mainstream, Pierre would commence to rebel against our own work. His attempts at other explanations and post-success rebellions has been pummeled by referees, using our own theories against his newer alternative explanations. I have personally witnessed this having a disheartening effect on Pierre, but he never has allowed criticism to quell his enthusiasm for a better explanation. During a recent lunch meeting, he paid his ultimate tribute to the success of some of our work by declaring it to be "too establishment", saying that "it is more fun to criticize and come up with something better" than to further build upon this work. I will therefore be presenting some of our successful models that more recently must endure scathing criticisms by Pierre himself.

1.0.1. *Relativistic few-particle scattering theory*

Quantum physics provides probably the most fundamental description of the nature of physical phenomena. Yet, most of our macroscopic observations appear to involve de-coherent, classical processes. Some enigmas of quantum physics, like coherence and entanglement, remain puzzling to this day. When examined in the context of few-particle dynamics, these effects can become enhanced. For instance, the eternal triangle effect discovered by Pierre using a three-body Schrödinger equation formulation was independently reformulated in terms of energy levels by V. Efimov, and is often referred to as the Efimov effect. The effect was experimentally verified, more than three decades after Pierre first discussed it.

We know that any fundamental description of physical phenomena must incorporate the kinematics of special relativity. The non-linear kinematics introduces peculiar issues with regards to cluster decomposability and the classical separation of systems that only very weakly interact. We developed a successful formulation of non-perturbative relativistic scattering theory that incorporates all of the non-relativistic phenomena described by the well accepted Faddeev equations, while demonstrating the appropriate kinematics of special relativity [1–4].

Upon our successful formulation of relativistic few-particle scattering theory, Pierre approached a respected colleague, Steven Weinberg (who is known not only for formulations of few particle scattering theory, but also for unification of electroweak interactions), with our results. Weinberg replied that until we could incorporate pair creation into a scattering theoretic formulation, he would not be satisfied that such an approach could accurately describe relativistic physics. As a response, we developed what we call symbar [5], a symmetric bar transformation on a unitary amplitude, that utilizes the analytic association of negative 4-momenta with antiparticles in the same manner as done by Feynman. In Section 2, we will explore some of the subtle manifestations of quantum coherence in few particle systems, including both non-relativistic features like the Efimov effect, as well as inherently relativistic features such as pair creation.

1.0.2. *Dark energy and cosmological fluctuations*

There is general agreement among physical cosmologists that the current expansion phase in the evolution of our universe can be extrapolated back toward an initial state of compression so extreme that we have neither direct laboratory nor indirect (astronomical) observational evidence for the laws of physics needed to continue that extrapolation. Examining the reverse evolution of the universe from the present, a period of gravitational coherence can be maintained at a density well under the Planck scale, suggesting that this pre-coherent phase has global coherence properties. We will use general arguments to examine the energy scales for which a quantum coherent description of gravitating quantum energy units is necessary.

The expected gravitational vacuum energy density at the scale that gravitational coherence is lost is expected to freeze out, and this process of decoherence can imprint this resulting dark energy upon the fluctuations from uniformity in the energy density of the universe.

We expect that microscopic interactions cannot signal with superluminal exchanges. However, we know that *coherent quantum systems* have supraluminal *correlations* which cannot be used for supraluminal *signaling*. We focus our examination of the early universe to the period for which the scale expansion rate satisfies $\dot{R} = c$, asserting that earlier than this time, it is a fully coherent quantum system. The scale of the amplitude of fluctuations produced during de-coherence of cosmological vacuum energy are found to evolve to values consistent with those observed in cosmic microwave background radiation and galactic clustering.

In Section 3 we will explore how the microphysics of vacuum energy during decoherence can affect cosmology. We presume a universality throughout the universe in the available degrees of freedom determined by fundamental constants during its evolution. In fact there is direct experimental evidence that quantum mechanics does apply in the background space provided by the Schwarzschild metric of the Earth due to experiments by Overhauser *et.al.* [6, 7] and others. These experiments show that the interference of a single neutron with itself changes as expected when the plane of the two interfering paths is rotated from being parallel to being perpendicular to the "gravitational field" of the Earth. Since quantum coherent objects have been shown to gravitate, we expect that during some period in the past, quantum coherence of gravitating systems should have qualitatively altered the thermodynamics of the cosmology. Note that the exhibition of quantum coherent behavior for the gravitating systems presented does *not* require the quantization of the gravitation field *itself*.

1.0.3. *Scientific eschatology*

Coincidental with the founding of ANPA, Freeman Dyson [8] published a paper with the objective of determining whether the then known models of cosmology allowed the perpetuation of complex biological forms. At that time he could take a geometric view of the problem, defined by whether the spatial curvature parameter k was positive (which results in a closed universe that eventually re-collapses), zero (which describes spatial flatness, resulting in a universe that barely never re-collapses), or negative (which results in an open universe that expands indefinitely). One of his objectives in the paper was to make a case for the possibility that rational, scientific, ecological communities of living organisms could look forward to a "time without end". He demonstrated that if the members of such communities depend on the biochemistry with which we are familiar, there is no hope that his objective can be reached within any class of cosmologies he considered. He then extended the definition of "biology" to include "organisms" that have the same *structure* as those we have become familiar with on our planet. This was done by postulating a scaling law for quantum mechanical interactions which gave precision to his meaning of *structure*. For open cosmologies he found a way to show that appropriate "biological" strategies could provide a (subjective) *time without end* for organisms and ecosystems of ever increasing complexity.

At the time Dyson wrote, observational evidence for a "cosmological constant" did not exist. Einstein's motivation for introducing this constant (i.e. to preserve a steady state in the infinite universe satisfying his *cosmological principle*) had evaporated with discovery of the red shifts of distant galaxies and Hubble's law. Thus there was no reason for Dyson to include in his discussion the de Sitter cosmologies produced by a positive cosmological constant. We now believe that until about 5-7 Giga-years ago the rate of expansion of the universe was *decreasing* due to gravitational attraction, while more recently the rate has been *increasing*. From very early times until the present ($\sim 13.8\ Gyr$) the observational data can be fitted by a positive cosmological constant Λ together with an evolving energy density which can be checked in other ways. The data are consistent with Λ being strictly constant throughout this period. If Λ is indeed a "constant of nature", this fact would render Dyson's analysis moot, since the analysis assumed gravitational *deceleration* of the universe.

A review of Dyson's paper will be briefly presented in Section 4, supporting the possiblity of a "time without end" for the type of "biology" proposed. We will then examine the consequences of having a cosmological constant fixed for all late times.

2. Fundamentals of Microscopic Physics

In this section, a summary of a formulation for microphysical interactions using parameters and concepts that are meaningful at every stage of any given calculation, and incorporates many of the ideals of modeling natural philosophy, will be given. Our approach has been to always use properly normalized boundary states parameterized by physical masses, charges, and measured quantum numbers. Since the equations are non-perturbative, the problem then becomes one of developing analytically relevant scattering functionals that are both calculable and intuitive enough to allow meaningful connections to the phenomena being described. Once analytic forms for the basic interactions have been developed, those forms can be consistently embedded into more complex processes in a manner consistent with both relativity and classical correspondence.

This approach has been appealing to Pierre because the use of scattering theoretic techniques allows a direct examination of how fundamental measured parameters, such as coupling constants and masses, enter into the models of a scattering process. Few particle scattering theory inherently incorporates the complicated kinematics of the scattering process independently of the specific dynamics of the interaction. Combinatorial calculations of the dynamical coupling constants can be directly incorporated into a structure that consistently represents the correct space-time kinematics of special relativity. This allows parallel paths of the exploration of the fundamentals of physical processes. One path involves understanding the hierarchy and nature of the fundamental interactions. The other involves understanding the nature of the arena upon which the actions and interactions are staged.

One should recognize that in any physically realized experiment, there will always be a finite number of particles in initial and final states. However, high energy phenomenology demonstrates the possibility of pair creation (particle-antiparticle production) and annihilation, so that the actual number of particle states in the final state can be different from the number in the initial state. Single quantum states can be emitted from an excited or pseudo-stable system (like excited atoms and radioactive nuclei), but particle states require that lepton and baryon number conservation be satisfied for observed phenomenology. In Figure 1, the dashed lines in the diagram on the left represent the antiparticle states associated with the nearest solid line on the diagram.

Fig. 1.   Pair creation in a fixed particle number scattering process.

Under the *symbar operation* we developed, the diagram on the left can be identified with that on the right, where two of the solid lines are identified with negative 4-momentum as espoused by Feynman. The figure demonstrates that in all physically realizable cases, the asymptotic particle states can be represented in terms of a fixed particle number scattering process that has incoming/outgoing particle states transformed into outgoing/incoming antiparticle states. Once the procedure has been established for mapping these related amplitudes, one only needs to calculate the fixed particle number scattering amplitude related to any given particle scattering process. That amplitude can then be unitarily incorporated into a multi-channel scattering process.

2.0.4. *Combinatorial input to dynamics*

Part of the allure that scattering theory had to Pierre was that the input and output parameters are all well defined in and of themselves. The question then becomes one of how combinatorial ideas can contribute to descriptions of the dynamics represented by a given scattering process. One possibility suggested by Pierre is that the dynamics of the fundamental interactions are to first order due to the purely combinatorial counting of a discrete set of states, which define the coupling constants. We refer to this description of dynamics as the *Noyes Combinatorial Conjecture*, which can be stated as follows:

Noyes Combinatorial Conjecture: Once a quantum interaction vertex forms, the likelihood of it re-opening into the specific final state with conserved quantum numbers is inversely proportional to the number of combinatorial states on that level of interaction.

For example, consider the 137 level of the combinatorial hierarchy developed by the founding members of ANPA (excluding kinematic factors). Figure 2 demonstrates that only one of the 137 possible final state outcomes satisfies the constraint of this conjecture, approximately describing the appropriate coupling due to an electromagnetic interaction.

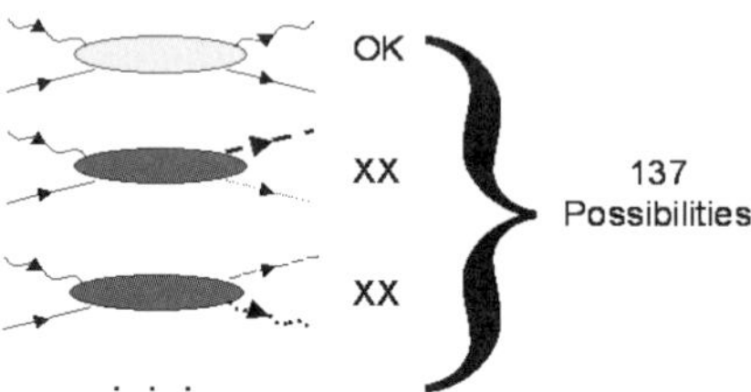

Fig. 2.   The Noyes Combinatorial Conjecture. Once the vertex forms, only one out of 137 possible final outcomes, the one labeled OK, satisfies the criterion for a viable scattering amplitude.

This conjecture gives a possibility of a direct input of dynamical parameters calculated using combinatorial techniques into relativistic scattering predictions.

2.1. *Fixed particle number relativistic scattering theory*

Perturbative approaches to scattering calculations involve sums over an infinite number of virtual states. There are some inherent flaws associated with naive perturbative approaches:

- it is difficult for one to make correspondences between the perturbative approaches in relativistic quantum field theory with those procedures developed in introductory courses in non-relativistic quantum mechanics;
- perturbative quantum field theory does not describe low lying bound states, since one cannot apply perturbative methods near the singular points of an amplitude;
- for quantum chromodynamics (QCD), certain asymptotic parameters, like the mass of the quarks, cannot be directly measured because of confinement;
- there do exist physical phenomena, like superconductivity, where there is no perturbative weak coupling limit because of an essential singularity at zero coupling in analytic forms involving the electron-electron coupling.

The formulation we developed [1–4] has well defined and straightforward non-relativistic limits and classical correspondences, allow bound states to be described in a straightforward manner, and produces analytic forms in cases for which perturbative series expansions would be invalid. However, it also brings certain non-trivial concerns, such as:

- one expects extremely complicated functional forms for kinematic dependencies;
- there is a need to construct unitary relativistic input amplitudes, which incorporate the conservation of probability fluxes and persistence of internal particle quantum numbers;
- the inclusion of the creation and annihilation of particles is not trivial.

2.1.1. *Faddeev's formulation of scattering theory*

Our formulation is a relativistic generalzation of the few body formalism developed by Faddeev [9]. The three body Faddeev equations for the transition amplitude involves decomposing this amplitude into channels associated with the possible clusters of the system, which can be diagrammatically represented as in Figure 3.

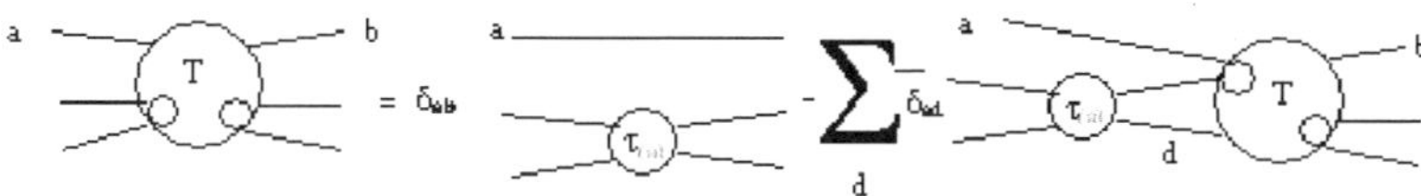

Fig. 3.   Diagrammatic representation of the three particle transition operator. The initial channel is labeled b, while the final channel is labeled a.

In this figure, the second term on the right involves a sum over all all possible intermediate channels d exclusive of channel a, as forced by the parameter $\bar{\delta}_{ad} \equiv \begin{cases} 1 \text{ if } a \neq d \\ 0 \text{ if } a=d \end{cases}$. This parameter excludes all self interactions, and eliminates the need for renormalizations. Self interactions are already included in the physical masses and couplings of the boundary state particles. Since there are no self interactions, there are no self-energy bubbles, or singularities due to disconnected clusters. Also, all particle legs represent physical particles. Several of the infinities associated with perturbative approaches to quantum scattering (such as self-energy bubbles/mass renormalization, charge renormalization at non-physical vertices, etc.) are absent in this approach. However, the full complication of the internal kinematics and dynamics are present in the intermediate states parameterized by channels d in the sum.

The primary complication in developing a relativistic finite particle number scattering formalism was in constructing a mechanism that properly embeds dynamical clusters in the global space-time. Cluster decomposability is essentially the requirement that the kinematics of distant, non-interacting systems should not in any way influence the dynamics of an interacting cluster. This is related to quantum dis-entanglement or de-coherence, and is necessary for the realization of the classical limit. Classical physics requires that systems can be described in some manner without quantum entanglement, and therefore, without cluster decomposability one cannot have correspondence with classical physics. There was considerable difficulty with incorporating cluster decomposability in a system whose kinematics was consistent with the non-linear nature of special relativity, yet did not change the energy spectra of the dynamical cluster. The successful description involved properly separating relativistic kinematics from quantum dynamics, and recognizing the difference between the off-shell quantum dynamics associated with the uncertainty principle and intermediate state propagation, versus the off-diagonal description of intermediate states in terms of a complete set of boundary states.

The key elements of the solution are as follows:

- The *velocity* describing the Lorentz frame of reference should be conserved, rather than the composite *momentum*, for off-diagonal energies. If momentum was conserved off-energy-diagonal, then the Lorentz frame of reference associated with the different energies would be different. This can be seen in terms of the generators of the Poincare group $[K_j, P_k] = i\delta_{jk}H$. If one chooses a unique frame of reference, one cannot have off-diagonal energies $E \neq E'$ and on-diagonal momenta $\mathbf{P} = \mathbf{P}'$:

$$\mathbf{u} = \frac{\mathbf{P}}{E} \neq \frac{\mathbf{P}'}{E'} = \mathbf{u}' \text{ unless } \mathbf{P} \neq \mathbf{P}'.$$

- Spectating cluster kinematics enter only parametrically. Otherwise for off-diagonal intermediate states, spectator kinematics would directly affect the energetics of the dynamical cluster, violating cluster decomposability.
- All amplitudes and parameters involved in any calculation have well defined non-relativistic limits, and the equations uniquely go to the Faddeev equations in that limit.

The solution for flat Minkowski space-time turns out to be quite instructive in solving the related problems in formulating dynamics in quantum gravity [10].

2.1.2. *The Efimov effect*

One of the most profound implications of three particle scattering was discovered independently by Efimov (the Efimov effect [11]) and Noyes (the Eternal Triangle effect). The effect predicts the existence of long range 3-body effects, despite the absence of long range pair-wise interactions. In particular, 3-body bound states can form in the absence of any 2-body bound states. This occurs if the phase shift has significant low energy modifications in the two-body interactions, allowing the two body long-range stationary states to generate three-body bound states near threshold energies. The effect was observed [12] in bosonic cesium atoms cooled to 10 nano-Kelvin. A magnetic field was then used to fine-tune the scattering length (which is inversely related to the pairwise energy scale) between pairs.

2.2. *Inclusion of particle-antiparticle symmetries*

We needed to expand our fixed particle number scattering theory into one that included antiparticles and the potential of pair creation. The natural parameters to describe a particle-particle scattering process can be expressed in terms of the 4-momenta in Figure 4.

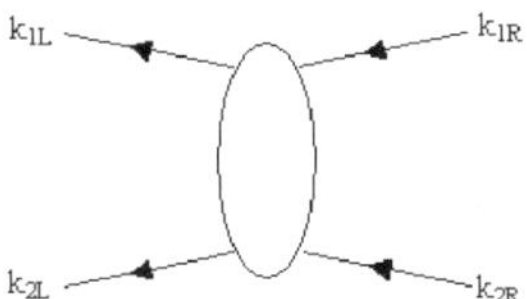

Fig. 4.   Particle-particle scattering in terms of incoming 4-momenta on the right, and outgoing 4-momenta on the left.

To determine the particle-antiparticle annihilation/pair creation amplitude, we first formulate the particle-particle amplitude in Figure 4. We then reverse the signs of the first particle's outgoing four-momentum (changing it to an incoming antiparticle) and the second particle's incoming four-momentum (changing it to an outgoing antiparticle), as indicated in the particle-antiparticle (bar) channel Figure 5.

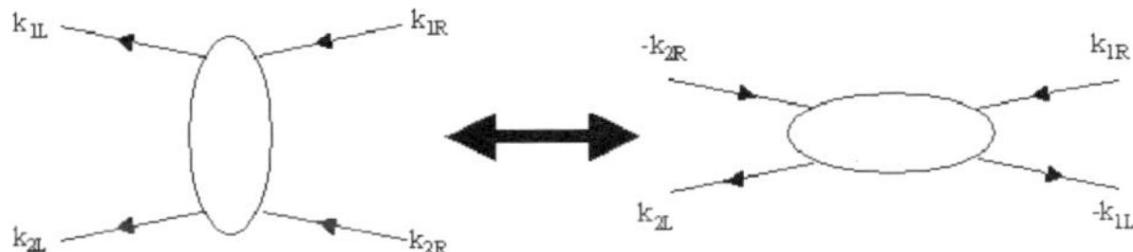

Fig. 5.   Annihilation/creation channel scattering.

This amplitude can then be incorporated into a general multichannel scattering amplitude in a manner that preserves unitarity. Therefore, once an analytic form has been established for the particle-particle scattering amplitudes in a fixed-particle-number scattering formalism, antiparticle scattering, annihilation, and pair creation amplitudes can be generated in a straightforward (though non-trivial) manner. New particle types will not be produced by this transformation (i.e., a left handed massless neutrino is associated only with a right handed anti-neutrino).

2.3. *Compton scattering*

To finish this section, we demonstrate the amplitude associated with the emission of a photon from an excited distant source, the subsequent scattering of that photon from a charged particle, and its final asymptotic absorption to place a detector in an excited state. The classical dis-entanglement of the source and detector requires that there be no direct interaction between these components during the scattering process as represented in Figure 6.

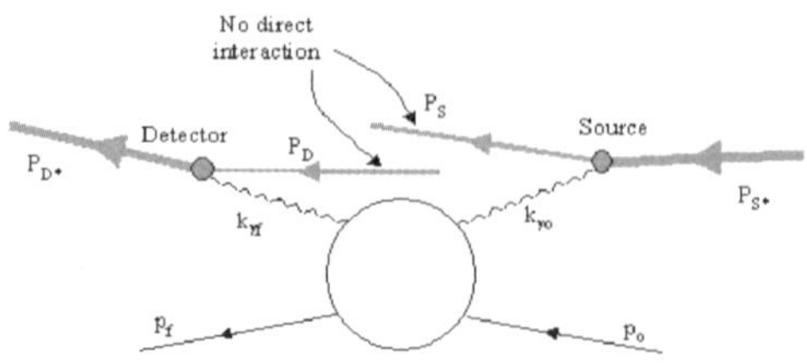

Fig. 6.   Compton Scattering, inclusive of the photon source and detector.

In this process, the source cluster and the detector cluster interacts with the particle via two-body unitary scattering amplitudes from which one can extract the photons in question. In order for the process to include the kinematic possibility of describing the emission of a quantum that can be interpreted as a boundary state, this scattering must be *anelastic* in the sense that the source changes mass when the quantum is emitted. Similarly, the detector must engage in a unitary anelastic two-body scattering with the quantum. The kinematic parameters labeling excited states are labeled with an asterisk in Figure 6.

2.3.1. *Quantum dis-entanglement of classical sources and sinks.*

The specific postulate needed to connect Compton scattering to a three-particle scattering process is that the source and detector be (classically) disentangled from each other, except through the interaction with the particle, which is possible using our formulation. Our demonstration that this non-perturbative calculation directly corresponds with the perturbative calculation of quantum electrodynamics using photons as asymptotic states proved the Wheeler-Feynman point of view that any description treating photons as particles can equivalently be described by treating them as implicit characteristics of particle-particle interactions including the sources and sinks. The apparent particulate behavior of the photons can be extracted from their interaction with the source/detector.

3. Cosmological Scale and Dark Energy

Besides the fundamentals of microphysics and quantum mechanics, Pierre has a considerable interest in the fundamentals of cosmology. Some of this interest was motivated by Ed Jones' early advocacy of dark energy before it became incorporated into the standard model of cosmology [13, 14]. In this section, we will review some insights into the interplay of microphysics with cosmology.

The horizon problem in standard cosmology ponders the reason for the large scale homogeneity and isotropy of the observed universe. The reason that this is somewhat an inigma is that the universe has expanded by a factor of about 1100 since the formation of the observed Cosmic Microwave Background (CMB) radiation. The ratio of the distance a photon could have traveled since the beginning of the expansion with that during recombination is about 100, which means that the subsequent expansion implies that light from the CMB comes from $100^3 = 10^6$ causally disconnected regions. Yet, angular correlations of the fluctuations beyond causally connected regions have been accurately measured by several experiments [15].

Furthermore, "standard candle" data examining the luminosities of distant Type Ia supernovae, which have a well understood time and frequency dependency, indicate clearly that the rate of expansion of the universe has been accelerating for several billion years [16]. This conclusion is independently confirmed by analysis of the detailed structure of the CMB radiation [17]. Both results are in quantitative agreement with a (positive) cosmological constant fit to the data. A fundamental question remains: What is this dark energy causing an *acceleration* in the expansion of the universe?

Our approach has been to incorporate quantum behaviors into the cosmological expansion. We expect quantum effects to be significant within regions of the Compton wavelength of a given energy unit. In particular, for relativistic gravitating mass units with quantum coherence within the volume generated by a Compton wavelength λ_m^3, the zero point momentum is expected to be of order $p \sim \frac{\hbar}{V^{1/3}} \sim \frac{\hbar}{\lambda_m}$. This gives a zero point energy of order $E_0 \approx \sqrt{2}mc^2$. If we estimate a mean field potential from the Newtonian form $V \sim -\frac{G_N m^2}{\lambda_m} = -\frac{m^2}{M_P^2}mc^2$ which has a magnitude much less than E_0, it is evident that the zero point energy will dominate the energy of such a system. A period of global quantum coherence solves the horizon problem, since quantum correlations are in some sense supraluminal. We should also be able to associate the dark energy with the quantum zero point energy during some period of gravitational coherence.

3.1. *Motivation*

Our initial examinations of micro-cosmology were motivated by the explorations of Ed Jones [14]. We will assert that there is some special cosmological scale $R(t)$ associated with the microscopic physics of the coherent cosmology that will be specified by the Robertson-Walker expansion scale $a_{micro-cosmology}(t) \to R(t)$. This scale will take on a well defined value related to the energy density when global gravitational coherence is lost, resulting in the hot big bang that describes observations so well.

3.1.1. *Jones' Microcosmology*

Jones envisaged an extremely rapid ("inflationary") expansion from the Planck scale (i.e. from the Planck length $L_P = \frac{\hbar}{M_P c} \cong 1.6 \times 10^{-33} cm$, where the Planck mass

$M_{Pk} = [\frac{\hbar c}{G_N}]^{\frac{1}{2}} \cong 2.1 \times 10^{-8} kg \cong 1.221 \times 10^{19} GeV/c^2$, and G_N is Newton's gravitational constant) to a length scale $R_\epsilon \sim \frac{\hbar c}{\epsilon}$, where ϵ is a single unit of virtual energy, of which there are a large number during the transition from gravitationally driven coherence. This expansion, whose details are not here examined, is characterized by the dimensionless ratio $\mathcal{Z}_\epsilon = \frac{R_\epsilon}{L_P} = \frac{M_P}{\epsilon}$.

After this expansion has occurred, the virtual energy which drives it makes a thermodynamic equilibrium transition to normal energy densities (i.e. dark matter, baryonic and leptonic matter, electromagnetism,...) at a mass-energy scale characterized by the mass parameter m_θ and length scale $\frac{\hbar}{m_\theta c}$. Jones uses the energy parameter ϵ as a unit of this energy which he calls one *Planckton*, defined as one Planck mass worth of energy distributed over the volume $V_\epsilon \sim \left(\frac{\hbar c}{\epsilon}\right)^3$. The virtual energy during the transition is assumed to consist of N_{Pk} Plancktons of energy ϵ giving total energy $N_{Pk}\epsilon$ in the region of coherence, corresponding to an energy density $\sim \frac{N_{Pk}\epsilon}{V_\epsilon}$. This virtual energy makes a transition to normal thermal energy, so that

$$N_{Pk}\epsilon^4 \sim (m_\theta c^2)^4. \tag{3.1}$$

By utilizing the Dyson-Noyes argument which follows in the next section, Jones then argued that the number of Plancktonic energy units in the region was related to the dimensionless ratio by $N_{Pk} = \mathcal{Z}_\epsilon^2$.

Finally, Jones assumed that a virtual energy unit ϵ spread throughout the region R_ϵ is left unthermalized (not transitioned to normal matter), and hence was interpreted as the cosmological constant density ρ_Λ for later times. It then can be measured as dark energy during the present epoch.

3.1.2. *Dyson-Noyes-Jones Anomaly*

Jones assumes that the dense state during decoherence can be specified by using an extension to gravitation [18, 19] of an argument first made by Freeman Dyson [20, 21]. Dyson pointed out that if one goes to more that 137 terms in the coupling $\alpha \equiv \frac{e^2}{\hbar c}$ in the perturbative expansion of renormalized quantum electrodynamics (QED), and assumes that this series is analytic in the coupling at zero coupling (and thus also applies to a theory in which e^2 is replaced by $-e^2$) , clusters of like charges will be unstable against collapse to negatively infinite energies. Schweber [21] notes that this argument convinced Dyson that renormalized QED can never be a *fundamental theory*. The case of replacing e^2 by $-G_N m^2$ corresponds to a classical theory for gravity. Noyes [18] noted that *any* particulate gravitational system consisting of masses m must be subject to a similar instability and could be expected to collapse to a black hole.

We identify $Z_{e^2} = 1/\alpha_{e^2} \cong 137$ as the number of electromagnetic interactions which occur within the Compton wave-length of an electron-positron pair ($r_{2m_e} = \hbar/2m_e c$) when the Dyson bound is reached. If we apply the same reasoning to gravitating particles of mass m (and if we are able to use a classical gravitational form), the parameter α_{e^2} is replaced by $\alpha_m = -\frac{G_N m^2}{\hbar c} = -\frac{m^2}{M_P^2}$. The parameter determining the Dyson-Noyes (DN) bound becomes the number of *gravitational* interactions within $\lambda_m \equiv \hbar/mc$ which will produce another particle of mass m, and is given by

$$\mathcal{Z}_m \sim \frac{M_P^2}{m^2}. \tag{3.2}$$

If this dense state with Compton wavelength λ_m, contains $\mathcal{Z}_m$ interactions within λ_m, then the Dyson-Noyes-Jones (DNJ) bound is due to the expected transition transition $\mathcal{Z}_m m \to (\mathcal{Z}_m + 1)m$ lowering the energy of the system, thereby exhibiting instability against gravitational collapse due to relativistic particle creation.

The production channel for the masses m due to $\mathcal{Z}_m$ interactions within λ_m is diagrammatically represented in Figure 7.

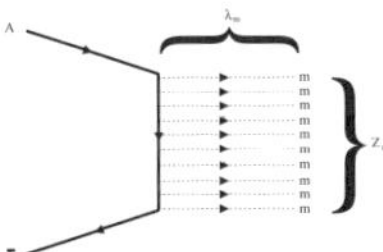

Fig. 7.   Noyes-Jones collapse of gravitating quanta.

If there are $\mathcal{Z}_m$ scalar gravitating particles of mass m within the Compton wavelength of that mass, a particle falling into that system from an appreciable distance will gain energy equal to mc^2, which could produce yet another gravitating mass m. Clearly, the interaction becomes anomalous. Generally, when the perturbative form of a weak interaction becomes divergent, it is a sign of a phase transition, or a non-perturbative state of the system (eg bound states). One expects large quantum correlations between systems of mass m interacting on scales smaller than or comparable to the Compton wavelength of those masses. We will assume that if the number of gravitational interactions of mass units m which can occur within a region of quantum coherence is greater than the DNJ limit, a phase transition into systems with quantum coherence scales of the Compton wavelength of those mass units will occur.

3.2. *Coherent Gravitating Matter*

Following Jones we have associated a Planck energy unit (cf. Sec.3.1.1) within a region that has quantum coherent energy of one Planck mass M_P. Within such a region, quantum gravitational effects are expected to be significant. There are expected to be many Planck units of energy within a scale unit of the universe $R(t)$. The Planck units will be envisioned as internally coherent gravitating units that become incoherent with each other during the period of de-coherence.

The localization of interactions has to be of the order $\hbar/mc$ in order to be able to use the DNJ argument. This allows us to obtain the number of gravitational interactions that can occur within a Compton wavelength of the mass m which provide sufficient energy for a phase transition, namely $\mathcal{Z}_m \sim \frac{M_P^2}{m^2}$. The process of de-coherence occurs when there are a sufficient number of virtual particles such that gravitational interactions of quantum coherent states of Friedman-Lemaitre (FL) cosmology could have a DNJ anomaly. A single Planck unit of coherent energy in a scale of R_ϵ will have $\mathcal{Z}_\epsilon$ partitions of an available Planck mass of energy that can constitute interactions of this type. This is shown schematically in Figure 8.

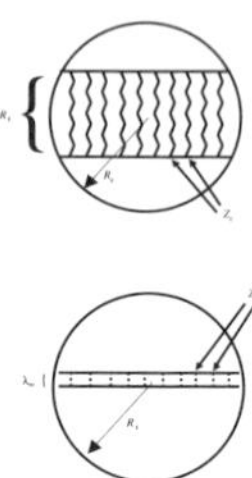

Fig. 8.   Counting of gravitating quanta during de-coherence

Since there is significant microscopic de-coherence at later times, the quantity $\mathcal{Z}_\epsilon$ can only be calculated during the pre-coherence.

When the number of partitions of a given Planck energy unit equals the DNJ limit, in principle there could occur a transition of the DNJ type involving interaction energies equal to a Planck mass unit for which gravitation is significant, thus allowing us to conclude that de-coherence defines a mass scale from the relationship

$$\mathcal{Z}_\epsilon = \mathcal{Z}_m \equiv \mathcal{Z}. \tag{3.3}$$

Expressed in terms of the energy scales, this gives the fundamental equation connecting the observable parameter ϵ (related to dark energy) to the mass scale at decoherence m

$$m^2 c^2 \approx \epsilon M_P. \tag{3.4}$$

This connection is a succinct summary of Jones' model; henceforth we will refer to it as the *Jones equation*.

3.3. *Correspondence with the Cosmological Constant*

We use the present day measurement of the cosmological density ρ_Λ to determine the energy scale m of cosmological de-coherence. The measured dark energy density is

$$\rho_\Lambda == \frac{\Lambda c^4}{8\pi G_N} \approx 4.10 \times 10^{-6} \frac{GeV}{cm^3} \approx 0.6 \, joules/kilometer^3 \sim \frac{\epsilon^4}{(\hbar c)^3} \tag{3.5}$$

We can immediately calculate the de-coherence scale energy, its corresponding cosmological scale, and the DNJ limit dimensionless parameter:

$$\epsilon \sim 10^{-12} GeV \quad , \quad R_\epsilon \sim 10^{-2} cm \quad , \quad \mathcal{Z}_\epsilon \equiv \frac{M_P}{\epsilon} \sim 10^{30} \sim \mathcal{Z}_m \qquad (3.6)$$

The equality of the interaction factors $\mathcal{Z}$ gives the Jones equation $m^2 c^2 = \epsilon M_P$ from which we calculate a value for the mass scale for quantum de-coherence

$$m \sim 5\, TeV/c^2, \qquad (3.7)$$

which is much smaller than the Planck mass $\sim 10^{16}\, TeV/c^2$. The number of Planck energy units per scale volume during de-coherence is given by

$$N_{Pk} \sim \mathcal{Z}^2 \sim 10^{60}. \qquad (3.8)$$

At this point we examined the Hubble rate equation during this transition. Using the standard equations for cosmological evolution, we obtain a rate of expansion given by

$$\dot{R}_\epsilon \sim R_\epsilon \sqrt{\frac{8\pi G_N}{3}(\rho_m + \rho_\Lambda)} \sim \frac{1}{\epsilon}\sqrt{\frac{1}{M_P^2}\mathcal{Z}_\epsilon^2 \epsilon^4} \sim c. \qquad (3.9)$$

This is a very interesting result, since it implies that the transition occurs at the time that the expansion rate of this important scale is the same as the speed of light.

3.4. *Gravitating massive scalar particle*

If the mass scale m represents a fundamental gravitating scalar particle, we expect the coherence length of interactions involving the particle to be of the order of its Compton wavelength with regards to couplings with other particulate degrees of freedom present. If the density is greater than

$$\rho_m \sim \frac{mc^2}{\lambda_m^3} \sim \frac{(mc^2)^4}{(\hbar c)^3} \qquad (3.10)$$

we expect that regions of quantum coherence of interaction energies of the order of mc^2 and scale λ_m will overlap sufficiently over the scale R_ϵ such that we will have a macroscopic quantum system on a cosmological scale.

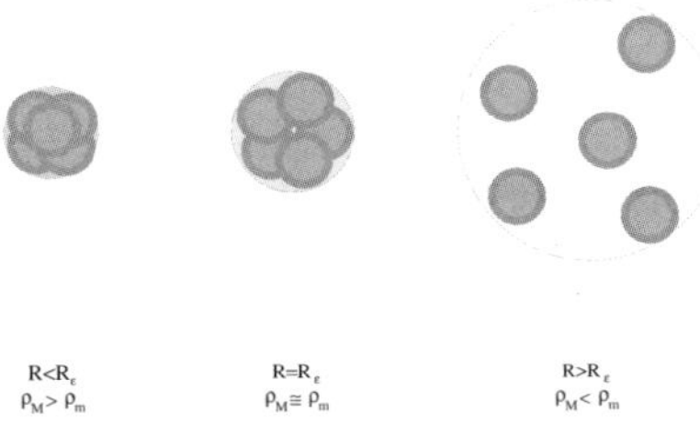

Fig. 9.   Overlapping regions of coherence during expansion.

As long as the expansion scale is less than the transition scale $R(t) < R_\epsilon$, cosmological vacuum energy is determined by this scale $R(t)$. However, once the energy density of FL energy becomes less than ρ_m, we expect that since the coherence length of the mass m is insufficient to cover the cosmological scale, and the energy density will break into domains of cluster decomposed (AKLN de-coherent [3]) regions of at most local quantum coherence. This phase transition should decouple quantum coherence of gravitational interactions at the cosmological scale R_ϵ. At this time (de-coherence), the cosmological (dark) vacuum energy density ρ_Λ is frozen at the scale given by

$$\rho_\Lambda \sim \frac{\epsilon}{R_\epsilon^3} = \frac{\epsilon^4}{(\hbar c)^3} \tag{3.11}$$

During de-coherence, we assume that the energy contained in a region of scale R_ϵ is given by N_{Pk} Planck mass units appropriately red-shifted to the de-coherence epoch. This gives an energy density of the form

$$\rho_{FL} \sim N_{Pk} \frac{\epsilon^4}{(\hbar c)^3}. \tag{3.12}$$

Since de-coherence is expected to occur when this density scale is the same as the quantum coherence density scale for the mass ρ_m, we obtain the following relationship between the de-coherence energy scale ϵ and the scalar mass m:

$$N_{Pk}\epsilon^4 \cong (mc^2)^4 \tag{3.13}$$

This allows us to consistently relate the number of Planck energy units in the region of coherence R_ϵ to the DN counting parameters:

$$N_{Pk} \cong \frac{(mc^2)^4}{\epsilon^4} = \frac{m^4}{M_P^4}\frac{(M_P c^2)^4}{\epsilon^4} = \frac{\mathcal{Z}_\epsilon^4}{\mathcal{Z}_m^2} = \mathcal{Z}^2, \tag{3.14}$$

which insures that all quantities relevant to our theory can be reduced to a single parameter in the Jones equation $m^2 c^2 \cong \epsilon M_P$.

3.5. *Subsequent thermal expansion*

The equations for cosmological expansion satisfy local energy conservation $T^{\mu\nu}_{;\nu} = 0$, which implies $\dot\rho = -3H(\rho + P)$ in terms of the Hubble rate $H \equiv \frac{\dot R}{R}$. The first law of thermodynamics then implies an adiabaticity condition on the expansion given by

$$\frac{d}{dt}\left(\frac{S}{V}R^3\right) = 0. \tag{3.15}$$

Assuming adiabatic expansion, we expect $g(T)\,(RT)^3$ to be constant far from particle production thesholds. Here, $g(T)$ counts the number of low mass particles contributing to the cosmological entropy density at temperature T. This gives a red shift in terms of the photon temperature during a given epoch

$$\frac{R_o}{R} \equiv 1 + z = (1 + z_{dust})\left(\frac{g(T)}{g_{dust}}\right)^{1/3}\frac{T}{T_{dust}}, \tag{3.16}$$

where z_{dust} is defined as the redshift at equality of radiation and pressure-less matter energy densities. In terms of photon temperature, we can count the average number of photons using standard results from black body radiation

$$\frac{N_\gamma}{N_{\gamma o}} = \frac{(RT)^3}{(R_o T_o)^3} \simeq \frac{g_o}{g(T)}. \tag{3.17}$$

This allows us to write a formula for the ratio of the numbers of gravitationally coherent mass units to the photons at the temperature of decoherence (T_{freeze}), in terms of its mass m and the measured baryon-photon ratio:

$$\frac{N_m}{N_\gamma} \simeq \frac{\Omega_m}{\Omega_{baryon}} \frac{N_{bo}}{N_{\gamma o}} \frac{m_N}{m} \frac{g(T)}{g_o}, \tag{3.18}$$

which gives $\frac{N_m}{N_\gamma} \simeq 2.9 \times 10^{-9} \frac{g(T_{freeze})}{g_o} \frac{m_N}{m}$.

To summarize, if there are $\mathcal{Z}_m$ coherent masses m within λ_m, there is high likelihood of the production of another scalar mass m. This interpretation *requires* us to be talking about *quantum coherent systems* when the Jones transition from microcosmology to a universe where we can use conventional physics and cosmology takes place. This line of reasoning suggested that this transition itself *must* in some sense correspond to *quantum decoherence.*

3.6. *De-coherence from Dark Energy*

In most of what follows we will assume flat spatial curvature $k = 0$. Prior to the scale condition $\dot{R}_\epsilon = c$, which we will henceforth refer to as the time of dark energy de-coherence, gravitational influences are propagating (at least) at the rate of the gravitational scale expansion, and microscopic interactions (which can propagate no faster than c) are incapable of contributing to cosmological scale equilibration. Since the definition of a temperature requires an equilibration of interacting "microstates", there must be some mechanism for the redistribution of those microstates on time scales more rapid than the cosmological expansion rate, which can only be gravitational.

3.6.1. *Dark energy*

As we have discussed in the motivation section, we expect that dark energy decoherence occurs when the cosmological scale is $R_\epsilon \sim 1/\epsilon$. The gravitational dark energy scale associated with de-coherence is given by ϵ, independent of the actual number of energy units N_ϵ in the scale region. Since the dark energy density must be represented by an intensive parameter which should be the same for the universe as a whole, we will express this density as the coherent vacuum state energy density of this macroscopic quantum system.

The energy levels associated with excitations on the quantum system should satisfy the usual condition

$$E_{N_\epsilon} = \left(N_\epsilon + \frac{1}{2} \right) \hbar\omega = (2N_\epsilon + 1)\,\epsilon. \tag{3.19}$$

In effect, the scale parameter R_ϵ provides the infrared cutoff for cosmological quantum coherent processes,

$$k_\epsilon = \frac{2\pi}{\lambda_\epsilon} = \frac{\pi}{R_\epsilon}. \tag{3.20}$$

3.6.2. *Rate of Expansion during De-Coherence*

Since we find the expansion rate equation $\dot{R}_\epsilon = c$ a compelling argument for the quantitative description of gravitational de-coherence, it is this relationship that we will use to determine the form for the energy density ρ_{FL} during de-coherence. This energy density is directly related to the number of gravitating excitations above vacuum energy in the pre-coherent state. The Hubble equation takes the form

$$H_\epsilon^2 = \left(\frac{c}{R_\epsilon}\right)^2 = \frac{8\pi G_N}{3}\left(\rho_{FL} + \rho_\Lambda\right) = \frac{8\pi G_N}{3}\left(2N_\epsilon + 1\right)\rho_\Lambda, \tag{3.21}$$

where we have written $\rho_\Lambda = \frac{\Lambda}{8\pi G_N}$. We see that $2N_\epsilon$ counts the number of Jones-Planck energy units per scale factor in the pre-coherent universe (referred to by Jones as $N_{Planckton}$). Using equations 3.21 and 3.11 the energy density during dark energy decoherence is therefore given by

$$\rho_{FL} = 2N_\epsilon \rho_\Lambda \cong \frac{3}{2\pi^3}M_P^2\epsilon^2. \tag{3.22}$$

This is a cosmological form of the Jones equation.

Upon substituting the phenomenological parameters, the dark energy scale and FRW scale given by $\epsilon \sim 6 \times 10^{-12}\,GeV$, $R_\epsilon \sim 6 \times 10^{-3}\,cm$. Using our calculation, the Planck energy partition $\mathcal{Z}_\epsilon$ and number of gravitational modes N_ϵ at dark energy de-coherence consistent with the Jones calculation:

$$\mathcal{Z}_\epsilon \cong 2.19 \times 10^{30} \quad , \quad N_\epsilon \cong 3.59 \times 10^{60}. \tag{3.23}$$

The coherent mass density scale $\frac{mc^2}{\lambda_m^3} = \frac{(mc^2)^4}{(\hbar c)^3}$ is given by $m \sim 2 - 4\,TeV/c^2$.

3.6.3. *Fluctuations*

Adiabatic perturbations are those that fractionally perturb the number densities of photons and matter equally. These fluctuations grow due to gravitational attraction in a well understood manner. This allows us to write an accurate estimation for the scale of fluctuations during de-coherence in terms of those at last scattering

$$\delta_\epsilon = \left(\frac{R_{dust}}{R_{LS}}\right)\left(\frac{R_\epsilon}{R_{dust}}\right)^2 \delta_{LS} \cong \frac{z_{dust}z_{LS}}{z_\epsilon^2}\delta_{LS} \tag{3.24}$$

if the fluctuations are "fixed" at dark energy de-coherence.

We expect the energy available for the fluctuations at de-coherence to be of the order of the vacuum energy:

$$\delta_{DC} \sim \left(\frac{\rho_\Lambda}{\rho_{FL} + \rho_\Lambda}\right)^{1/2} \cong \left(\frac{1}{2N_\epsilon}\right)^{1/2}, \tag{3.25}$$

regardless of the specifics of the pre-coherent state. Indeed, we obtain the correct order of magnitude for fluctuations at last scattering:

$$\delta_{LS} \cong 2.8 \times 10^{-5}. \tag{3.26}$$

A more general and precise calculation is published in reference [22].

4. Scientific Eschatology

One of Pierre's favorite topics of conversation involves the future of our cosmology and the civilizations, both terrestrial and extra-terrestrial, that inhabit it.

4.1. *Dyson's "Time without end"*

We begin examining late time cosmology by reviewing Dyson's pioneering paper [8]. As already noted he rapidly concludes that a "big crunch" universe does not provide enough scope for an optimistic view of the far future. Even in an open (or flat) universe, Dyson's arguments close the door on the continuation of "flesh and blood" biology of a complexity comparable to that currently experienced on our planet for a *physical* time without end. This focuses his interest on the biology of "...sentient black clouds [23], or sentient computers [24]...".

To make his argument plausible and quantitative, he proposes a "biological scaling hypothesis" that has as its first consequence the conclusion that the "...appropriate measure of time as experienced subjectively by a living creature is not physical time t but ..." an integral from zero to t of the temperature function defined by his scaling law ([8], Eq.56). This he calls *subjective* time. "The second consequence of the scaling law is that any creature is characterized by a quantity Q which measures its rate of entropy production per unit of subjective time." For a single human being Q works out to be about 10^{23} bits, and for our species overall $\sim 10^{33}$ bits. This sets a fixed lower bound for the temperature θ, which is ([8], Eq. 73) $\theta > (Q/N)\, 10^{-12}\,{}^\circ\text{K}$, where N is the number of electrons available to the society of complexity Q. For our current biosphere $N = 10^{42}$, so our current social complexity cannot be maintained at temperatures lower that $10^{-23}\,{}^\circ\text{K}$. This bound is arrived at by asserting that the rate of energy dissipation (i.e. use of energy by the society) must not exceed the power that can be radiated away into space. At that point it can still dissipate the energy which it must expend to keep on operating. Below that point it must decrease its complexity or increase its controlled number of electrons. Since the supply of energy available to the society is assumed finite, it must reach this point at a finite time, and Dyson remarks (ref. [8] p. 456):

> We have reached the sad conclusion that the slowing down of metabolism described by my biological scaling hypothesis is insufficient to allow a society to continue indefinitely.

Here Dyson examines a possible way to insure "time without end" for entities who "live" in this cold and forbidding future. This is, quite simply, the biological strategy of *hibernation*. Life can metabolize at a higher temperature and then hibernate at a much lower temperature to stretch out its *subjective* time. To quote Dyson again (ref. [8] p 4550])

> Suppose then that a society spends a fraction $g(t)$ of its time in its active phase and a fraction $[1 - g(t)]$ hibernating. The cycles of activity and hibernation should be short enough so that $g(t)$ and $\theta(t)$ [Here $\theta(t)$ is a function Dyson has already assumed technologically available subject to explicit thermodynamic constraints] do not vary appreciably during any one cycle. Then [the previous constraints] no longer hold.

Hence the constraints which led to his dismal conclusion no longer applied. In this way he was able to achieve a system with an infinite subjective time in an expanding universe, while expending only a finite amount of energy.

One matter where the finite energy limitation is serious is in the storage of memory. As Dyson remarks (ref. [8] p. 456)

> I would like our descendants to be endowed not only with an infinitely long subjective lifetime but also with a memory of endlessly growing capacity. To be immortal with a finite memory is highly unsatisfactory; it seems hardly worth while to be immortal if one must erase all trace of one's origins in order to make room for new experience.

Digital memory with the finite energy resources available to a periodically hybernating society is not an option. Dyson turns to analog memory and claims that the angles between a finite number of structures (e.g. stars) in an expanding universe can be used for an *analog* memory storage of ever increasing capacity. In Dyson's expanding universe scenario, even though the energy available to the society is finite, the unbounded expansion of the volume over which this energy is distributed means that there is, indeed, no *a priori* upper limit on the entropy. Consequently analog storage might meet the requirement he imposes.

4.2. *Late Time de Sitter Universes*

Recent observational results have convinced most of the experts that the energy density of our universe is currently partitioned into approximately 73% dark energy, 23% dark matter and 4% ordinary matter and radiation, in a space that is flat rather than either spherical or hyperbolic. The simplest way to fit the data is to assume that the various interlocking pieces of evidence which lead to this picture constitute an actual *discovery* of Einstein's cosmological constant Λ as a new universal constant *and* a measurement of that constant to a couple of percent.

The specific consequences of interest here which follow from this assumption are that:

(a) The energy density that drives the cosmological expansion in the dynamical Friedman-Lemaitre (FL) equation exclusive of dark energy, which we call ρ_{FL}, will eventually become insignificant compared to the cosmological constant density ρ_Λ.

(b) Subsequently, the rate equation is replaced at late times by $[\frac{\dot{R}}{R}]^2 \simeq \frac{8\pi G_N}{3c^2}\rho_\Lambda = \Lambda c^2/3$, which implies that $R(t) = R_E e^{\sqrt{\frac{\Lambda}{3}}ct} \equiv R_E e^{\frac{ct}{R_\Lambda}}$. Here R_Λ is sometimes called the de Sitter radius or horizon, and R_E is set at a time when ρ_{FL} is negligible compared to ρ_Λ. Although objects of the scale of the gravitationally bound super-cluster have dynamics which are determined by local matter densities, the late time exponential expansion defines a cosmological de Sitter horizon $R_\Lambda = \sqrt{\frac{3}{\Lambda}} \simeq 16.6\,Glyr$ which serves as a causal boundary. This means that anything which crosses this boundary can never re-establish luminal contact with our region of the universe. Our galaxy appears to be close to the edge of, and probably bound to a super-cluster with a radius of about 50 mega-parsecs $\approx 0.16\,Gly$. We expect that all other galaxies not in our local super-cluster will have crossed the de Sitter horizon after $100\,Gyr$. The consequence already precludes useful discussion of Dyson's analog memory storage and far-ranging communications strategies if Λ is indeed a universal constant. Both of these strategies rely on scaling laws that assume causal (i.e. luminal) contact can always exist if the society waits long enough.

(c) Since the de Sitter horizon has a finite area, the causal region of the de Sitter space will have a finite entropy [25] $S_\Lambda = k_B \frac{\pi R_\Lambda^2}{L_P^2}$, where k_B is Boltzmann's constant.

(1) Because finite entropy implies finite information storage capacity, using simple counting arguments this fact precludes any way of realizing Dyson's analog storage method for constructing an indefinitely extendable memory. As a counting argument, this holds for quantum-coherent systems (e.g. quantum computers) as well as for digital computers.

(2) Systems with finite entropy undergo Poincaré recurrences [26]. Such recurrences are due to the finite number of configurations (microstates) available to a system with finite entropy. Because there are only a finite number of configurations that the system fluctuates among, the system will eventually return to any given initial configuration. These recurrences have only to do with the counting of states. Such recurrences are maximally destructive of prior information content.

(3) One strategy of despair that has sometimes been suggested is to find a way to pass clues about our experience in this universe through any recycling to provide information that could prove useful to new societies evolving in the cycle that might emerge after we are consumed. Any total destruction of information precludes placing any hope in this possibility.

There might still be a finite strategy arising from the fact that in such universes the de Sitter horizon maintains a finite temperature. The existence of a horizon creates an information deficit within the space bounded by that horizon. The subtle quantum correlations in states near the horizon must be described statistically in terms of the degrees of freedom and parameters accessible in the causal region. Whenever there are such statistical degeneracies in ways to describe a particular physical state, the concept of temperature becomes a meaningful tool to describe macroscopic states. There are many states across any horizon which can describe any given measured state within the causal region, thus associating an entropy and temperature with that horizon.

4.2.1. *Are societal strategies futile?*

To explore possible strategies, if the mass density ρ_M is uniformly distributed throughout the region, the horizon scale satisfies

$$r_H = \sqrt{\left(\frac{3c^4}{8\pi G_N}\right)\frac{1}{\rho_m + \rho_\Lambda}}. \tag{4.1}$$

The entropy associated with this horizon is given by $S = k_B \frac{\pi r_H^2}{L_P^2}$, whereas the temperature is given by $T = \frac{\hbar c}{2\pi k_B r_H}$. Any societal activity which is capable of increasing the energy density ρ_m within the causal region would be expected to utilize processes that preserve information, converting entropy from the larger horizon (at a cooler temperature) into the increased local energy densities. From (4.1), an increase in energy density will result in a decrease of the horizon scale, a decrease in the horizon entropy, and an increase in the horizon temperature. If such processes can occur, the society has access to increasing energy supplies, increasing temperatures, and the associated increasing rates of subjective time as defined by Dyson.

As the horizon scale shrinks, the temperature associated with the horizon increases. Using Dyson's association of the subjective time rate with the temperature of the organism/society, any evolving "biological" organisms would have increasing rates of subjective time. However, the trade off is the decrease in the horizon entropy, *decreasing* the Poincaré recurrence time, which remains large exponentially in the entropy. In such an environment, Dyson's strategy of hybernation would not be useful, since it would only serve to slow down relative subjective time as the exponential expansion persists during the periods of hibernation.

Alternatively, placing energy into an increasing black hole will increase the size and entropy of the de Sitter horizon, cooling that surface as well as the surface of the black hole. Any such society must balance its perilous existance between those two horizons defining its ecosystem.

We have been unable to come up with any *a priori* reasons against such societal interventions on a cosmological scale. Note that in contrast to Dyson's scenario which accommodates biology to a changing cosmological environment, such a society must alter both itself *and* the cosmology to manipulate its subjective time.

4.3. *Is the "Cosmological Constant" Really Constant?*

The scenario discussed so far infers that Λ be constant. However, in some scenarios, this parameter is essentially constant only during well defined periods [27].

General relativity is a local theory of gravity. This means that we expect local geometry to be determined by local energy densities as described using Einstein's equation $G_{\mu\nu}(x) = \frac{8\pi G_N}{c^4} T_{\mu\nu}(x) + \Lambda g_{\mu\nu}(x)$. For dynamically significant periods of time, it is clear that the homogeneity and isotropy assumptions inherent in a Friedman- Lemaitre cosmology do not hold on the scale of galactic clustering. This means that the local geometry generated, $G_{\mu\nu}(x)$, is neither pure (cosmological) FRW-Lemaitre nor the gravitationally bound region of an isolated galactic cluster in Minkowski space (which would have negligible space-time expansion from dark energy). For instance, our local gravity is primarily the Schwarzschild space-time generated by Earth, with negligible influence from the overall cosmological acceleration due to the dark energy (or else we would all be leaving the surface of the Earth!). This means that our local space-time is not undergoing the exponential expansion associated with a cosmological constant, despite our presence in an accelerating cosmology. We expect the evolution of our local scales to be determined by our local energy (and dark energy) densities, appropriately matching boundary conditions. Likewise, on scales for which the cosmological matter inhomogeneities are important, the local densities are expected to have significant influence on the behavior of the geometry relative to cosmological dynamics. As the scale of relevance to galactic clustering crosses the de Sitter scale radius, one must take care in describing the de Sitter scale as a true horizon. It is not unreasonable to suggest that the association of a given scale distance with supra-luminal rates of expansion could be only a temporary phase in the evolution of a cosmology that contains radiation, matter, and dark energy.

One conjecture is that the de Sitter scale radius, having no physical system to support it, need not remain a "horizon". More precisely, this *apparent* horizon could only be a temporary loss of luminal contact as objects cross into a region which recede at supra-luminal rates. After that *trapping region* evaporates, objects which achieve (necessarily sub-luminal) escape velocity from the locally bound system would presumably continue to spread out in space. They would remain in (eventual) luminal contact. This conjecture raises eschatalogically important considerations and issues which we briefly explore. In this case, there would be no information horizon limiting the potential complexity of memory. This means that the information content of the local system need no longer be bounded, and there would be no "big crunch" due to a Poincaré recurrence. However, hibernation remains a very bad idea until *after* causal contact begins to be re-established with the remaining accessible parts of the universe.

In any case, our position on the fringes of the bound galactic super cluster is advantageous for the transitional stages of eschatology. Regions in the bound cluster nearest the outer orbits are well positioned with regards to communications, access to external information, and energy required for transportation. It looks as though we are destined to be on the dynamic frontier.

5. Conclusions and Discussion

We have developed well defined methods that allow the correspondence of the transition amplitudes in quantum processes with those that can be obtained using suitably renormalized perturbative techniques. One of the problems posed to us was how might combinatorial results be incorporated into physically testable models or formulations. We have demonstrated successful calculations of verified experimental results (such as Compton scattering) using calculational methods consistent with those utilized by ANPA approaches to natural philosophy. Our approach has been to clearly separate the analytic forms needed to describe the dynamics (which vary amongst the hierarchy of fundamental interactions) from the analytic behavior needed to embed that dynamics into the space-time kinematics associated with multiple frames of reference. The space-time kinematics need only be consistent with quantum energies and momenta that satisfy the principles of special relativity on the usual coarse scales associated with experiment.

We have also demonstrated that the size of cosmological fluctuations that ultimately result in galaxy, star, and planet formation, can be related to a cosmological scale when the thermal energy densities of normal matter de-coheres from the present dark (vacuum) energy. This de-coherence occurs when $\dot{R} = c$, which is only consistent with a spatially flat cosmology. When the cosmology has global coherence, the gravitational vacuum state is expected to evolve with the contents of the universe. When global coherence is lost, there remains only local coherence within causally independent clusters, and only the prior vacuum state maintains global scale coherence. The resultant dark energy scale is frozen out as a cosmological constant of positive energy density satisfying $\rho_\Lambda = \frac{\Lambda}{8\pi G_N} = \frac{\epsilon k_c^3}{(2\pi)^3}$ in terms of the present day cosmological constant.

Finally, we noted that if the cosmological constant is indeed constant, the prospects for the manipulation and retention of late-time information are dire. To determine the ultimate fate of the universe, one needs a fundamental understanding of the nature of the quantum vacuum. The Wheeler-Feynman interpretation of the propagation of quanta irreducibly binds those quanta to their sources and sinks. According to Lifshitz and others [28], the zero temperature electromagnetic field in the Casimir effect can be derived in terms of the zero-point motions of the sources and sinks upon which the forces act. In the absence of a causal connection between those sources and sinks, one has a difficult time giving physical meaning to a vacuum energy or Casimir effect. The same should be true of gravitational radiations. In summary, from our explorations we can draw one definitive conclusion: there remains much left to be discovered.

References

[1] J. Lindesay, A. Markevich, H.Pierre Noyes, G. Pastrana, A self consistent, Poincaré invariant and unitary three particle scattering theory SLAC-PUB-3661, Phys.Rev.D33:2339 (1986)

[2] A. Markevich, Angular momentum and spin within a self consistent, Poincare invariant and unitary three particle scattering theory SLAC-PUB-3715, Phys.Rev.D33:2350(1986). Also, see Relativistic three particle scattering theory, A. Markevich (Stanford U., Ph.D.Thesis), UMI 85-22194-mc (microfiche), Jun 1985. 62pp.

[3] M.Alfred, P.Kwizera, J.Lindesay, H.P.Noyes, A non-perturbative, finite particle number approach to relativistic scattering theory http://www.slac.stanford.edu/cgi-wrap/pubpage?slac-pub-8821, Foundations of Physics, 34 (2004) 581. See also, A test of the calculability of a three body relativistic, cluster decomposable, unitary, covariant scattering theory M. Alfred (Howard U., Ph.D. Thesis) May 2000.

[4] J.V.Lindesay and H.P.Noyes, "Non-Perturbative, Unitary Quantum-Particle Scattering Amplitudes from Three-Particle Equations", hep-th/0203262.

[5] J.V.Lindesay and H.P.Noyes, "Construction of Non-Perturbative, Unitary Particle-Antiparticle Amplitudes for Finite Particle Number Scattering Formalisms", nucl-th/0203042.

[6] A.W.Overhauser and R.Colella, Phys.Rev.Lett. **33**, 1237 (1974).

[7] R.Colella, A.W.Overhauser and S.A.Werner, Phys.Rev.Lett. **34**, 1472 (1975).

[8] F.J.Dyson, "Time without end: Physics and biology in an open universe", *Rev.Mod.Phys* **51**, 447-460 (1979).

[9] L.D.Faddeev, Zh.Eksp.Teor.Fiz.39,1459 (1960), Sov.Phys.-JETP 12, 1014 (1961). See also L.D.Faddeev, Mathematical Aspects of the Three-Body Problem in Quantum Scattering Theory (Davey, New York, 1965). For the extension to the four-body problem, see e.g. O.A.Yakubovsky, Yad.Fiz. 5, (1967), Sov.J.Nucl.Phys. 5,937 (1967).

[10] J. Lindesay, *Foundations of Quantum Gravity*, Cambridge University Press (2013).

[11] V. Efimov, Phys. Letters 33B, 563 (1970). V. Efimov, Nuclear Phys. A210, 157 (1973). V. Efimov, Sov. J. Nucl. Phys. 17, 589 (1971).

[12] Hanns-Chistoph Nageri of Innsbruck University, Nature 440, 315 (2006)

[13] E.D.Jones, private communication to HPN c. 1997. The result of Jones' calculation which gave $\Omega_\Lambda = 0.6 \pm 0.1$ is quoted in *Aspects II (Proc. ANPA 20)*, K.G.Bowden, ed. (1999), p. 207, and cited again in [19], p. 561.

[14] Private presentation by E.D. Jones at SLAC to L.H.Kauffman, W.R.Lamb and HPN on 29 April 2002.

[15] C.L. Bennett, et.al., Astrophys.J.Supp. 148, 1 (2003)

[16] A.G.Riess et al, Astron.J. 116, 1009 (1998); P. Garnavich et al, Astrophys.J. 509, 74 (1998); S. Perlmutter et al, Astrophys.J. 517, 565 (1999)

[17] Particle Data Group, **Astrophysics and cosmology**, as posted (2004).

[18] H.P.Noyes, "Non-Locality in Particle Physics" prepared for an unpublished volume entitled *Revisionary Philosophy and Science*, R.Sheldrake and D.Emmet, eds; available as SLAC-PUB-1405 (Revised Nov. 1975); reprinted as Ch. 1 in [19].

[19] H.P.Noyes, *Bit-String Physics*, World Scientific, Singapore 2001, Vol. 27 in *Series on Knots and Everything*.

[20] F.J.Dyson, *Phys.Rev.*, **85**, 631 (1952), and a seminar at about that date.

[21] S.S.Schweber, *QED and the Men Who Made It*, Princeton, 1994, Sec. 9.17, pp. 564-65.

[22] J.V.Lindesay, H.P.Noyes and E.D.Jones, "CMB Fluctuation Amplitude from Dark Energy Partitions", Physics Letters **B** 633 (2006) 433-435, astro-ph/0412477v2 11

Feb 2005.

[23] F.Hoyle, *The Black Cloud*, Harper, New York (1957).

[24] K.Capek, *R.U.R.*, translated by Paul Selver, Doubleday, New York (1923).

[25] L.Suskind and J.Lindesay, *AN INTRODUCTION TO BLACK HOLES, INFOR-MATION and the STRING THEORY REVOLUTION: The Holographic Universe*, World Scientific, Singapore, 2005, p. 121, Eq. 11.5.26.

[26] L.Dyson, J.Lindesay, and L.Suskind, "Is There Really a de Sitter/CFT Duality", JHEP 0208 (2002) 045; hep-th/0202163.

[27] J.V.Lindesay and H.P.Noyes, Scientific Cosmology, *Proc. ANPA 26*, 2004; arXiv:astro-ph/0407535v1 26 July 2004.

[28] E.M. Lifshitz, Soviet Phys. JETP 2, 73 (1956). I.D. Dzyaloshinskii, E.M. Lifshitz, and I.P. Pitaevskii, Soviet Phys. Usp. 4, 153 (1961). I.D. Landau and E.M. Lifshitz, *Electrodynamics of Continuous Media*, pp368-376 (Pergamon, Oxford, 1960)

Brief Biography

James Lindesay, a native of Kansas City, Kansas, received his SB in physics from MIT, having done research in scattering theory with Francis Low, designed and built drift chambers with Ulritch Becker and Samuel C.C. Ting, and written a (published) thesis on macroscopic quantum fluids working with Harry Morrison. He received his MS from Stanford University while studying the phenomenology of photo-production of hadrons with Stan Brodsky. He received his PhD developing the theory for few particle relativistic dynamics working with H. Pierre Noyes at the Stanford Linear Accelerator Center (SLAC). During his tenure as a graduate student, he received Stanford University's highest teaching honor (Gores Award), as well as being given the honorary faculty position Acting Instructor by the faculty of the Stanford Physics Department.

He received a Chancellor's Distinguished Postdoctoral Fellowship from the University of California, Berkeley, where he worked on the applications of abelian and non-abelian local gauge theories to problems in quantum fluids. In addition, he received a National Research Council / Ford Foundation Fellowship, where he worked at the Stanford Linear Accelerator Center (SLAC) to develop the first relativistically covariant cluster decomposable unitary few particle scattering theory. During 1985-87, he worked as a Peace Corps Volunteer as a Lecturer in Physics at the University of Dar Es Salaam, Tanzania. This service resulted in his nomination by the agency in Tanzania for International Volunteer of the Year (1986).

He has held a faculty appointment at Howard University since 1988, where he founded the Computational Physics Laboratory. To this point, during his academic career he has supervised or co-supervised 5 post-doctoral associates, 24 graduate students, and 41 undergraduate students, resulting in 7 PhDs, 6 MS, and 8 BShonors and BS theses. He has been a visiting professor at Hampton University, Stanford University, and a visiting faculty scientist at MIT.

He has received significant funding at Howard University as a principal investigator in the Condensed Matter Research Laboratory, the Computational Science and

Engineering Research Center, the Materials Science Research Center of Excellence, the Computational Mathematical Sciences Collaborative Laboratories, the AT&T Multimedia Laboratory, and additionally as the Director for Student Development of the Center for the Study of Terrestrial and Extra-Terrestrial Atmospheres. Other awards and honors include a Commitment to Higher Learning Award from the University of the District of Columbia, induction as an honorary member into the Golden Key International Honor Society, and induction as an honorary member into Phi Beta Kappa.

He has been on the Executive Council of the Alternative Natural Philosophy Association, is currently the Secretary of the International Association of Relativistic Dynamics, is a Board member of the Dr. Beth A. Brown Science Foundation, and has been on the Scientific Advisory Boards of Vicus Biosciences, Inc., and NutriGene, Inc. He has more than 90 journal and technical publications, has recently authored a book commissioned by Cambridge University Press titled "Foundations of Quantum Gravity" (2013), has co-authored 2 books (including the critically acclaimed "An Introduction to Black Holes, Information, and the String Theory Revolution: The Holographic Universe" co-authored with Lenny Susskind, which was in the top 5 of the Scientific American Main Selection Book Club during the World Year of Physics (2005), and has been a World Scientific Press best seller for more than 8 years) as well as 4 chapters in books, has 3 patent disclosures, and holds a patent for "Quantum Optical Methods of and Apparatuses for Writing Bragg Reflection Filters". His present research interests include theoretical physics, cosmology, biophysics, and foundations of physics.

Ordering Operators:
Towards a Discrete, Strictly Finite and Quantized Interpretation of the Tensor Calculus

David McGoveran

POB 2097 Boulder Creek, CA 95006
Principal, Alternative Technologies, Deerfield Beach, Florida E-mail:
mcgoveran@AlternativeTech.com

The Ordering Operator Calculus (1982) provided a discrete, strictly finite foundation for differential geometry and the calculus. This abstract mathematical formulation was subsequently incorporated in Foundations of a Discrete Physics (1988), which applied it to a reformulation of relativistic quantum mechanics, vis-à-vis the combinatorial hierarchy and Prof. Noyes' Bit String Physics. The present paper provides a review of ordering operators, discusses principles for applying the ordering operator calculus to physics, and provides a high-level introduction to an OOC notational calculus – the tensor calculus – so that tensor equations can be interpreted as a special case of ordering operator equations.

Preface

I first met Prof. Noyes ("Pierre") in the spring of 1978 at a weekly seminar I co-sponsored with Dr. Hewitt D. Crane at Stanford Research Institute (now SRI International) and initiated by my friend and co-author Eddie Oshins. Although I was familiar with the work of Dr. E. W. "Ted" Bastin, it was there Pierre brought us up-to-date on the Combinatorial Hierarchy and I first introduced him to the beginnings of the ordering operator calculus. Subsequently, ANPA was founded at Pierre's instigation, and he invited me to attend the second annual meeting at King's College, Cambridge University, where I had the honor of meeting Dr. Ted Bastin, Prof. Clive Kilmister, Dr. John Amson, Dr. Fredrick Parker-Rhodes, and others. I began working with Pierre at SLAC under his sponsorship in the early 1980s. Over the years, Pierre has been friend, mentor, teacher, and inspiration. In all likelihood, had it not been for Pierre, I would have abandoned my work in physics altogether and might well have abandoned the ordering operator calculus.

Thank you Pierre, and Happy 90th Birthday!

I. Introduction

The ordering operator calculus (OOC) is a purely discrete and finite mathematics, and provides a covering theory for certain theories of continuum mathematics. Proper subsets of OOC are isomorphic to those theories up to their requirements for infinities, infinitesimals, and unbounded procedures, the OOC having constructs that remove those requirements. Previous publications on the subject established a general foundation for the ordering operator calculus. Ordering operators were defined and the formalisn shown to be powerful enough to provide support for the apparatus of topology and differential geometry (The Ordering Operator Calculus, 1982–1985). This was done by recasting and redefining the foundations of finite differences, incorporating insights from recursion theory and combinatorics. Subsequent publications such as Foundations for a Discrete Physics ("FDP"), (SLAC-Pub 4526, June 1989) added conceptual discussion of ordering operators and explored applications, especially to physics and, in particular, special relativity, relativistic quantum mechanics and quantum field theory.

With the exception of the original OOC paper which introduced the abstract (*i.e.*, unapplied) mathematical concepts, previous papers have addressed results derivable when particular classes of ordering operators are used. In most of those publications pertaining to physics, the discussion was restricted to a special class of ordering operators that would reproduce Prof. Noyes' Bit String Physics in which bit strings partitionable into label and address portions are generated. In FDP, the conception of the input or output of ordering operators as being labels with unspecified structural complexity was a convenience not intrinsic to ordering operators in the general case. Many results obtained in FDP were already obtained in the more general case.

In prior publications, a unified apparatus for symbolic computation has not been given. That omission is corrected with this paper. As a by-product, OOC obtains application to general relativity and that application is born quantized. Mathematical details will be addressed in future papers as time permits.

II. Conceptual Beginnings

Many systems encountered in both theory and practice are a priori discrete, finite, and intrinsically process oriented. By discrete, I mean that there is no intrinsic reason to import or assume the properties of the continuum[1]. By finite, I mean that there is no reason for assuming any infinities (completed infinities) or infinite extensibility (*e.g.*, infinite recursion). With Leibnitz, we consider these to be fictions leading to computational shortcuts. Furthermore, certain properties may depend on the cardinality of the system and its subsystems. By intrinsically process oriented I mean that there is a inherent notion of the system evolving recursively in that certain

---

[1]One might argue, however, that the OOC approach is consistent with an Archimedean continuum based on ratios (as contrasted with the Wierstrass version based on limits).

properties depend on the generation or on comparison of generations. Examples are physical systems that display characteristic numbers of the Fibonacci series, Combinatorial Hierarchy, and the like.

In representing such systems, it is important to be very reluctant to introduce any continuum properties in the mathematical description, and to do so only deliberately and with full cognizance of their effects. Preferably, these effects can be contained or bounded so that they do not pervade the representation and erode (*e.g.*, contradict) the finite, discrete, and process properties. The received foundations of mathematics and logic introduce continuum properties in many ways, both within the object language and the metalanguage[2].

III. Physical Motivations

A. Problems with the Computational Apparatus

There exists a seemingly irresolvable tension between mathematics and science that is never more apparent than it is in the foundations of physics, where it becomes a violent collision. Many physicists choose to ignore these problems, treating them as inconsequential artifacts. A science should be questioned when its mathematics must be circumvented by procedures that have no physical motivation except to avoid absurdities. Problems such as:

1. the inherent incompatibility of quantization and geometrodynamics while both quantum theory and geometrodynamics give correct empirical results,

and

2. the appearance of infinities that require ad-hoc renormalization, especially when that renormalization leads to astoundingly accurate predictions as in quantum field theory,

should not be dismissed. These incompatibilities do not exist in the "real world" – whatever that might be, it is necessarily a self-consistent entity.

Although renormalization procedures are motivated by vague physical requirements (*e.g.*, bounding momentum or dimensionality), the choice of renormalization procedure is really little more than trial and error (*i.e.*, picking one that yields correct results) with heuristic motivation. How baldly embarrassing all this would be if it were it not for the astounding accuracy that obtains. Of course, all this occurs simply because some of our mathematical entities (*e.g.*, operators and variables) are ill-defined. When the complexity of a physical process like particle interaction is

---

[2]Indeed, Abraham Robinson's non-standard analysis takes this infection to the extreme and makes it inherent in the mathematics. I disagree with Robinson's historical perspectives, including the notion that infinitesimals are intuitive – that perspective came into being only in modern times after decades of indoctrination and, I dare say, is not held by the mathematically unsophisticated.

recursively extensible without bound (because you don't know when the recursion should halt based on any physical understanding), then no finite quantities characterizing the process have any meaning. As is well-known, this fact raises its ugly head precisely because we can't normalize probabilities in the first place. To put it bluntly, we can wave the magic wand of renormalization in QED or QCD to get something that works, but we really don't know what we are talking about.

I'm not being critical of my physicist and mathematician friends. These are very serious, very difficult problems. Most practicing physicists just want to get on with the business at hand and assume that all this will eventually sort itself out – or that it is anomalous in some way.

I submit that these problems arise in the first place because we have the wrong mathematical model on several counts: the continuum. To be clear, my position is that continuum mathematics as currently practiced by physicists is an inadequate tool for modeling the foundations of physics. This position has been held by numerous physicists and mathematicians including Weyl, Gödel, and Wheeler. According to Wheeler, Weyl believed the continuum of the natural numbers an idealization and that the lesson of Gödels incompleteness theorems was that we commit a folly when we construct or believe in completed infinities (*i.e.*, infinity as number).

B. Conflicts Between Ontology and Epistemology

Bohr seems to have taken an epistemological view of physics, choosing to understand the task of physics as being to describe the information we can have about reality, and going so far as to deny any reality. By contrast, Einstein's ontological view of physics poses the task of phyisics as being to describe reality. Neither can be completely correct nor are they strictly contradictory.

On the one hand, we are surely trapped in informational theories and can never directly perceive some assumed objective reality. All we can demand is epistemological consistency. In this sense, Bohr was right. However, he went too far in claiming that the limits of a particular epistemological theory (*e.g.*, Copenhagen quantum mechanics) were necessarily inviolate or that that epistemological theory could be uniquely correct. In particular, an assertion that reality is inherently probabilistic is simply unprovable – about as helpful as asserting that an omnipotent god is responsible for everything.

On the other hand, Einstein's desire to answer the ontological question is clearly more in line with our intuitive understanding of the task of physics. However, simultaneously postulating a continuum, infinite reality and a complete, deterministic descriptive theory is contradictory, while denying epistemological limitations, is contradictory. In this sense, Einstein was wrong. At best all physics can do is identify that class of theories powerful enough to describe our (presumed objective) experience of reality and eliminate those theories that contradict that experience. That is, the task of physics (and, in general, of any experimental science) is to tell us what reality is not.

I see the primary task of physics as being, first and foremost, to provide a unified approach to our understanding of information[3]. Then and only then can we consider information as an explanation of the causal structures we call reality. What is needed is an abstract mathematical apparatus for constructing abstractions that (a) is rich enough to model physical properties and their measurement and (b) can do so without a priori or imported abstractions or limitations.

IV. Ordering Operators

A. Graph Representation

For pedagogical purposes, an ordering operator is a recursive generator of a particular directed acyclic graph (DAG) defined on an ensemble of nodes of cardinality N. The graph is not embedded in any space. We call this particular DAG the ordering operator's canonical DAG (CDAG). There are two elements to this notion: (1) the canonical directed graph and (2) its generation. It is convenient to think of a specific ordering operator as a dedicated purpose computer which contains its canonical CDAG in internal memory and instructions for recursively providing a walk of that CDAG.

As is well known, a directed graph G containing n nodes (a.k.a. vertices) can be represented by an $n \times n$ adjacency matrix A which shows which node of the graphs are connected to which other nodes. Generally, the a_{ij} entry is the number of connections (a.k.a. arcs or edges) from the ith node to the jth node. Every arc has an initial node n_i and a final node n_j. A binary adjacency matrix A or connection matrix restricts the number of arcs from the ith node to the jth node to Boolean values of either 0 or 1. Thus, a connection matrix satisfies the first part of an ordering operator by representing the reachability relation for the graph. If the entries for the ith column of A are all zero, then we say the ith node is an *initial node* of G. If the entries for the jth row of A are all zero, then we say that the jth node is a *terminal node* of G.

Define an adjacency sub-matrix A_k for the binary adjacency matrix A as an $n \times n$ matrix in which some of the arcs represented in A are disallowed; that is, the corresponding entry is 0 in A_k where in A it was a 1 (or more).

The generation of a directed graph has a dual representation, either as a totally ordered set of states of the graph or as a totally ordered set of transition matrices.

An ordering operator may be conveniently thought of as a particular walk of a pre-existing DAG, although this does not capture the ontological content of the OOC and can be misleading[4]. An adjacency matrix contains insufficient information

---

[3]I do not mean information theory.

[4]Care must be taken not to ignore the generative aspects of the computation, which are essential to understanding the process aspects of an OOC application.

to capture more than one walk of the CDAG. Walking a graph consists in visiting each node of the graph in such a way that every arc is traversed at least once, subject to the requirement that the initial node of the arc is either an initial node of G or is a final node in the previous step. Every walk is either a walk on a directed graph or may be understood as inducing a direction on each arc of an undirected graph.

A *state* can be defined as a matrix showing which nodes have been most recently visited, beginning with those nodes of the graph that have only outbound arcs (*i.e.*, have only 0s in corresponding column of the adjacency matrix) and ending on those nodes which of the graph that have only inbound arcs (have only 0s in the corresponding row of the adjacency matrix). For a walk of graph G, define a state vector at step S_m of the walk as the n-element vector for which the ith element is zero unless the most recent step has followed at least one arc that terminates at the corresponding node. By convention, we will express state vectors as column vectors.

The initial state of G is represented by the state vector S_i in which the jth element is 1 if the jth node is an initial node of G and is 0 otherwise. The final state of G is represented by the state vector S_f in which the jth element is 1 if the jth node is a terminal node of G and is 0 otherwise. In general, the nth state of the ordering operator is represented by an n-vector in which every node that has been visited has a value of 1 and every node that has not been visited has value 0.

A *transition matrix* is a projection of the adjacency matrix that delineates a single step in generating the adjacency matrix. Combining the set of transition matrices in order yields the adjacency matrix. In order to capture the evolution of the DAG, the ordering operator must be represented by an ordered set of states or an ordered set of transition matrices, analogous to a tensor. Indeed, when we consider generalized ordering operators in which the DAG is literally modifed by the interaction of multiple ordering operators, we see that this analogy is precise and allows us to reinterpret the abstract symbolic tensor calculus in discrete, finite, process terms.

In general, note that each directed acyclic graph may be associated with a large number of ordering operators, just as there are many DAGs for n labeled nodes. In particular, for n labeled nodes the number of DAGs is given by the recurrence relation:

$$a_n = (k-1)\, 2^{k(n-k)}\, a_{n-k}\,.$$

The same numbers count the (0,1) matrices in which all eigenvalues are positive real numbers[5]. The proof is bijective: a matrix A is an adjacency matrix of a DAG if and only if the eigenvalues of the (0,1) matrix $A + I$ are positive, where I denotes the identity matrix.

Representing an ordering operator as a sequence of transition matrices enables us to apply a specific ordering operator to a class of DAGs in addition to the

---

[5]See McKay, B.D., *et al* (2004).

CDAG. For example, multiple subnets may be isomorphic to the ordering operator's canonical DAG or a subnet of it. Alternatively, a DAG may be isomorphic to the canonical DAG up to some transformation T. That is, if the adjacency matrix that results when all subnets having a particular adjacency matrix are replaced by a new subnet is isomorphic to the adjacency matrix of the canonical DAG, then the DAG may be said to be isomorphic to the canonical DAG up to the transformation T. A transformation of particular interest is one that replaces the given subnet with a single node having the inbound and outbound arcs of the subnet. Such a transformation will be called a *reduction* and the given subnet the *reduction subgraph*[6].

A specific walk of G can be understood as an ordered sequence of pairs of state vectors and adjacency sub-matrices:

$$S_c^{n+1} = \sum S_c^n \times A_{r,c}^n.$$

B.   Properties and Objects

Following Leibnitz (identity of indiscernables), we treat properties as fundamental and objects are being a confluence of properties rather than having *a priori* existence. OOC represents a discrete, finite, process system as an ordering operator O with a specific CDAG and each occurrence of a specific property P in that system as an occurrence of a subgraph in the CDAG. Thus, every ordering operator generates a set of one or more properties and, for sufficiently complex CDAG, there may be many occurrences of a property.

The occurrences of each property P in a CDAG may be partially ordered and in a variety of ways, only one of which corresponds to the partial ordering of generation of P by the ordering operator O. Every other partial ordering corresponds to a walk of the graph, but may violate the acyclicity provided by O, effectively inducing cycles.

The co-occurrence of collection of properties having the same partial ordering for some subset of the occurrences of each property in a graph define multiple occurrences of an object defined by those properties. In set theoretic terms, we would say that the collection of properties are the defining or required properties of the set (a.k.a. the meaning criteria). A predicate corresponding to the requirement of these properties is then the membership function of the set. An instance of an object may be inferred ("exists" in an ontological sense) only by virtue of the co-occurrence of the defining properties for that object.

By suitable graph reduction, the occurrences of an object may be seen to be ordered in various ways. If they are totally ordered, we say the ordering is "time-like" and if partially ordered, then "space-like". When we say that an object is

---

[6]The problem of finding all similar subgraphs among two given graphs is computationally NP-complete, but we will not have need of that computation here. The problem is approached differently: we generate a hierarchy of graphs, each of which is a reduction of the previous, wherein the reduction subgraph is specified in advance.

moving through spacetime, we mean that its invariant, defining properties can be found in multiple subgraphs and that each instance is associated with properties that satisfy some definition of spacetime (such as a metric) across some ordering of those instances.

C. Metrics and Probability

Measures and metrics are inherent in the combinatorial constructions of OOC. In the generation of a CDAG, the occurrences of a particular subgraph may be counted. Since the graph is finite, the total number of occurrences is known in advance. Each ordered occurrence provides a ratio which can be interpreted as a relative frequency or, equivalently, as a distance measure on the graph. Thus, OOC unifies probability and metrics, and provides a multi-connected topology with many metrics, each specific to the property (subgraph) or properties being measured. This unification has many useful applications in quantum theory (*e.g.*, understanding EPR), but requires finite constructions.

D. Composition via Tensor Product

For graphs of sufficient complexity, there are many possible decompositions. The resulting decomposition graphs need not be acyclic. When the decomposition graphs are independent, we call them *projections*. Consider a CDAG corresponding to an ordering operator O defined as the tensor product of n independent graphs $P_1 \otimes P_2 \otimes \cdots \otimes P_n$, each with corresponding ordering operator. The tensor product then corresponds to a tensor operator in n-dimensional space and, simultaneously, to an ordering operator.

V. Correspondences to Physics

A. General Considerations

The principle of property confluence (or "co-occurrence") outlined above is similar to how we would recognize macroscopic objects. It is even how we recognize particles between two particle events in particle physics: if the track matches a feasible trajectory and the events obey the appropriate conservation laws, we assume particle identification and continuity. As an ontological principle, objects have existence only in terms of a confluence of defining properties. In physics, those properties may be quantum numbers and other invariants as represented by conservation laws. This is a kind of Wheeler-Feynman rule combined with the Eddington idea that particles are conceptual carriers between events of properties (quantum numbers) that satisfy invariants (conservation laws). A key insight is that causal structure – and therefore our notions of time and space – are defined in this way *via* Lorentz invariance.

The generation order of the directed acyclic graph generated by an ordering operator is not to be identified with time. Neither is the distance (by whatever metric) between two subgraphs A and B to be identified with spatial separation *per se*.

Instead, we find instances of the confluence of defining properties as subgraphs and establish an ordering on these subgraphs that satisfies spatio-temporal properties such as Lorentz invariant metric. The requirements for such metrics have been shown to be intrinsic to OOC[7]. In a sufficiently complex graph there will be many spatio-termporal "paths" between object instances.

B. Feynman's Discrete Path Integrals

If the spatio-temporal paths are given a representation in action-time, the graph ceases to be acyclic so that some paths are generated "backwards in time". Furthermore, it is clear that there will be a path that corresponds to a minimum of the summed action along the path – that is, a classical trajectory. We can characterize alternative paths as being separated by a "phase" that characterizes the difference in action. This representation is coordinate free and contravariant – hence the corresponding ordering operator representation is an ordered set of adjacency matrices.

Feynman and Hibbs showed how to construct the one-dimensional discrete sum over all paths for the two slit. They treated this construction as being in one spatial dimension with steps occurring in time, and no one has been able to extend it to three spatial dimensions so that the correct results are obtained. If instead of taking the Feynmann and Hibbs construction as being one-dimensional and needing to be extended, we can take the construction to be on a spatial dimension along the classical trajectory (*i.e.*, in the preferred coordinate frame of the "particle"). A different strategy now becomes apparent – we want to decompose Feynman's construction into three independent generators, *i.e.*, the projections of the paths into three coordinate bases. To put it another way, the ordering operator generates the phase space representation in "action-generation" space. The coordinate spacetime represenation must be derived from this *via* graph decomposition and reduction.

By definition, such decomposition must be a possible physical representation. Feynmans paths in phase space become real: it is the imposition of classical spacetime that is artifactual. In essence, Feynman solved the three dimensional discrete sum over all paths problem without realizing it. Furthermore, when viewed through the classical spacetime lens, the generative order on the graph appears to have an intrinsic Zitterbewegung. As explained in FDP, this Zitterbewegung satisfies the requirements for a metric with Lorentz invariance. The construction is discrete, has the required quantum mechanical properties, and is born relativistic. Even more important, it is born without infinities.

---

[7]See, for example, FDP.

This notion can be illustrated by considering a hypothetical ordering operator that generates a DAG in which nodes represent an interaction event among elementary particles. To be a pair of events associated with a classical particle trajectory, the pair of events must conserve certain physical properties such as energy and momentum and must satisfy certain relations such as the Lorentz transformation. These conditions allow us to "reverse engineer" spacetime on the DAG. However, it should be clear that neither space nor time are then directly generated by the ordering operator. Instead, it is imposed in hindsight as a way of organizing the information carried by the DAG. In some ways, it is closer to a discrete version of Feynman's sum over all paths and, in fact, any recursive generator of Feynman's discrete sum over all paths in 1+1 dimensions (see Feynman's and Hibbs' derivation of the solution to the 1 | 1 Dirac equation) can be seen as a special case of an ordering operator. The ordering operator calculus allows us to derive the propagator (see FDP), and to generalize the discrete sum over all paths into four dimensional spacetime as outlined here.

Note that the ordering operator representation provides a deterministic history, but a probabalistic future in which there are a finite (though possibly very large) number of multiple possible trajectories. As each observable event is generated, these multiple potential trajectories are reduced to a single trajectory which satisfies the relevant conservation laws (constraints) in such a way that is consistent with the classical trajectory constructed so far. In physics, we infer that a particle "carries" the conserved quantities between events. An inexact and incomplete way of characterizing this representation in causal terms might be to consider observation to "collapse" the state of this inferred particle.

A similar process occurs in the generation and recognition of linguistic events. Within a given corpus, the recipient (hearer or reader) of linguistic signals can from time to time predict with certainty the next signal, whether that signal is a phoneme, a word, a phrase (noun, verb, adverbial, adjectival), or a sentence. In between these fully determined events, there are multiple possibilities. Note that a hierarchy of overlapping "state waves" are being co-generated.

C. Interpretation of Tensor Calculus

A tensor may have either covariant or contravariant components. These correspond to the ordering operator representations as a sequence of state vectors *vs.* a sequence of adjacency matrices. As with tensors, a composition may be mixed. As representations of physical systems, the tensor calculus deals with basis transformations (including coordinate bases). These are special cases of ordering operators (as prescribed in FDP), most often derivable as projections and reductions of an ordering operator having a graph in an abstract action-generation space.

If we understand the notation of the tensor calculus in terms of state vectors and adjacency matrices having an n-dimensional basis and a metric based on occurrences of properties (subgraphs), it is obvious that the tensor calculus can represent the much more general ordering operators with similar rules of interpretation. Although the underlying state vectors and adjacency matrices will be binary, property metrics introduce non-binary measures and both state vectors and adjacency matrices in the resulting representation become the more familiar tensors. Characteristic numbers of the underlying combinatorial structures become connection coefficients. Inasmuch as it has been previously shown (FDP) that the OOC is a covering theory for differential geometry, where differential geometry is understood as approximating the high cardinality, finite combinatorics of OOC, we can reasonably anticipate a precise correspondence between ordering operators and tensors.

The tensor calculus may be understood as a sub-theory of OOC, with OOC being capable of capturing relationships outside conventional basis representations. While the tensor calculus encourages representations in a topology which is singly-connected, OOC provides representations in a multi-connected topology. As such, relationships outside the Lorentzian causal structure become not only possible, but natural.

VI. Conclusion

According to James Gleick, John Archibald Wheeler left behind "an agenda for quantum information science". We repeat this agenda here, annotated with comments relating the steps of the agenda to progress in OOC:

1. Go beyond Wootters and determine what, if anything, has to be added to distinguishability and complementarity to obtain all of standard quantum theory.

Comment: OOC provides a combinatorial theory of distinguishability and explains complementarity as a natural property of such finite and discrete process systems.

2. Translate the quantum versions of string theory and of Einstein's geometro-dynamics from the language of continuum to the language of bit.

Comment: Although we eschew much of string theory as an unnecessary and obfuscating complication, previous work has addressed much of relativistic quantum theory, combinatorially deriving the fine structure constant, propagator, uncertainty, and the relativtistic Schrödinger for the hydrogen atom, and the present paper makes connection with geometrodynamics.

3. Sharpen the concept of bit. Determine whether "an elementary quantum phenomenon brought to a close by an irreversible act of amplication" has at bottom (1) the 0-or-1 sharpness of definition of bit number in a string of binary digits, or (2) the accordion property of a mathematical theorem, the length of which, that is, the number of supplementary lemmas contained in which, the analyst can stretch or shrink according to his convenience.

Comment: TBD. However, it is clear that physics is more a theory about information and its representation of knowledge, and less about some ontological reality.

While I find no reason to accept the constraints of the Copenhagen interpretation, it does seem likely that information is more context dependent than not, and so a bit more like (2) in this regard.

4. Survey one by one with an imaginative eye the powerful tools that mathematics – including mathematical logic – has won and now offers to deal with theorems on a wholesale rather than a retail level, and for each such technique work out the transcription into the world of bits. Give special attention to one and another self-referential deductive system.

Comment: Many of those mathematical tools rely upon features that are inconsistent with finitism, discreteness, and process-orientation and so, from my perspective, are incompatible with "the world of bits".

5. From the wheels-upon-wheels-upon-wheels evolution of computer programming dig out, systematize and display every feature that illuminates the level-upon-level-upon level structure of physics.

Comment: This is precisely the importance of the Combinatorial Hierarchy, and of OOC decomposition and reduction. There are other combinatorial and hierarchical relationships of importance as well, too numerous to go into here.

6. Capitalize on the findings and outlooks of information theory, algorithmic entropy, evolution of organizisms, and pattern recognition. Search out every link that each has with physics at the quantum level. Consider, for instance, the string of bits 1111111...and its representation as the sum of the two strings 1001110...and 0110001...Explore and exploit the connection between this information-theoretic statement and the finding of theory and experiment on the correlation between the polarizations of the two photons emitted in the annihilation of singlet positronium and in like Einstein-Podolsky-Rosen experiements. Seek out, moreover, every realization in the realm of physics of the information-theoretic triangle inequality recently discovered by Zurek.

Comment: These relationships have been explored in previous papers and OOC has been shown to accommodate if not explain (previously characterized as "simulate") EPR results. Prof. Noyes' Bit String Physics provides a bit string representation of the standard model of particle physics and a bit string can be understood as a projection of an ordering operator with respect to quantum numbers. I believe that the greater richness of graphs is necessary for a complete model, especially if quantum general relativity is to be incorporated. Much remains to be done.

7. Finally. Deplore? No, celebrate the absence of a clean clear definition of the term 'bit' as elementary unit in the establishment of meaning. We reject "that view of science which used to say, 'Define your terms before you proceed'. The truly creative nature of any forward step in human knowledge," we know, "is such that theory, concept, law, and method of measurementforever inseparableare born into the world in union." If and when we learn how to combine bits in fantastically large numbers to obtain what we call existence, we will know better what we mean both

by bit and by existence.

Comment: Here we disagree with Wheeler in part and insist that our terms be defined before we proceed. On the other hand, Wheeler may well have been concerned that such a stricture was too rigid to accommodate learning and refinement. Our solution (see FDP) is to engage in a methodology that permits definitions of terms and their relationships to be refined iteratively, mimicking the idealized scientific method. Interestingly, this process can be modeled using OOC and so is consistent with it. Our goal is like Wheeler's – in OOC, "theory, concept, law, and method of measurementforever inseparableare born into the world in union."

All the foregoing point to much work to be done, putting flesh on the skelaton as it were and refining the interpretation of tensor calculus notation. However, the results obtained to date and the ability of OOC to represent both intrinsically discrete theories such as quantum mechanics and intrinsically continuum theories such as geometrodynamics is encouraging.

References

Feynman, R. P., and Hibbs, A. R., *Quantum Mechanics and Path Integrals: Emended Edition*, (Daniel F. Styer - Editor), Dover Publications , July 21, 2010.

Gleick, J. 'The Information: A History, A Theory, A Flood', Vintage Books (c) 2011.

McGoveran, D., and Noyes, H. P., *Foundations for a Discrete Physics*, SLAC-Pub 4526, June 1989.

McGoveran, D., 'The Ordering Operator Calculus', ANPA Annual Proceedings, 1984.

McKay, B. D., Royle, G. F., Wanless, I. M., Oggier, F. E., Sloane, N. J. A., Wilf, H. 'Acyclic digraphs and eigenvalues of (0,1)-matrices', *Journal of Integer Sequences* 7, Article 04.3.3. (2004)

Wheeler, J.A., *Information, Physics, Quantum: the Search for Links*, in: *Proceedings 3rd International Symposium on the Foundation of Quantum Mechanics*, Tokyo, 354–368 (1989)

Zurek, W.H. (Ed.), *Complexity, Entropy and the Physics of Information*, New York, 3–28. (1990)

. . . .

Brief Biography

David McGoveran is founder and principal of Alternative Technologies, a consulting firm specializing in emerging software technologies and theoretical contributions to computer science. Educated in the foundations of physics, mathematics, and logic, he has periodically held visitor positions with the Theory Group, SLAC under the sponsorship of Prof. Noyes (since the mid-1980s) and has joyfully collaborated whenever possible with Prof. Noyes over the last thirty years.

Information, Entropy, and the Combinatorial Hierarchy : Calculations

Michael Manthey

E-mail: manthey@acm.org

Douglas Matzke

E-mail: doug@quantumdoug.com

We provide exact calculations of the information content (and its transformation) of expressions in the geometric algebras $\mathcal{G}(n,0) = \mathcal{G}_n$ over $\mathbb{Z}_3 = \{0, 1, -1\}$. Being fundamentally combinatorial, the results are theory-neutral.

The overall picture is a "Bit Bang" modelled as the algebraic expansion $\mathcal{G}_0 \to \mathcal{G}_1 \to \mathcal{G}_2 \to \mathcal{G}_3 \to \mathcal{G}_4$, which expansion is driven by entropy creation via the conversion of information from space-like (non-Shannon) to time-like (Shannon) form. This generates the Standard Model in $\mathcal{G}_3$. $\mathcal{G}_4$ further yields a bridge from quantum mechanics to general relativity in the form of time-like quaternion isomorphs called tauquernions, plus dark matter structure [5]. In the end, driving the entropic expansion of the cosmos, emerging like the Cheshire cat, we find the Combinatorial Hierarchy.

Introduction

We provide exact calculations of the information content (and its transformation) of expressions in the geometric algebras $\mathcal{G}(n,0) = \mathcal{G}_n$ over $\mathbb{Z}_3 = \{0, 1, -1\}$. Being fundamentally combinatorial, the results are theory-neutral.

The overall picture is a "Bit Bang" modelled as the algebraic expansion $\mathcal{G}_0 \to \mathcal{G}_1 \to \mathcal{G}_2 \to \mathcal{G}_3 \to \mathcal{G}_4$, which expansion is driven by entropy creation via the conversion of information from space-like (non-Shannon) to time-like (Shannon) form. This generates the Standard Model in $\mathcal{G}_3$. $\mathcal{G}_4$ further yields a bridge from quantum mechanics to general relativity in the form of time-like quaternion isomorphs called tauquernions, plus dark matter structure [5]. In the end, driving the entropic expansion of the cosmos, emerging like the Cheshire cat, we find the Combinatorial Hierarchy.

Section §1 calculates the numerical information-theoretic skeleton of our $\mathbb{Z}_3$ $\mathcal{G}_4$ algebra, which is possible because of its finiteness and relatively small size: $3^{16} \approx 43$ million expressions. Section §2 then describes the computational mechanisms that define and create the aforementioned space-like, non-Shannon information. [Our use of the term "non-Shannon information" is distinct from, but consistent with, a like-sounding entropy-related term, "non-Shannon-type inequalities".]

We distinguish between $\mathcal{G}$ and $G = \{1, a, b, c, \ldots, ab, ac, \ldots, abc, \ldots, \ldots\}$, in that G denotes *actual instantiated elements*, though these of course also belong to the *abstract* geometric algebra $\mathcal{G}$. The elements of G, being mutually orthogonal, form a space with inner product, whence Parseval's Identity applies:

Theorem *(Parseval, 1799)*. Let F be an n-dimensional single-valued function over S, an orthogonal space with inner product $\cdot$. Then $F \cdot S$ is the Fourier decomposition of F on S (and wave-particle duality in a nutshell).

Since the algebra therefore *is* the phase space, the *exact* (!) numerical skeleton that results from this analysis has cosmological implications that we pursue in §3, arriving naturally at the Combinatorial Hierarchy in §4. In general, see [5], from which the present article is extracted.

1. Information Content and Kind

The formal concept of *information* is due to Claude Shannon (1948), who defined the information content $\mathcal{I}$ of an event x as

$$\mathcal{I}(x) = -lg\ p_x$$

where p_x is the probability of occurrence of the event x, and lg is the logarithm to the base 2. Thus, as is well known, the more improbable the event, the greater its information content. The import of this definition is best understood with the example of an ***if-then-else**-type* decision. The form

$$X(\,1 + \langle -1 - a, \pm\mathsf{a}\rangle\,)\ +\ Y(\,1 + \langle -1 + a, \pm\mathsf{a}\rangle\,)$$

describes the computation *if* a *then* X *else* Y, where the brackets $\langle\,,\rangle = \pm1$ indicate the inner product of the idempotent measurement probe $-1 \pm a$ with the process-entity $\pm\mathsf{a}$ in the surround, and $+$ indicates the concurrency of the processes $X, Y \in G$. Here we see that a *static* bit of information — encoded in the $\pm$ state of a — is converted into the *motion* [state change] of *one* of the processes X or Y, since one of the two expressions *will* yield zero and the other minus one (minus because X (or Y) now changes state). Note particularly that *the information is consumed*: a has been changed by the measurement and no copies made. One correctly concludes that a binary decision *costs* one bit of information.[1]

Applying this to G_n, this means that a measurement sequence that would locate some entity $\in G_n$ having an information content of m bits would require m such nested *if*'s. Furthermore, this decision process <u>transforms</u> the static *space-like* information contained in the current state of G_n into dynamic *time-like* information

---

[1] Notice, by the way, how the 1-dimensional 180° opposition between X and Y as coded in $\pm\mathsf{a}$ becomes a conjugate (90°) opposition, $-1 - a$ vs. $-1 + a$, in the translation from the sequential to the concurrent view.

at an exchange rate of 1 : 1. It is this transformation (on a massive scale) that constitutes our expanding time-like universe.

This transformation is fundamentally entropic in character. Because the algebra is finite, we can calculate the probability of occurrence of an expression, and so we can know its information content. Knowing that, we can follow the entropy trail — loss of information - and make predictions about what further transformations will occur.

We therefore now embark on the calculation of the information content, measured in *bits*, of every element of G_n , $n = 0, 1, 2, 3, 4$. This is an *exact* calculation, since it is based on pure combinatorics and resulting integer ratios.

The binary nature of our algebra allows us to fully expand the combinatorial content of any given expression in the fashion of a "truth table". Below we show the tables for ab, abc, and $abcd$. Beneath the tables are vectors of the respective *result* (rightmost) columns; *these result vectors are the basis for our information content analysis.*

As an example, we take the result vector for abc and add to it $+\mathbf{1}$ and $-\mathbf{1}$:

$$abc = [-\ ++-+--+]\qquad abc = [-++-+--+]$$
$$+\mathbf{1} = [+++++++++]\qquad -\mathbf{1} = [--------]$$
$$[\ \cdot\ --\ \cdot\ -\ \cdot\ \ \cdot\ -]\qquad\qquad [+\ \cdot\ \ \cdot\ +\ \cdot\ ++\ \cdot\ ]$$

where $\cdot$ denotes zero = "does not occur". Note that the *pattern* of symbols is the same for abc and the two sums, the only difference being that in abc, the two symbols that appear are $+$ and $-$, whereas in the sums the two symbols are $\cdot$ and $-$, and $+$ and $\cdot$, respectively. But the *ratios* are the same: here, four of each and *no* third symbol. And if you think about it, this proportionality will always hold — all that happens with the summing of abc with $\pm\mathbf{1}$ is that one so-to-speak rotates a three-symbol mapping vector $[\cdot ,+,-]$ first to $[- ,\cdot ,+]$ and then to $[+,- ,\cdot]$: *the proportions will therefore always be the same.*

a	b	ab
−	−	+
−	+	−
+	−	−
+	+	+

a	b	c	abc
−	−	−	−
−	−	+	+
−	+	−	+
−	+	+	−
+	−	−	+
+	−	+	−
+	+	−	−
+	+	+	+

a	b	c	d	$abcd$
−	−	−	−	+
−	−	−	+	−
−	−	+	−	−
−	−	+	+	+
−	+	−	−	−
−	+	−	+	+
−	+	+	−	+
−	+	+	+	−
+	−	−	−	−
+	−	−	+	+
+	−	+	−	+
+	−	+	+	−
+	+	−	−	+
+	+	−	+	−
+	+	+	−	−
+	+	+	+	+

$$[+--+]\quad [-++-+--+]\quad [+--+-++--++-+--+]$$

Argument. A *pattern encoding* consists of a 3-tuple (#0's, #1's, #-1's), which forms a *signature* of the vector's structure. Suppose we have the pattern vector $(2, 2, 4)$ and imagine a (minimal) decision tree — think nested *if*'s — that identifies any expression having this pattern. Then the amount of information embedded implicitly in the tree's decision points is the measure of the tuple's information content. The three symbols are interchangeable because the tree's *form* (the structure of the search space) is indifferent to which symbols lie at its leaves. Since the ratios are invariant under exchange of symbols, the counts can appear in any order, so we just sort the tuples numerically.

This symbol-invariance implies that $\pm abcd$ and $\pm 1 \pm abcd$ all have the same information content. Since the latter form defines a measurement on the former, and these two therefore *should* be the same, this is comforting. For this and similar reasons, we think that any classification scheme (cf. *binning*, below) must subscribe to the collapsing of $0, 1, \tilde{1}$ into one signature.[2]

This all means that we can classify *every* expression in the algebra in terms of its result-vector's signature. We will soon see that these informational classifications exactly match the particle structures of the standard model, plus dark matter and the Higgs , cf. Table 1 below and [5], Appendix I.

Since a polynomial $\in \mathcal{G}_n$ has maximally $|G_n| = 2^n$ mutually orthogonal terms, and their coefficients can be one of $0, 1, -1$, we get the set S, of size $|S| = 3^{2^n}$, which covers all of the possible expressions in G. With S in hand, we can count how many times k each pattern X occurs, and we can then divide k by $|S|$ to get the probability p of X's occurrence:

$$p_X = \frac{k}{3^{|G_n|}}$$

If $k = 1$, then is there is but one single occurrence of X in S, so p_X would be minimal, but this actually can't happen — the best you can do is the three scalar constants, $0, 1, \tilde{1}$, where $k = 3$.

From the other end of the microscope, a minimal X requires the full measure of the information in S in order to be identified and isolated. That is, the information content $\mathcal{I}$ of an expression $X \in G$ is

$$\mathcal{I}(X) = -lg \; p_X = -lg \; \frac{k}{3^{|G_n|}} = lg \; \frac{3^{2^n}}{k} \quad bits$$

X's information content is thus a function of how many other X's share its signature, and the size of the space it occurs in.

An obvious application of this is to ask, What is the information content of some particle P, having in mind the fact that 1 *bit* = 4 *Planck areas* $/ln\,2$ ($\approx 10^{-66} \; cm^2$).

---

[2]*Void* cannot have its own category because, by definition, it has no properties by which it might be so categorized, including the property of having no properties. *Void* can first become manifest in the distinction $[1, \tilde{1}]$.

Thus, for example, a single Higgs boson $\mathcal{H} = (1 + wxyz)(xy + xz + yz) = xy + xz + yz + wx + wy + wz$ exists in 16 states out of the 64 possible in the form. Its information content is therefore

$$\mathcal{I}(\mathcal{H}) = lg\,\frac{3^{16}}{16} = 21.3594000\ bits \qquad ^3$$

The next step, the conversion of *bits* to *Gev*, turns out to be unexpectedly complicated, and is our current focus.

Of interest equal to individual particles, however, is the picture painted with the broader brush of the signatures and bin counts themselves.

Table 1 lists the information content, calculated in this broader way, of relevant elements of G_n. Because rarity/information is relative to the size of the space, the measure of (say) ab is 2.17 bits in G_2, 7.29 bits in G_3, and 18.9 bits in G_4. But at the same time, all of a, ab, abc, and $abcd$, at any given level, have the *same* measure, since their uniqueness stays proportional to n; note that namely these *also* have the highest information content after $0, 1, \tilde{1}$. In general, the lower the bit value, the larger the family of entities having that count, and oppositely, the higher the count, the smaller the family. We now explore this a little more.

The function $bits\mathsf{N}(X) = lg\,\frac{3^{2^N}}{count(X's)}$ calculates the information content of $X \in G$ relative to G_N.

Then, *re* G_0, the three scalar constants $0, 1, -1$ are all *known* and occupy the entire space, which is of size $3^{2^0} = 3$ states, one each for $\{0, 1, -1\}$,

- So each occurs with probability $p = \frac{1}{3} \mapsto lg\,3 = 1.58$ bits, *but*
- *Known* means $bits0(0) = bits0(1) = bits0(-1) = lg\,\frac{3}{3} = lg\,1 = 0$
- So G_0 actually contains no information.

In G_1 there are $3^{2^1} = 9$ states, three for G_0's scalars, $\in (0, 0, 2)$, and $2 + 4 = 6$ more for $\pm a$ and $\pm 1 \pm a$, both $\in (0, 1, 1)$:

- The scalar constants are *known*, and so they contain no information, but nevertheless occupy three slots in the state space $\Rightarrow bits1(1) = lg\,3 = 1.58$ bits.[4] [It is a mod-3 coincidence that the numbers for G_0 and G_1 are the same.]
- 1-vectors occupy the remaining states in G_1, so $bits1(\pm a) = lg\,\frac{9}{6} = 0.58 = bits1(\pm 1 \pm a)$.
- The net result is that exactly 1 (classical) bit of information is encapsulated in the structure a: $bits1(1) - bits1(\pm a) = 1.00$.

For G_2, the algebra of pure $q$bits:

---

[3]This result differs from Table 1 (below) because we have ignored the other members of its bin, (4,4,8). Also, we don't know what the experimentalists are actually measuring — perhaps we should have calculated $\pm 1 \pm H$, etc.

[4]Recall that $0, 1, -1$ all map to the same pattern, whence 3 and not 1.

- The scalar constants are *known*, but occupy state space: $bits2(1) = \lg(\frac{81}{3}) = 4.75$ bits.$^{(ditto)}$
- Here is a smallest addressable state: $(1-a)(1-b) = 1-a-b+ab \in (0,1,3)$ $\mapsto bits2((1-a)(1-b)) = lg(\frac{81}{24}) = 1.75$ bits, corresponding to a single row of the form's "truth table". The 24 count comes from the 2^4 sign variants of $1-a-b+ab$ plus the 2^3 sign variants of $a+b+ab$.
- The next section shows how simple concurrency, $a+b$, *mere concurrent existence*, contains and encodes information. Here we just calculate: $bits2(a) = 2.17 = bits2(b)$, $bits2(ab) = 2.17$, $bits2(a+b) = 1.17$ bits,
- Whence $bits2(ab) - bits2(a+b) = 1.00000000$, where we show in the 0's the number of significant digits that actually are available in these (exact) calculations; we show rounded values otherwise.

In G_4:

- Let $m = a+b+c+d$, whence $D = m + m\,abcd$ and $D^2 = 0$. As shown in Table 1, $\mathcal{D}_0 \in (4,4,8)$ and each D contains 5.53 bits of information.
- But $abc\,D = -1 + ab - ac + ad + bc + bd + cd + abcd$ computes to 6.87 bits (not shown). One does not expect a reversible operator like abc to change the information content of an entity.
- The explanation is that the rotation by abc changes the *signature bin* that the expression falls into, and the new bin, namely $(2,6,8)$, has fewer members, and so the information content is higher. "It's not the rotation's fault." [We will exploit this phenomenon in our Bit Bang story in 2.]
- In Table 1, there are two examples of binnings that further differentiate the 3-signature — $(a+b+c)d$ and $\mathcal{M}_2$ are both $\in (4,6,6)$, yet their bit-measures differ, 12.1 vs. 7.08, and again $a+bcd$ and $\mathcal{D}_0$ are both $\in (4,4,8)$, but their measures are 15.1 vs. 5.53.

These examples show that information content values like those in Table 1 are sensitive to the binning algorithm that is used. Fortunately, whatever the binning, the results will always be consistent because the underlying population is the same.

Our general-purpose binning algorithm (used in Table 1) first applies the 3-pattern signature, and then further bins together only those expressions having the same number of non-scalar terms.[5]

---

[5]This increases the number of bins from 10 to 14 for G_3, and from 30 to 86 for G_4. For example, in the text just above, $(a+b+c)d \in ((4,6,6),3)$, $\mathcal{M} \in ((4,6,6),6)$, $a+bcd \in ((4,4,8),2)$ and $\mathcal{D}_0 \in ((4,4,8),8)$. All 43 million expressions were binned.

Particle/Form		Vector ($\mathcal{G}_3$ and $\mathcal{G}_4$ samples)	$\mathcal{G}_1$	$\mathcal{G}_2$	$\mathcal{G}_3$	$\mathcal{G}_4$
$\mathcal{G}_0$						
$Void \mapsto \mathbf{0}$	*is*	$[\cdot\;\cdot\;\cdot\;\cdot\;\cdot\;\cdot\;\cdot\;\cdot\,] \in (0,0,8),0$	1.58	4.75	11.1	23.8
± 1	*are*	$[\pm\,\pm\,\pm\,\pm\,\pm\,\pm\,\pm\,\pm] \in (0,0,8),0$	1.58	4.75	11.1	23.8
$\mathcal{G}_1$						
a	*±exist*	$[-\;-\;-\;-\;+\;+\;+\;+] \in (0,4,4),1$	0.58	2.17	7.29	18.9
$1 - a$	*(measure)*	$[-\;-\;-\;-\;\cdot\;\cdot\;\cdot\;\cdot\,] \in (0,4,4),1$	0.58	2.17	7.29	18.9
$Row0\;(1-w)\dots(1-z)$		$\in (0,1,1),(0,1,3),(0,1,7),(0,1,15)$	0.58	1.75	7.09	18.8
$\mathcal{G}_2$						
ab	*±spin*	$[+\;+\;-\;-\;-\;-\;+\;+] \in (0,4,4),1$	–	2.17	7.29	18.9
$a + b$	*co-occ*	$[+\;+\;\cdot\;\cdot\;\cdot\;\cdot\;-\;-] \in (2,2,4),2$	–	1.17	4.70	15.1
$a + b + ab$	ν	$[-\;-\;-\;-\;-\;-\;\cdot\;\cdot\,] \in (0,2,6),3$	–	1.75	5.29	15.6
$a + ab$	$W, Z\;\dagger$	$[\cdot\;\cdot\;+\;+\;\cdot\;\cdot\;-\;-] \in (2,2,4),2$	–	1.17	4.70	15.1
$1 + ab$		$[-\;-\;\cdot\;\cdot\;\cdot\;\cdot\;-\;-] \in (0,4,4),1$	–	2.17	7.29	18.9
$\mathcal{G}_3$						
abc	*±charge*	$[-\;+\;+\;-\;+\;-\;-\;+] \in (0,4,4),1$	–	–	7.29	18.9
$a + bc$	*quark*	$[\cdot\;+\;+\;\cdot\;-\;\cdot\;\cdot\;-] \in (2,2,4),2$	–	–	4.70	15.1
$ab + ac$	e	$[-\;\cdot\;\cdot\;+\;+\;\cdot\;\cdot\;-] \in (2,2,4),2$	–	–	4.70	15.1
$a + b + c + ab + ac$	p	$[-\;-\;-\;-\;\cdot\;+\;+\;-] \in (1,2,5),5$	–	–	2.70	11.5
$a + b + c$	γ	$[\cdot\;-\;-\;+\;-\;+\;+\;\cdot\,] \in (2,3,3),3$	–	–	3.29	12.1
$ab + ac + bc$	*3-space*	$[\cdot\;-\;-\;-\;-\;-\;-\;\cdot\,] \in (0,2,6),3$	–	–	5.29	15.6
$\mathcal{G}_4$						
$abcd$	*+mass*	$[+\;-\;-\;+\;-\;+\;+\;-\;-\;+\;+\;-\;+\;-\;-\;+]$	$\in (0,8,8),1$	–	–	18.9
$1 - abcd$		$[\cdot\;-\;-\;\cdot\;-\;\cdot\;\cdot\;-\;-\;\cdot\;\cdot\;-\;\cdot\;-\;-\;\cdot\,]$	$\in (0,8,8),1$	–	–	18.9
$q_A q_B$	*2 qbits*	$[\cdot\;\cdot\;\cdot\;\cdot\;\cdot\;+\;-\;\cdot\;\cdot\;-\;+\;\cdot\;\cdot\;\cdot\;\cdot\;\cdot\,]$	$\in (2,2,12),4$	–	–	14.1
$a + b + c + d$		$[-\;+\;+\;\cdot\;+\;\cdot\;\cdot\;-\;+\;\cdot\;\cdot\;-\;\cdot\;-\;-\;+]$	$\in (5,5,6),4$	–	–	10.1
$(a + b + c)d$		$[\cdot\;\cdot\;+\;-\;+\;-\;-\;+\;+\;-\;-\;+\;-\;+\;\cdot\;\cdot\,]$	$\in (4,6,6),3\ddagger$	–	–	12.1
$\mathcal{M}_1\;(16/64)$ *proto-mass*		$[\cdot\;\cdot\;\cdot\;+\;\cdot\;+\;+\;\cdot\;\cdot\;+\;+\;\cdot\;+\;\cdot\;\cdot\;\cdot\,]$	$\in (0,6,10),6$	–	–	13.1
$\mathcal{M}_2\;(32/64)$ *proto-mass*		$[+\;+\;-\;\cdot\;-\;\cdot\;-\;+\;+\;-\;\cdot\;-\;\cdot\;-\;+\;+]$	$\in (4,6,6),6\ddagger$	–	–	7.08
$\mathcal{H}\;(16/64)$	*Higgs*	$[-\;\cdot\;+\;+\;\cdot\;-\;+\;-\;-\;+\;-\;\cdot\;+\;+\;\cdot\;-]$	$\in (4,6,6),6$	–	–	7.08
$Bell = ab + cd = T'$		$[-\;\cdot\;\cdot\;-\;\cdot\;+\;+\;\cdot\;\cdot\;+\;+\;\cdot\;-\;\cdot\;\cdot\;-]$	$\in (4,4,8),2$	–	–	15.1
$Magic = ab - cd = T$		$[\cdot\;-\;-\;\cdot\;+\;\cdot\;\cdot\;+\;+\;\cdot\;\cdot\;+\;\cdot\;-\;-\;\cdot\,]$	$\in (4,4,8),2$	–	–	15.1
$B_0 = -ac + bd$		$[\cdot\;+\;-\;\cdot\;+\;\cdot\;\cdot\;-\;-\;\cdot\;\cdot\;+\;\cdot\;-\;+\;\cdot\,]$	$\in (4,4,8),2$	–	–	15.1
$M_0 = ad - bc$		$[\cdot\;+\;-\;\cdot\;-\;\cdot\;\cdot\;+\;+\;\cdot\;\cdot\;-\;\cdot\;-\;+\;\cdot\,]$	$\in (4,4,8),2$	–	–	15.1
$a + bcd$	*dark*	$[+\;\cdot\;\cdot\;+\;\cdot\;+\;+\;\cdot\;\cdot\;-\;-\;\cdot\;-\;\cdot\;\cdot\;-]$	$\in (4,4,8),2\ddagger$	–	–	15.1
$\mathcal{D}_0$	*dark*	$[-\;\cdot\;-\;\cdot\;\cdot\;-\;\cdot\;+\;-\;\cdot\;+\;\cdot\;\cdot\;+\;\cdot\;+]$	$\in (4,4,8),8\ddagger$	–	–	5.53
$\mathcal{D}_q$	*dark*	$[+\;+\;-\;\cdot\;+\;+\;-\;-\;+\;+\;-\;-\;\cdot\;+\;-\;-]$	$\in (2,7,7),8$	–	–	6.87
$\mathcal{D}_u$ *(80/96)*	*dark*	$[-\;-\;\cdot\;\cdot\;\cdot\;\cdot\;-\;+\;-\;+\;\cdot\;\cdot\;\cdot\;\cdot\;+\;+]$	$\in (4,4,8),8$	–	–	5.53
$\mathcal{D}_u$ *(16/96)*	*dark*	$[+\;\cdot\;\cdot\;\cdot\;\cdot\;\cdot\;\cdot\;\cdot\;\cdot\;\cdot\;\cdot\;\cdot\;\cdot\;\cdot\;\cdot\;-]$	$\in (1,1,14),8$	–	–	15.9

Table 1: Information content (in *bits*) of principal $\mathcal{G}_n$ forms [5]   † Tentative ‡ See text.

Therefore, co-occurrences/qbits $x + y$, electrons $xy + xz$, and quarks $x + yz$, which already have the same signature, will still bin together. Thus, the numbers in Table 1 and its cousins will always be indicative rather than definitive, since how one bins is determined by which interaction-classes one is interested in.

There are other interesting things in Table 1: the information content of 3-space as described by classical quaternions is 3.3 bits smaller than that of matter ($15.6-18.9$). Photons ($a+b-c$) and their confounding ($a+b-c)*d$ have the same measure, 12.1, which is rather larger than $\mathcal{H}$iggs 7.08, which contains them. There are apparently *two* forms of proto-mass $\mathcal{M}$ (13.1 vs. 7.08), and we note that the former is a sparse +1 variant. Singletons always have the highest bit value after the scalars, even more than two classical $q$bits $q_A q_B$. But then, given their spin, they *are* bits, yo. Finally, note that the Bell/Magic states, $\mathcal{D}$ark matter, quarks, and electrons all have the same measure, 15.1, only slightly less likely than light, 12.1; and versus the rather more likely $\mathcal{H}$ and $\mathcal{M}$ at 7.08 bits.

Howsoever, as the expansion proceeds - - $\mathcal{G}_1 \rightarrow \mathcal{G}_2 \rightarrow \mathcal{G}_3 \rightarrow \mathcal{G}_4$ in Table 1 - - Ψ's information content shrinks as the information in $3 + 1$d gets denser and denser. For example, the two classical bits q_A, q_B use 4 spinors and 14.1 bits to encode 1 *ebit* — time-like stability costs! Matter itself is only slightly denser at 18.9 bits per: frozen potential (because actualized), robbed of its variability through loss of degrees of freedom. This is the fate of the space-like non-Shannon information that is converted, as the expansion of the universe, into time-like Shannon information.

We pursue this entropic expansion in a cosmological setting in 2. Before doing so, we introduce and define the concept of non-Shannon information, and show how this builds structure.

2. Non-Shannon information

There is a subtle paradox — concerning *kinds* of information - that we must deal with before going further. Shannon's concept of information, as we have seen, can be viewed as a descent into a binary tree from root to leaf, where at each branch point, one bit is consumed in the choosing of one path versus the other. Two points should be noted: (1) the (*information represented by the*) bits are(*is*) *consumed* and converted into the motion/advance of the descent-process; and (2) *action* is what this is all about... this *sequential* process is blind to context, and sees only its own (namely causal) point of view. The process concept, here exemplified, *is* sequence and action, combined. Thus Shannon's view of information is purely *time-like*.

It is difficult to see how Shannon's definition misses anything out, and yet ... it does. There *is* a kind of information that falls beyond it, namely the information of *concurrent existence*, what we call *non-Shannon information*. The following *Coin Demonstration* makes the argument.

Act I. *A man stands in front of you with both hands behind his back. He shows you one hand containing a coin, and then returns the hand and the coin behind his back. After a brief pause, he again shows you the same hand with what appears to be an identical coin. He again hides it, and then asks, "How many coins do I have?"*

Understand first that this is not a trick question, or some clever play on words — we are simply describing a particular and straightforward situation. The best answer at this point then is that the man has "at least one coin", which implicitly seeks *one bit* of information: two possible but mutually exclusive states: *state1* = "one coin", and *state2* = "more than one coin".

One is now at a decision point — *if* one coin *then X else Y* — and only one bit of information can resolve the situation. Said differently, when one is able to make this decision, one has *ipso facto* <u>received</u> one bit of information.

Act II. *The man now extends his hand and it contains two identical coins.*

Stipulating that the two coins are in every relevant respect identical to the coins we saw earlier, we now know that there are *two* coins, that is, *we have received one bit of information*, in that the ambiguity is resolved. We have now arrived at the dramatic peak of the demonstration:

Act III. *The man asks, "Where did that bit of information come from?"*

Indeed, where *did* it come from??! The bit originates in the *simultaneous presence* of the two coins — their **co-occurrence** — and encodes the now-observed *fact* that the two *processes*, whose states are the two coins, respectively, do not exclude each other. [6]

Thus, there is information in (and about) the environment that *cannot* be acquired sequentially, and true concurrency therefore cannot be simulated by a Turing machine. Penrose correctly concluded in *The Emperor's New Clothes* that Turing machines cannot simulate quantum mechanics. Both Turing and Penrose consider the case $f \parallel g$, meaning execute the non-interacting processes f and g in parallel (and harvest their results when they end). Clearly one gets the same results whether one runs f first ($f; g$) or g first ($g; f$), or simultaneously, $f \parallel g$. In this *functional* view of computation, the only difference is wall-clock time. The Coin Demonstration is not about these cases *at all*, but rather asks, Can f *exist* simultaneously with g, or do they *exclude* each other's existence? This is the fundamental distinction that we draw.

More formally, we can by definition write $a + \tilde{a} = 0$ and $b + \tilde{b} = 0$, meaning that (process state) a excludes (process state) $\tilde{a}$, and similarly (process state) b excludes (process state) $\tilde{b}$. [7] Their *concurrent* existence can be captured by adding these two equations, and associativity gives two ways to view the result. The first is

$$(a + \tilde{b}) + (\tilde{a} + b) = 0$$

which is the usual excluded middle: if it's not the one (eg. that's $+$) then it's the other. This arrangement is convenient to our usual way of thinking, and easily encodes the traditional *one/zero* (or $1/\tilde{1}$) distinction. [8] The second view is

$$(a + b) + (\tilde{a} + \tilde{b}) = 0$$

---

[6] Cf. Leibniz's indistinguishables, and their being the germ of the concept of space: simultaneous events, like the presence of the two coins, are namely indistinguishable in time.

[7] This is the logical bottom, and so there are no superpositions of $a/\tilde{a}$ and $b/\tilde{b}$: they are 1d exclusionary distinctions . Superposition first emerges at level 2 with ab via the distinction *exclude* vs. *co-occur*.

[8] Since $\tilde{x}$ is not the same as $0x$, an occurrence $\tilde{x}$ is meaningful; in terms of sensors, $\tilde{x}$ is a *sensing* of an externality x, not x itself.

which are the two superposition states: either both or neither.

The Coin Demonstration shows that *by its very existence*, a 2-co-occurrence like $a + b$ contains one bit of information. Co-occurrence relationships are *structural*, ie. *space-like*, by their very nature. Such bits, being space-like, are the source of non-Shannon information.

[Cf. Table 1, this information is twice that of a or b alone *in* G_1, but $2.17 - 1.17 = 1$ bit less than a, b or ab in G_2.]

Act IV. *The man holds both hands out in front of him. One hand is empty, but there is a coin in the other. He closes his hands and puts them behind his back. Then he holds them out again, and we see that the coin has changed hands. He asks, "Did anything happen?"*

This is a rather harder question to answer. To the above two concurrent *exclusion*ary processes we now apply the *co-exclusion inference*, whose opening syllogism is: *if a excludes $\tilde{a}$, and b excludes $\tilde{b}$, then $a + \tilde{b}$ excludes $\tilde{a} + b$* (or, conjugately, $a + b$ excludes $\tilde{a} + \tilde{b}$).... This we have just derived.

The inference's conclusion is: *and therefore, ab exists*. The reasoning is that we can logically replace the two one-bit-of-state processes a, b with one two-bits-of-state process ab, since what counts in processes is sequentiality, not state size, and exclusion births sequence (here, in the form of alternation). That is, the existence of the two co-exclusions $a + \tilde{b} \longrightarrow \tilde{a} + b$ and $a + b \longrightarrow \tilde{a} + \tilde{b}$ contains sufficient information for ab to be able to encode them, and therefore, logically and computationally speaking, ab can rightfully be instantiated. We write $\delta(a + \tilde{b}) = ab = -\delta(\tilde{a} + b)$ and $\delta(a + b) = ab = -\delta(\tilde{a} + \tilde{b})$. A fully realized ab is, we see, comprised of two *conjugate* co-exclusions, a sine/cosine-type relationship.

We can now answer the man's question, *Did anything happen?* We can answer, "Yes, when the coin changed hands, the state of the system rotated 180°: $ab(a + \tilde{b})ba = \tilde{a} + b$." We see that one bit of information ("something happened") results from the alternation of the two mutually exclusive states.

With the co-exclusion concept in hand, we can now add a refinement to the idea of co-occurrence. Recall that S is the space of all imaginable expressions in $\mathcal{G}$. But, thinking now computationally, this means that they are all "there" at the same time! That is, *S is the space of superpositions*, of all *imaginable* co-occurrences of elements of $\mathcal{G}$ *all at the same time*; whereas G is the space of *actually occurring* (but still space-like) entities, which means no co-exclusionary states allowed. When things move from S to G, superposition is everywhere replaced by reversible alternation, ie. G is a sub-space of S.

Co-exclusions, being superpositions, thus live exclusively in S, whereas co-occurrences can exist in both S and G, though their objects are slightly different. Co-occurrences in $\mathcal{T}$auQuernion-space have yet another flavor. Each of the transitions $S \rightarrow G$ and $G \rightarrow \mathcal{T}$ is entropically favored. We now look at the former, the latter being the standard theory of quantum measurement.

As a first example, consider the scalar distinction $[1, \tilde{1}]$, an element of S, which is mapped to the vector a, an element of G, and therewith encapsulates one bit, cf. Table 1. $[1, \tilde{1}] \in S$ because both 1 and $\tilde{1}$ must be simultaneously present if the idea of their distinction is to be meaningful. Thus, what is a *superposition* of 1 and $\tilde{1}$ in S becomes an *alternation* between 1 and $\tilde{1}$ in $a \in G$. A degree of freedom has been lost.

A second example: the co-exclusions $(a + \tilde{b} \,|\, \tilde{a} + b)|(a + b \,|\, \tilde{a} + \tilde{b})$ induce the formation of ab. What happens is that the *superpositions* in S represented by the co-exclusions — three of them — have been replaced by their actualized *alternations*, $(a + \tilde{b} \leftrightarrow \tilde{a} + b) \leftrightarrow (a + b \leftrightarrow \tilde{a} + \tilde{b})$ in G. [9] That is, the superpositions in S are replaced by space-like exclusions in G, which is, again, a *reduction* in the number of states. In the next step, this reversible alternation in G is replaced by *before/after*, that is, it becomes a time-like (irreversible) exclusion in $\mathcal{T}$.

The overall movement of information is thus from superposition in S to space-like exclusion ("alternation") in G to time-like exclusion ("before-after") via projection/measurement in $\mathcal{T}$. Each of these steps increases entropy by (further) compartmentalizing information, which reduces correlation, ie. increases noise, which is entropy.

The information that Shannon defined is namely *time-like,* and is exactly modelled by a binary decision tree descent from root to leaf. In contrast, what δ does is to *build* that tree from the leaves (detailed co-occurrences like $a + \tilde{b}$) first to ab, ie. $\delta(a + \tilde{b}) = ab$, and from there up to the root $abc \ldots z$. In doing so, it reduces the information content of S by turning its superpositions into exclusionary distinctions in G, which in turn, at level 4, are projected into $3 + 1$d tauquernion spacetime. The Bit Bang explosion is much like the irresistible salesman who argues that owning one cow *after* the other is really just as good as owning two cows at the *same* time. (Although it isn't, as we know.)

When we calculate the information content of $G = \Psi$, we are counting non-Shannon information. And yet, the conceptual *basis* for this counting up of non-Shannon information is Shannon's time-like information, information you can use to locate and identify things in a space, cf. the binary tree descent! This is a subtle paradox.

We resolve the paradox by viewing the entropic expansion $\mathcal{G}_0 \to \mathcal{G}_1 \to \mathcal{G}_2 \to \mathcal{G}_3 \to \mathcal{G}_4$ as the conversion of the space-like information in S and G into time-like information in $\mathcal{T}$-space — ebits, mass, 3d space, gravity, entropy, and time. That is, causal *potential* is converted into causal *actuality,* and it is in this conversion that the Shannon encoding of non-Shannon information is rendered meaningful, as namely Shannon information.

The continuation of this entropically-favored process of increasing encapsulation

$$a \xrightarrow{\delta} ab \xrightarrow{\delta} abc \xrightarrow{\delta} abcd \xrightarrow{\delta} \ldots \xrightarrow{\delta} abc \ldots z$$

---

[9]Note that the co-exclusion form sums to 0, and so holds no contradiction.

would seem to lead to the conclusion that black holes are to be described by pseudo-scalars of grade $4n$, where n is *very* large, and "4" because this is a gravitational phenomenon, and the algebra cycles semantically $mod\,4$ (and more subtly, $mod\,8$). We are namely looking at (ie. inside) the interior of a gigantic gravitationally res-onantly bound particle with 2^{4n} dimensions. At this extremely high level of gravi-tational organization (read heavily entangled), everything is so intensely correlated with everything else that, in the limit, all entities become indistinguishable from each other. In this way, the stage is set for a new expansion. [10]

3. Cosmological Evolution

The preceding section dilineated the information content of elements of the algebra, and thereafter how these elements are stitched together computationally and mathe-matically (namely with co-exclusion $\mapsto \delta$) to create ever more actualized structures. Left unaddressed however, is how exactly these algebraic elements come to be in the first place.

Metaphysics aside, we rely on two pillars of support in this telling of this story:

- The structure of the algebra itself, without questioning whether this is putting too much in by hand.
- The entropic propensity, ie. the truth of the 2nd Law of Thermodynamics.

These are the governing principles in what follows.

The overall story arc is that the information creation via co-occurrence (cf. the Coin Demo), which is both dominant and non-Shannon, can be sustained using reversible mechanisms. The result is an exponentially expanding space-like information space, namely $G = \Psi$. This information is then bled off by its conversion into its time-like form, which we experience as $\mathcal{H}, \mathcal{M}, \mathcal{D}$, the Big Bang, and its aftermath.

The primitive mechanisms that contribute to the creation of bits of information are

- Distinctions: scalar 1 vs. $\tilde{1}$, and (multi-)vector $XY = -YX$
- Products, XY
- Co-occurrences, $X + Y$

The last of these dominates the information content of both S and $G = \Psi$ because the number of co-occurrences grows hyper-combinatorially. The two distinctions are clearly proto-bits. "Products" get their own line because, if co-occurrence is the steam locomotive, then products - being the generators of novelty — are the coal car, without a constant supply of which, the train will grind to a halt. This is detailed below.

---

[10]We note that the Pythagorean relationship $B_1^2 + B_2^2 = (B_1 + B_2)^2$ for the total entropy of the merge of two black holes B_1, B_2 [21] is satisfied by any two tauquernions so long as they anti-commute. See also the discussion of $\mathbb{Z}_n$ arithmetics in Appendix I of [5].

Since we are dealing chiefly with co-occurrences, all "information" is non-Shannon unless otherwise noted. We are dealing only with *extant* elements of S, that is, with the elements of G as so far constructed. Abusing combinatorial notation, we are generating the set $\{\binom{\{1,a,b,\ldots\}}{m}\}$ of all the possible forms in $G = \{1, a, b, \ldots\}$. This generates $\Sigma\binom{n}{m} - 1 = \sum_1 \binom{n}{m} = 2^n - 1$ elements.

The reason that the formula is $\sum_1 \binom{n}{m}$, ie. leaving out one possibility $(m = 0)$, is that $\mathcal{V}oid$ cannot be a party to a co-occurrence. This is because by definition, 0 means "does not occur", in the sense that $\mathcal{V}oid$ does not "happen", does not "take place", in either space or time, as opposed to the mis-understanding "not there at all". Thinking back to the Coin Demonstration, it simply cannot be performed when there is $\mathcal{N}oThing$ in the man's hand, but this does not deny $\mathcal{V}oid$'s presence.

We begin our construction with the scalars, G_0. These are $\mathcal{V}oid \mapsto 0$ and the primitive distinction $[1, \tilde{1}]$ that emerges from $\mathcal{V}oid$. The scalars have no dimensionality but *can* represent a primitive distinction if one has two of them. Dimensionally, $\mathcal{V}oid \mapsto 0$ represents a point, and the two-valued distinction ± 1 is the prototype of a line.

Including $\mathcal{V}oid$, G_0 has three distinctions $[\neg\mathcal{V}oid, \neg 1, \neg\tilde{1}]$ leading to $lg\,3 = 1.58$ bits; counting just the two non-zero states, this represents $lg\,2 = 1.00$ bit. These two different bit-measures express the difference between the space S of *possibilities*, and the space G of *extant* (in Ψ) entities, ie. those that have actually been constructed out of the possibilities.

The transition from G_0 to G_1 maps the scalar distinction $[1, \tilde{1}]$ to a 1-vector, a. This is an entropically favorable transition, according to Table 1, because a has one bit less information than the scalars from which it is formed. This mapping reifies into an exclusion what previously was only a potential to be 1 or $\tilde{1}$. Both scalars and vectors are now present, and Table 1 shows that they always have the highest information content of all.

The forms $\sum_1 \binom{2}{m}$ of G_1, which we might also write as $\sum_1 \binom{\{1,a\}}{m}$, yield the set $\{1, a, 1 + a\}$, but $\delta(1 + a) = a$, which we already have, so no novelty is generated.

The expansion must therefore seek another route ... which is (to await) the co-occurrence $a + b$, wherein we imagine the parallel existence of many G_1's (this is all an idealization, of course). Once a and b co-occur, they can co-exclude, whence ab, a *new entity*, is added to G. This is the coal car feeding the steam engine: every time a new entity is added to G, the number of co-occurrences, the size of G, doubles. [11]

Note that even though the multiplication $a + b \mapsto ab$ is reversible (eg. $a(ab) = b$), information is nevertheless created when ab is created (2.17 vs. 1.17 bits). As noted in 1.1, what is going on is that bins — of *possibility* — are simply being visited.

---

[11] Strictly speaking, we should not count 1-vectors and pseudo-vectors, the $\binom{n}{1}$ and $\binom{m}{m}$ terms of the Σ, since we're counting co-occurrences, and these are singletons. On the other hand, including $\mathcal{O}(n)$ singletons has negligible impact on $\mathcal{O}(2^n)$.

Addition (co-occurrence) is doing most of the work of the expansion — it's always entropically favored. But multiplication supplies a vital piece, namely the step from $a + b$ to ab. This being a crucial step, we reason that ab has the same information content as a and b, so in multiplying the latter together, it's $1 \times 1 = 1$ so to speak: we are simply combining things of the same measure and nothing is being "manufactured". Nevertheless, ab is still *novel*, so in the context of S and G and their basis in co-occurrences, we still harvest an information windfall from ab's appearance, because this gives (entropically favored) birth to a whole new generation of co-occurrences.

This may sound dodgy — something for nothing is always suspect - but the mathematics speaks clearly. It is non-Shannon (ie. space-like) information that becomes available via (though not because of) space-like rotation, $G = \Psi$ is expanding (because of addition), and there is no time-like context here.

This reasoning applies to all co-occurrences and products, and thus the expansion of Ψ is a general free-for-all application of co-occurrence $+$ and action $\times$ over and between all extant entities, biased in the general direction of entropy generation. But we are ahead of the story, and now must back up.

Eventually, all the elements of our G_1, call it $G_1^a = \{1, a\}$, will have been generated, so we must await a co-occurrence with a new entity, call it $b \in G_1^b$, and we then can generate $G_1^a + G_1^b$. Recall that co-occurrences always have a lower information content than the singletons composing them, so $G_1^a + G_1^b$ is entropically favored.

Once there is co-occurrence, there can be action: G_2 is created by $G_1^a \times G_1^b = \{1, a, \} \times \{1, b\} = \{1, a, b, ab\} = G_2^{ab}$. Besides $q$bits, this produces, in particular, the high-information bivector ab, and thence W/Z and neutrinos.

Nevertheless, at some point, the combinatorial possibilities of G_2^{ab} too will be realized, whence we await a co-occurrence with an entity belonging to another G, say $G_2^{ab} + G_1^c$, leading to the product $G_2^{ab} \times G_1^c$:

$$G_3^{abc} = \{1, a, b, ab\} \times \{1, c\} = \{1, a, b, c, ab, ac, bc, abc\}$$

With G_3^{abc} we get photons, electrons, quarks, protons, neutrons, mesons, gluons — all the familiar members of the Standard Model.

Similarly, cf. [5], $G_2^{ab} \times G_2^{cd}$ and $G_1^d \times G_3^{abc}$ together generate G_4^{abcd} - giving us $\mathcal{H}, \mathcal{M}, \mathcal{D}$, $3 + 1$ spacetime, mass, gravity, and entropy — at which point we leave quantum mechanics. $G_{4n} \times G_{4n}$ describe higher-order gravitational structures.

However, we have again gotten ahead of our story. In generating G_2 from $G_1 \times G_1$, we can further imagine the co-occurrence and subsequent product of several (say four) G_1's (over, say, a, b, c, d), which will then produce the six bivectors ab, ac, ad, bc, bd, cd.

Once again we recall that co-occurrences always have a lower information content than the singletons that compose them, so entities like $ab + cd$ will again be entropically favored. These are, of course, $\mathcal{T}$'s ($\Rightarrow$ Bell/Magic states and $e$bits), and so we see that there is an entropically favored route to $\mathcal{H}$ and $\mathcal{M}$. [The same applies to $xy + xz$ (electrons) and $x + yz$ (quarks).] Since, all else seeming equal, there are three times as many $\mathcal{M}$ states as $\mathcal{H}$ states, the tendency here will be for the formation of normal matter.

Similarly, G_1+G_3 will produce co-occurrences like $w+xyz$, the atoms of dark matter, so $\mathcal{D}$ is also an entropically favored outcome. Note that with the exception of $^{16}\mathcal{D}_u$, dark matter will be formed preferentially to normal matter, cf. $5.53, 6.87, 5.53$ versus 15.9 in Table 1.

4. The Combinatorial Hierarchy

Finally, we arrive at the ***Combinatorial Hierarchy***, whose investigation was the constant and prescient focus of ANPA and its founders, Frederick Parker-Rhodes, Ted Bastin, Clive Kilmister, John Amson, and the beloved recipient of this festschrift, ***H. Pierre Noyes***, from 1979 and before.

The *Combinatorial Hierarchy (CH)* concerns the generation of G_{i+j} from $G_i \times G_j$. Let $A = \{1, a\}$, whence we are in G_1. A-space is $\pm 1 \pm a \Rightarrow 2^2 = \mathbf{4} = 2^{2^1}$ states. Now let $B = \{1, b\}$. Then

$$A \times B = \{1, a, b, ab\}$$

and the resulting space is of size $2^4 = \mathbf{16} = 4^2 = 2^{2^2}$.

The next step is

$$\{1, a\}\{1, b\}\{1, c\} = \{1, a, b, c, ab, ac, bc, abc\}$$

which is of size $2^8 = \mathbf{256} = 16^2 = 2^{2^3}$.

Next is $\{1, a\}\{1, b\}\{1, c\}\{1, d\} =$

$$\{1, a, b, c, d, ab, ac, bc, ad, bd, cd, abc, abd, acd, bcd, abcd\}$$

which is of size $2^{16} = \mathbf{256^2} = 65536 = 2^{2^4}$.

$Lvl = n = generators \rightarrow$	G_1	G_2	G_3	G_4
# terms	$2^1 = 2$	$2^2 = 4$	$2^3 = 8$	$2^4 = 16$
Full state contents G	$\pm 1 \pm a$	$\pm 1 \pm a \pm b \pm ab$	$\pm 1 \pm a \pm b \pm c \pm$ $\pm ab \pm ac \pm bc \pm abc$	$\pm 1 \pm a \pm b \pm c$ $\pm d \pm ab \pm ac \pm bc$ $\pm ad \pm bd \pm cd \pm abc$ $\pm abd \pm acd \pm bcd \pm abcd$
$|G| = 2^{2^n} = (2^{2^{n-1}})^2$	$2^{2^1} = 2^2 = 4$	$2^{2^2} = 4^2 = 16$	$2^{2^3} = 16^2 = 256$	$2^{2^4} = 256^2 = 65536$
Occurrences $S_G = \sum_1 \binom{k}{j}$	$\sum_1 \binom{2}{j} = 3$	$\sum_1 \binom{3}{j} = 7$	$\sum_1 \binom{7}{j} = 127$	$\sum_1 \binom{127}{j} = 2^{127} - 1 \approx 10^{38}$
	$\{1, a, 1 + a\}$	$\{a, b, a + b \xrightarrow{\delta} ab,$ $a + ab, b + ab, a + b + ab\}$	$\{a, b, c, ab, ac, bc,$ $a + ab, a + ac, \ldots\}$	$\{a, b, c, d, ab, ac, ad, bc, bd,$ $cd, a + ab, a + ac, a + ad, \ldots\}$

Table 2:   *The Combinatorial Hierarchy,   CH* $[1, 2]$

The sequence of space-sizes increases as the square, $4 \to 16 \to 256 \to 256^2$, because of course $2^{2^n} = 2^{2^{n-1}} 2^{2^{n-1}}$. At the same time, the number of elements in these spaces (a subset of S) is growing even faster, and these two sequences are related. Table 2 shows the generation process, and the intertwining of the two sequences is visible in the related powers of 2 that appear.

In the early ($\mathbb{Z}_2$) analysis [1] of this construction — the original *CH* — it was understood in terms of the state vectors of one level being stacked to make square matrices, which matrices had to be capable of mapping the resulting next-level space onto itself. The intriguing aspect then is that while the matrix, being a stack of basis vectors, exists for $n = 1, 2, 3$, at $n = 4$ the number of co-occurrences explodes, and the $((256)^2)^2 = 2^{32}$ basis vectors are completely swamped by the 2^{127} co-occurrences they should map among. That is, 4 covers 3, 16 covers 7, and 256 covers 127, but then it's over. So the construction halts, or must begin anew, or, at least, something *new* has to happen, seemed to be the message back then.

[Three brief comments: (1) S_G is that part of S that corresponds to G's alternations; (2) the bottom two rows of the table show only + variants because the signature collapses all sign variants to the same bin; and (3) the base of the combinatorics, 2-ary distinctions, is the one that generates the most structure: 3- and 4-ary distinctions cut off sooner, and 5-ary doesn't even get off the ground [1].]

The present ($\mathbb{Z}_3$) perspective sees something *new*: the line that is crossed is the one that separates localizable effects from distributed ones, ie. weak, strong, and electromagnetic from EPR and gravity. Either way, the cut-off occurs with consistent and physically meaningful interpretations, and it seems clear that the two instances of the *CH* ($\mathbb{Z}_2$ and $\mathbb{Z}_3$) are both isomorphic and being imbued with the same physical import.

Finally, the observations that $3 + 7 + 127 = 137 \approx \frac{1}{\alpha}$, α being the fine structure constant, and that $3 + 7 + 127 + 2^{127} \approx 10^{38}$ roughly approximates the electromagnetism : gravity ratio, plus the above-described interpretation, led Bastin and Kilmister to refine this purely combinatorial approach to $\frac{1}{\alpha}$. Their most recent result [2] calculates this to 137.036011393. vs. the measured 137.035999710(96). We note that Bagdonaite *et alia.* report [4] that the proton-electron mass ratio has not varied in the past 7 billion years.

In both cases, the expansion is hyper-exponential, and, being prior to the actual formation of 3+1d spacetime via the $\mathcal{T}$'s , is also not limited by the speed of light. Thus this combinatorial expansion presumably models the initial inflationary episode of standard cosmology.

Summarizing the cosmological development, both graphs in Figure 1 show the two major pathways to space/mass creation: upward on the left, the creation of $3 + 1$d space and normal matter, $\delta(\mathcal{H} \cup \mathcal{M}) = abcd$, via the pathway $\delta(\delta(a + b) + \delta(c + d)) = abcd$; and upward on the right, dark matter, via the pathway $\delta\mathcal{D} = \delta(d + \delta(c + \delta(a + b))) = abcd$, but then also for the latter, a "back door" down to $\mathcal{H} \cup \mathcal{M}$ via $\mathcal{D}_q^2$, $\mathcal{D}_u^2$, and $abc\mathcal{D}$.

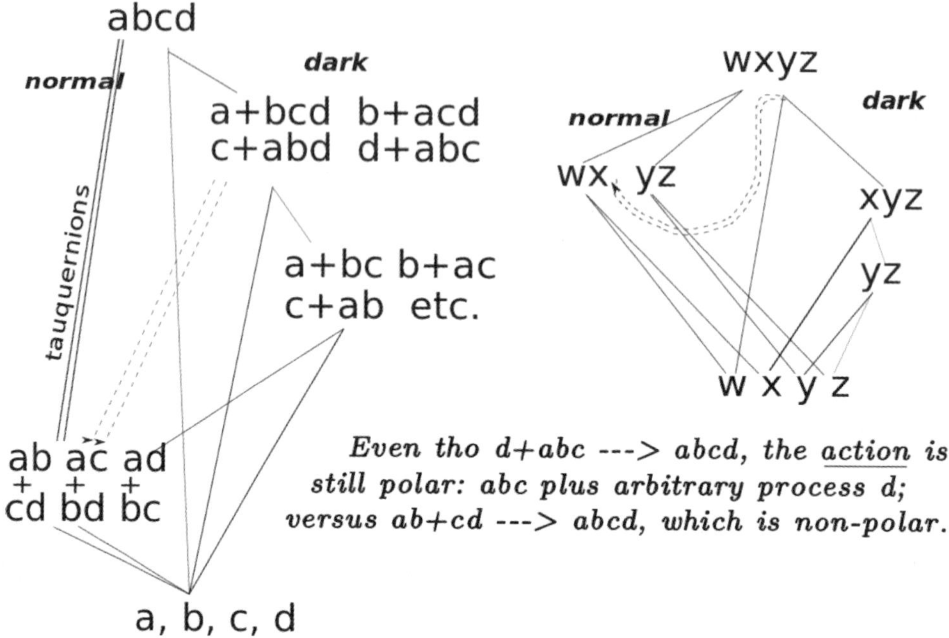

Figure 1: Two equivalent graphs of normal & dark matter creation.
Growth (δ) is upward, as the ambient energy falls.
The dotted lines symbolize the indirect tauquernion creation
from dark-dark interactions.

We thank Pierre (and Clive, Ted, John, Frederick, and ANPA)
for showing us where our path would ultimately lead.

References

1. Amson, J. C. (with A.F. Parker-Rhodes), 'Essentially Finite Chains',
   International Journal of General Systems, Vol.27, No.1-3, pp. 81-92
   1998, original paper dated 1965).
2. Bastin, T. and Kilmister, C. *The Origin of Discrete Particles.*
   World Scientific, Series on Knots and Everything, L. Kauffman, Ed., v. 42, 2009.
   Isbn-13: 978-981-4261-67-8.
3. Bastin, T. *The London Times* obituary. Nov. 17, 2011.
   http://trove.nla.gov.au/work/159129448.
4. Bagdonaite, J. *et alia.* 'A Stringent Limit on a Drifting Proton-to-Electron Mass
   Ratio from Alcohol in the Early Universe'. Science v. 339 4 Jan 2013.
5. Manthey, M. and Matzke, D. 'Tauquernions τ_i, τ_j, τ_k : 3+1 Dissipative Space
   out of Quantum Mechanics'. TauQuernions.org, v1.2 April 2013.

Brief Biographies

Michael Manthey taught computer science — architecture, operating systems, networking, distributed systems — for 25 years before retiring to pursue his research full time.
He has been affiliated with Aarhus University, Denmark; SUNY/Buffalo; University of New Mexico, Albuquerque; New Mexico State University, Las Cruces; Aalborg University, Denmark.

Doug Matzke, received his Ph.D. in Electrical Engineering from the University of Texas at Dallas in 2002 and earned his MSEE in from the University of Texas in Austin. Dr Matzke has been interested in limits of computing and quantum computing for over 30 years and was the chairman of two workshops *PhysComp '92* and *PhysComp '94* that brought over 100 people to Dallas to discuss this topic. He published the lead–off paper entitled "Will Physical Scalability Sabotage Performance Gains?" in the special issue of *Computer* Vol 30, No 9 in Sept 1997. Dr Matzke has 7 granted patents and over 50 papers and presentations. He is a software developer and consultant.

Spacetime, Dirac and Bit-Strings

G. N. Ord

Department of Mathematics
Ryerson University, Toronto, Canada
gord@ryerson.ca

Interleaved bit-strings provide a simple mechanism for the encoding of relativistic periodic processes underlying both spacetime and the Dirac equation. We illustrate this using simple model clocks.

Introduction

In a beautiful paper [1], Kauffman and Noyes brought bit-string physics [2] to the Dirac equation through an analysis of the Feynman Chessboard model. In that paper (henceforth DPDE) the paths in the Chessboard model are discrete and map onto stochastic sequences of the letters R and L (for Right and Left), the sequences being equivalent to bit strings. By superimposing a periodic sequence of signs onto the bit string one obtains the weight of a chessboard path and these appropriately combined yield a discrete version of the Dirac propagator. Among other things, this insight into the propagator illuminated the significance in discrete physics of the unit imaginary i.

Although there are many papers on the Chessboard model, there can be little doubt that DPDE is the most insightful work on the subject. It has initiated much exploration, particularly through ANPA, on the physics that it illuminates. This paper reviews some of the author's own excursions in the area. These have been strongly influenced by DPDE and ANPA meetings, but of course only the author is responsible for the views expressed here.

The connection between bit-strings and the Dirac equation in DPDE is very direct and as a result bit-strings are implicated in both quantum propagation and relativity. Since Minkowski spacetime is more accessible from a physical standpoint than is quantum mechanics, it is of interest to see if the bit-string perspective can be used to explore spacetime along with the Dirac equation. We do this here using some recent results on 'intrinsic timekeeping' [3].

We can enter the arena of spacetime by asking ourselves how a single clock in empty space might keep time through some signal. All the precise clocks in current use rely on periodic processes and are of finite precision, so this is where we shall start. We consider a discrete periodic process and to make contact with

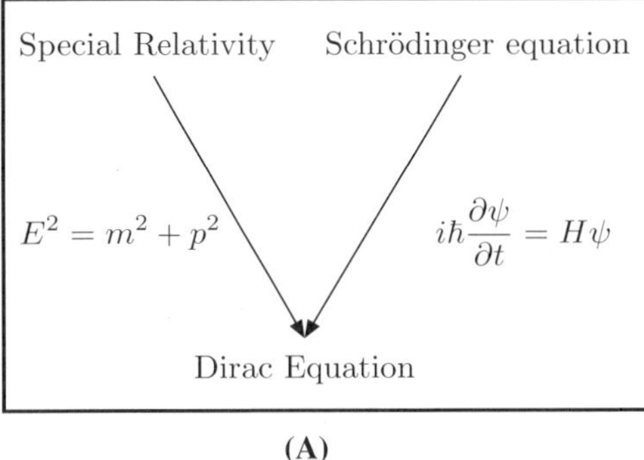

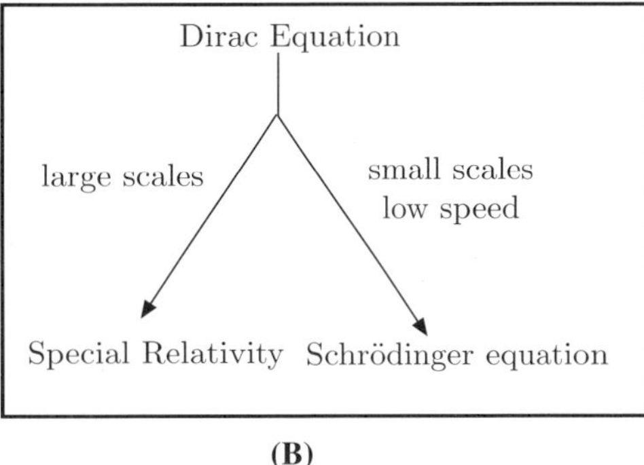

(A) **(B)**

Fig. 1. (A) The Dirac equation is conventionally obtained by merging special relativity and quantum mechanics. The argument that the foundations of quantum mechanics may be revealed by investigations in the Schrödinger regime assume that nothing essential to quantum mechanics is gained in the merger. (B) From the perspective of the domain of applicability, the Dirac equation is more fundamental than either classical special relativity or the Schrödinger equation. We expect special relativity to hold on large scales due to the correspondence principle. The Schrödinger equation holds on small scales where speeds are small compared to the velocity of light.

spacetime we take continuum limits first of all assuming a Galilean world and then a relativistic one. In the latter case two different continuum limits distinguish classical and quantum physics. In the former case the bit-string picture is 'submerged' in a limit that forces Nature to draw smooth curves between events. In the latter, the bit-string picture is preserved as a form of wave propagation. It forms a physical basis for quantum propagation.

Since the implications of the models are very general, we start with a prolog that argues for the consideration of discrete time in physics. In section 1 we review a simple model clock to look at the relationship between periodic processes, spacetime and scale. Section 2 extends clock models to the Dirac equation and the last section discusses the relation to bit-string physics.

Prolog: Spacetime, the Unit Imaginary and Discrete Origins

Currently, quantum mechanics and its use in high-energy physics give us our most precise probe of space and time. However, despite near universal acceptance of quantum mechanics as *the* fundamental theory of the micro-world, there is still no general consensus on the ontology of wavefunctions, the primary calculational object within the framework of the theory.

The most successful initial contact between relativity and quantum mechanics arose with the Dirac equation. There, special relativity and quantum mechanics were united in an enigmatic but seemingly fundamental way. On the grounds of the domain of applicability, the Dirac equation is more fundamental than either the Schrödinger equation or special relativity. However, the complexity of the equation tends to inhibit its consideration in the foundations of quantum mechanics, primarily because it is regarded as a *merger* of the Schrödinger equation and spe-

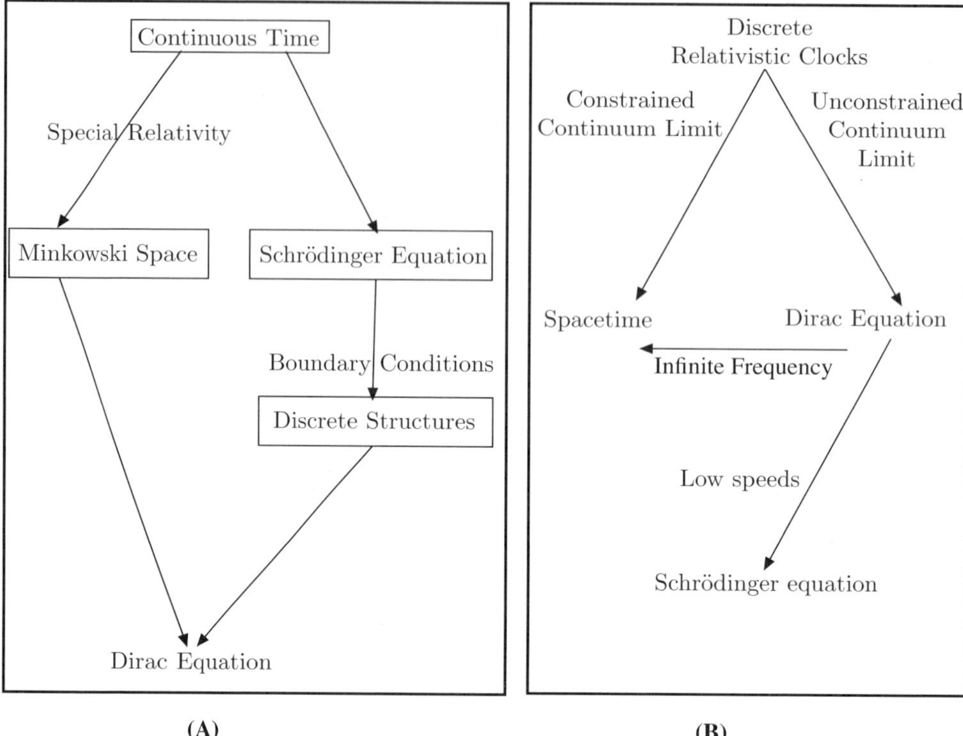

Fig. 2. (A) Both special relativity and quantum mechanics are conventionally formulated with the direct assumption that time is continuous and can be parameterized by a real number. Both theories employ a form of formal analytic continuation so that $t \to it$ is a useful replacement that either induces a pseudo-Euclidean space with odd signature (Spacetime Algebra, left path) or takes a classical partial differential equation and quantizes it (Schrödinger equation, right path). In the latter case, discrete structures appear as a result of boundary conditions. The Dirac equation inherits those discrete structures from Schrödinger. (B) In this paper we start with discrete relativistic periodic clocks. Two types of continuum limit present themselves. One leads directly to spacetime algebra and classical special relativity. The other leads to the Dirac equation. In this picture, classical spacetime as a construct is an infinite frequency limit.

cial relativity, rather than the progenitor of both. Thus, so the argument goes, the Schrödinger equation may be assumed to contain the fundamental essence of quantum mechanics, uncluttered by the additional details of special relativity, Fig. 1.

We shall adopt a perspective in which both spacetime and quantum propagation emerge from a discrete form of special relativity via two different continuum limits, Fig. 2. There are several general observations that encourage this emergent view.

(1) Both conventional quantum mechanics and special relativity start with a continuum formulation that ultimately produces discrete structures. In the former case, wave propagation is invoked in order to obtain *discrete* energy levels via boundary constraints. In the latter case, a discrete structure is built into the odd signature of spacetime. This is signalled in older versions of special relativ-

ity through the fourth coordinate $x_4 = ict$. The presence of the unit imaginary is a harbinger of a discrete four-state process. In more modern versions this periodicity is produced by the odd signature of spacetime.

(2) There are classical analogs of the Dirac and Schrödinger equations that have well-known origins in classical statistical mechanics. For example, the coupled equations:

$$\frac{\partial U}{\partial t} = c\,\sigma_z \frac{\partial U}{\partial z} + m\,\sigma_x U \tag{0.1}$$

are a form of the Telegraph equations. Here the σ_k are Pauli matrices. The Kac model [4] provides a classical microscopic statistical mechanics for these equations in which the continuum limit is an ensemble average over discrete processes. On long time scales the coupled equations may be replaced by the diffusion equation. For both the Telegraph and Diffusion cases, the existence of the underlying kinetic theory model means that the equations themselves are idealizations of discrete processes in which a finite number of configurations may be partitioned and counted. Ultimately, nothing more complicated than a frequency interpretation of probability is needed to understand the emergence of equation(0.1) from statistical mechanics.

For comparison the coupled equations:

$$\frac{\partial U}{\partial t} = c\,\sigma_z \frac{\partial U}{\partial z} + im\,\sigma_x U \tag{0.2}$$

are a representation of the Dirac equation in a two dimensional spacetime. The only difference between equations (0.1) and (0.2), apart from scale, is the presence of the unit imaginary in the second term in (0.2). The imaginary unit has two notable effects on the equations. The first effect is to induce a Lorentz covariance that is not found in (0.1). The second is to produce a form of wave propagation that ultimately allows a persistant discrete energy spectrum to arise from boundary constraints. The simultaneous initiation of Lorentz covariance and wave propagation is suggestive that both may be linked to a common antecedent.

A further similarity illustrated by these equations is that the solutions are, qualitatively, clocks. Solutions of (0.1) measure time through a characteristic decay e^{-mt}. One can in principle measure the exponential decay and invert the exponential to establish the time through the density U. Such solutions are examples of 'thermodynamic clocks' in which entropy provides the link between time and the state of the system. Radioactive dating is a common example of a practical use of this.

Solutions of (0.2) are reversible clocks based on a periodic process. The characteristic thermodynamic decay e^{-mt} is replaced by the periodic oscillator $e^{\pm imt}$. The lack of a kinetic theory basis for (0.2) is, in part, explained by this shift from thermodynamic to periodic time measurement.

(3) A general feature that suggests we look for progenitors of relativity and quantum mechanics is the fact that both theories essentially correct for the transition

from a Newtonian world in which absolute time is *assumed,* to a world in which the ageing process is path dependent. In special relativity this is accomplished through the Lorentz transformation which itself is often handled algebraically by invoking a spacetime of odd signature. In non-relativistic quantum mechanics, the deviation from absolute time is handled through path-dependent phase as occurs explicitly in the path-integral formulation. The link between these two 'corrections' to absolute time can be seen through examination of the Dirac equation [5–8].

The above heuristic arguments suggest that we look beneath the assumption that time intervals can always be directly modeled by real numbers. Both special relativity and quantum mechanics seem, in their own way, to have a predilection for the combination *it*. At its simplest, i represents a four-state process and we shall take this as a starting point when constructing a clock.

1. Clocks, Transfer Matrices, and Newton's Absolute Time

In this section we build a simple discrete clock that allows us to examine various continuum limits. The objective is to understand the Newtonian picture that regards time as a dimension separate from space. This gives the **space-time as a container** view in which all events essentially position themselves in physical space via a single universal sequencing parameter t that can be assumed to be real. To make contact with subsequent sections, we shall use concepts such as 'inertial frame', 'light cones' and 'causality' despite the fact that these only assume conventional meaning in section 2. We do this so that we can verify the Newtonian picture from a model that, when subjected to Einstein's light-speed postulate, gives us the Minkowski picture. We shall assume an initial lab frame in which our model clock is stationary and in which the speed of light is isotropic.

To begin with, consider a two dimensional space-time in which to build a simple digital clock [9]. The clock will be assumed periodic and for reasons that shall appear later, will be based on a period four process.

Figure 3A represents a periodic sequence of events at the origin in an inertial frame. (B) shows the past and future light cones of these events and (C) the causal areas that lie between the sequential events. Here the scale is such that the speed of light is 1 and the 'causal' areas are simply those areas between events that are in the future of the earlier event and the past of the later event. The significance of these areas is that regardless of the actual clock mechanism, the areas represent the maximum time-like domain of influence that successive ticks have in common. In (D) we colour the area boundaries to show that we can imagine the boundaries to be drawn as a single continuous curve from t=0 to the last event and thence back to the origin. Drawing the boundary this way gives the causal areas an alternating orientation and places the events at intersections of smooth segments of the two curves. If we think of the boundary curve as a path, there are two kinds of events. The outside corners represent impulsive accelerations (direction changes) while the

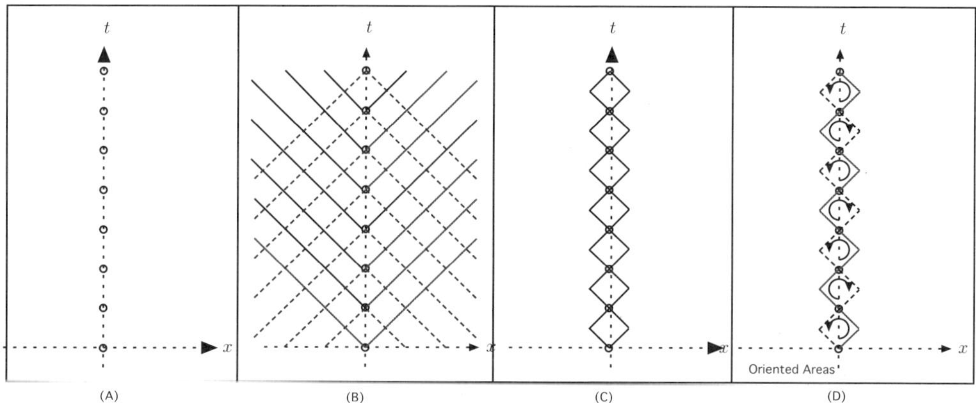

Fig. 3. (A) Periodic events in a two dimensional *spacetime*. (B) Past and future light-cones for the events. Units are chosen so that $c = 1$. (C) The causal areas between events. (D) The area boundary is chosen to provide a single traversal. This results in an alternating orientation of causal areas.

path-crossing points themselves are simply points where the curve crosses itself. We call the latter intersections on-worldline events and the former off-worldline events. From a path perspective, our clock is equivalent to a couple of featureless photons confined between two walls that generate a 'tick' at crossing points.

Notice that in terms of paths, the first on-worldline event after the origin has 2 corners in the two paths to the event, the second on-worldline event has four such accelerations, etc. We can think of the digital clock as counting unit impulses along the area boundaries. For simplicity, we can split the boundary in half and just use one side of it, as in Fig. 4.

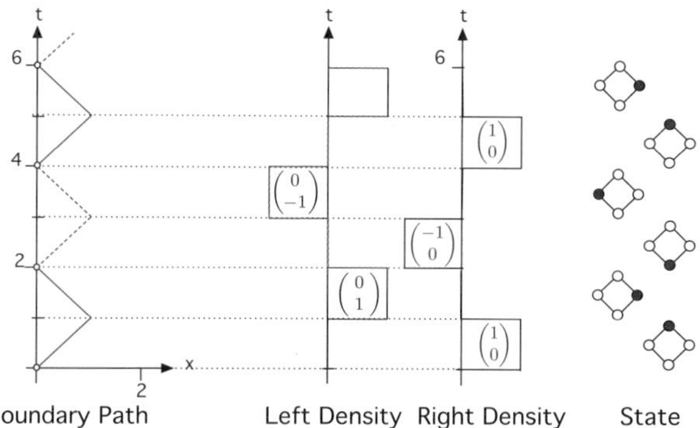

Fig. 4. The right boundary of the chain of causal areas can be in any one of four states. The boundary densities and column vector description are illustrated. On-worldline events occur at the even integers. The states correspond to four corners of a square centred at the origin. Each component when present is a binary sequence of ± 1. The two components interleaved form a period four process.

In this figure, a two-component 'state vector' records projections onto the left and right light cones and because the area boundaries are themselves null segments, successive states are orthogonal. The states themselves yield a period 4 sequence

$$s_k \in S_0 = \left\{ \begin{pmatrix} 1 \\ 0 \end{pmatrix}, \begin{pmatrix} 0 \\ 1 \end{pmatrix}, \begin{pmatrix} -1 \\ 0 \end{pmatrix}, \begin{pmatrix} 0 \\ -1 \end{pmatrix} \right\} \tag{1.1}$$

which is itself two interleaved binary processes

$$\left\{ \begin{pmatrix} 1 \\ 0 \end{pmatrix}, \begin{pmatrix} -1 \\ 0 \end{pmatrix} \ldots \right\} \text{ and } \left\{ \begin{pmatrix} 0 \\ 1 \end{pmatrix}, \begin{pmatrix} 0 \\ -1 \end{pmatrix} \ldots \right\}. \tag{1.2}$$

The two interleaved bit strings seem to suggest we have redundant information since either could be used to represent a clock. However we shall see later that the interleaved process is necessary to allow for special relativity.

If we think of the column vectors as the active display of the digital clock, the clock mechanism corresponds algebraically to left multiplication by the transfer matrix

$$T_0 = \begin{pmatrix} 0 & -1 \\ 1 & 0 \end{pmatrix} \tag{1.3}$$

where we note that $T_0^2 = -I_2$. The 'tick' matrix is a square root of minus the identity and a fourth root of the identity matrix. In terms of the Pauli matrices, $T_0 = -i\sigma_y$. If the clock is in state s_k at time k then the subsequent state, after the next event is

$$s_{k+1} = T_0 s_k. \tag{1.4}$$

Since

$$s_k = T_0^k s_0 \tag{1.5}$$

the power of the transfer matrix corresponds to the discrete displacement in time.

To make this clock analog, we have to construct a process that agrees with the clock at the integer events where the digital clock ticks, but provides extra events between ticks. The simplest way to do this is to construct a clock, similar to the original, but running at a higher frequency. Figure 5 shows an example of the original clock with similar clocks running at higher frequencies. As the clock frequency is altered to increase the density of on-worldline events, the causal areas between individual events decreases. As illustrated in the figure, the off-worldline events are brought closer to the on-worldline events as the frequency increases and the inter-event areas go to zero in the continuum limit. The dimension of the causal chain of areas collapses to one from its value of two for the discrete clock.

In terms of the transfer matrix T_0, a clock running at n times the original frequency, with n an integer, will have a transfer matrix T_n such that $T_n^n = T_0$ so that we can take T_n as the n-th root of T_0 via an eigenvalue expansion, thus:

$$T_n = \begin{pmatrix} \cos\left(\frac{\pi}{2n}\right) & -\sin\left(\frac{\pi}{2n}\right) \\ \sin\left(\frac{\pi}{2n}\right) & \cos\left(\frac{\pi}{2n}\right) \end{pmatrix}. \tag{1.6}$$

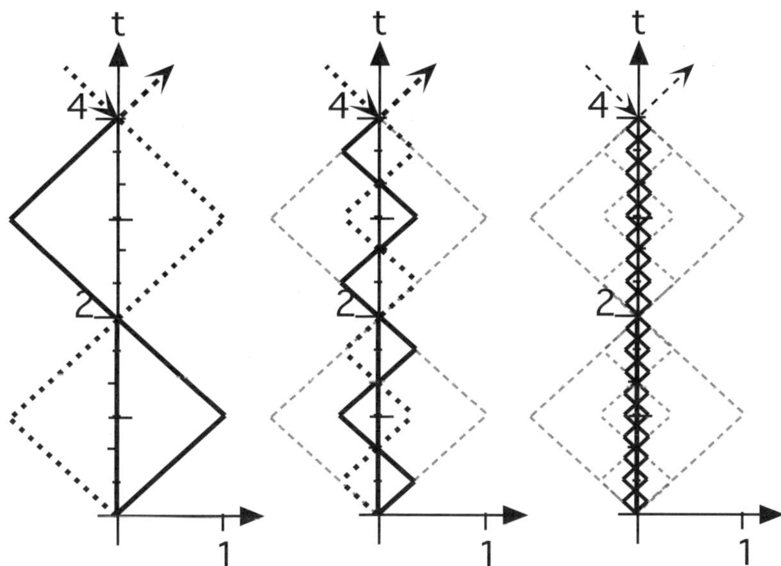

Fig. 5.   A full clock cycle for the original clock with similar clocks running at triple and nine times the original frequency. The higher the clock frequency, the closer the off-worldline events are to the t-axis and the denser the on-worldline events on the t-axis. In the continuum limit the distance between off-worldline events and the t-axis goes to zero as do the areas between events.

We can use this to take the limit as $n \to \infty$, in which case we get:

$$T(t) = \lim_{n \to \infty} T_n^{nt} = \begin{pmatrix} \cos\left(\frac{\pi t}{2}\right) & -\sin\left(\frac{\pi t}{2}\right) \\ \sin\left(\frac{\pi t}{2}\right) & \cos\left(\frac{\pi t}{2}\right) \end{pmatrix}$$

$$= \cos\left(\frac{\pi t}{2}\right) I_2 + T_0 \sin\left(\frac{\pi t}{2}\right). \tag{1.7}$$

Notice that $T(n) = T_0^n$ for integer n so that our 'analog clock' agrees with the digital clock at the integers where the original digital clock has events. In fact, we can recover the digital clock sequence equation (1.5) by using the greatest integer function $\lfloor t \rfloor$. Equation (1.5) could be replaced by

$$s_t = \left(\cos\left(\frac{\pi \lfloor t \rfloor}{2}\right) I_2 + T_0 \sin\left(\frac{\pi \lfloor t \rfloor}{2}\right)\right) s_0. \tag{1.8}$$

We have adapted the index on the states s_k to be real, however we still have only four states and it is apparent that we could not invert the right hand side of (1.8) to recover t to arbitrary precision. Equation (1.8) simply shows our original digital clock as an approximation to an ideal, perfectly periodic (PP) clock (1.7).

From equation (1.7) we can see that the analog transfer matrix $T(t)$ is just a rotation matrix and that the continuum limit replaces the set of four state vectors S_0 with the set of all column unit vectors $S = \left\{ \begin{pmatrix} u \\ v \end{pmatrix} : u^2 + v^2 = 1, u, v \in \mathbb{R} \right\}$. To see a simple connection to complex numbers we can consider a change of variables $\begin{pmatrix} x \\ y \end{pmatrix} = \frac{1}{\sqrt{2}} \begin{pmatrix} i & -i \\ 1 & 1 \end{pmatrix} \begin{pmatrix} u \\ v \end{pmatrix}$. This diagonalizes the original transfer matrix T_0 which becomes $i\sigma_z$ and equation(1.7) becomes

$$T(t) \rightarrow \begin{pmatrix} \exp[i\left(\frac{\pi t}{2}\right)] & 0 \\ 0 & \exp[-i\left(\frac{\pi t}{2}\right)] \end{pmatrix}$$

$$= \exp\left[\left(\frac{i\pi\sigma_z t}{2}\right)\right]. \tag{1.9}$$

From equation (1.9) we see that the diagonal form of the transfer matrix has two unimodular complex numbers on its diagonal, both synchronized but rotating in opposite directions as t increases. As a result, both components of the resulting state vector are unimodular complex numbers rotating in opposite directions. Either component could be inverted over half periods to extract t as a function of the component itself. Since we only need one component of the state vector to extract t, it seems again that the 2×2 transfer matrix has more information than we need. As noted earlier the matrix encodes two interleaved bit-strings, not just one. The necessity of the two complex components will surface in the next section where we require the clock to be relativistically correct.

In a Newtonian world where the Galilean transformation holds, the above light clock would work in the same way, regardless of the inertial frame or the initial frequency. The continuum limit would likewise be motion-independent and the above exercise would motivate the following argument:

> *We cannot physically build clocks with arbitrary precision to complete the limit in eqn.(1.7), however provided that clocks moving in uniform motion tick at the same rate as stationary clocks, we can safely assume that the limit (1.7) is an idealization that all real clocks approximate. This being the case, we can ignore the period and phase information in the transfer matrix and parameterize time by a real number.*

Accepting the above argument allows us to reserve a separate independent dimension for time through which *all* objects move in a synchronized fashion. This makes explicit an ontological connection between the intrinsically frequency-limited clocks that one might physically build, and an independent absolute time Fig. 6 .

2. Minkowski's Clock and the Transition to *Spacetime*

In the previous section, the digital to analog transition depended on the fundamental period of the clock. However, while the period was assumed invariant under a change of inertial frames and was a persistent feature of individual clocks, *all* clocks shared the feature that they had a continuum limit in which time, as a real parameter, was the eventual result. This feature seemed to make the details of period and phase irrelevant.

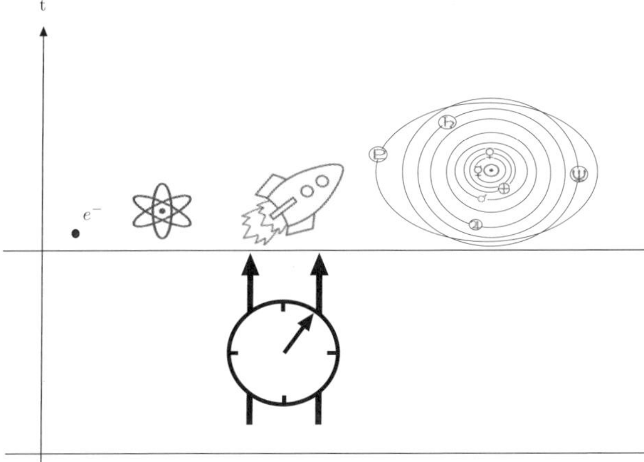

Fig. 6. In the Newtonian picture of absolute time, all objects, regardless of scale or relative speed, age in a synchronized fashion. It is as if a universal clock pushes all objects through time. The global nature of time makes plausible the assertion that natural digital clocks produce approximations to absolute time so that any 'timekeeping machinery' is irrelevant to dynamics. In particular the fundamental period (4 in our model) is irrelevant once the continuum limit is taken.

In a relativistic world, clocks need to *automatically* satisfy Lorentz transformations. The invariance of the speed of light in all inertial frames fixes the appearance of the graphs of clocks in moving frames. In Fig. 7, the graph of the first cycle of a digital clock is sketched. The causal inter-event areas have the same magnitude and orientation as the stationary clock but projections onto the left and right cones are different [8]. To see how to handle this algebraically, consider the original digital to analog conversion equation (1.6) that approximates T_0 by the n-th power of T_n. What is happening is that a single tick of the original clock is replaced by n smaller ticks of equal size to bring the higher frequency device to the same state. That is, the first step from state 1 to state 2 becomes:

$$T_0 \rightarrow \underbrace{T_0\, I_2\, I_2\, \cdots I_2}_{n-\text{terms}} \rightarrow \underbrace{T_n\, T_n\, \cdots T_n}_{n-\text{terms}} \tag{2.1}$$

where replacement by n copies of T_n is an interpolation, I_2 being the identity matrix. The application of either T_0 or its replacement switches $\begin{pmatrix} 1 \\ 0 \end{pmatrix}$ to $\begin{pmatrix} 0 \\ 1 \end{pmatrix}$ with $T_0 \rightarrow T_n^n$ being a smoothing operation that allows a continuum limit. In the original Newtonian clock the transition to the subsequent state would be accomplished by the same sequence as equation (2.1) regardless of the state of motion of the clock. However, relativistically a clock moving at constant speed v in the x direction has to stay in states 1 and 3 longer than states 2 and 4, Fig. 7. We might expect something like

$$T_0^2 \rightarrow \underbrace{T_0\, I_2\, I_2\, \cdots I_2}_{n-\text{terms}}\underbrace{T_0\, I_2\, I_2\, I_2\, \cdots I_2}_{m-\text{terms}} \rightarrow \underbrace{T_{m+n}\, T_{m+n}\, \cdots T_{m+n}}_{m+n-\text{terms}} \tag{2.2}$$

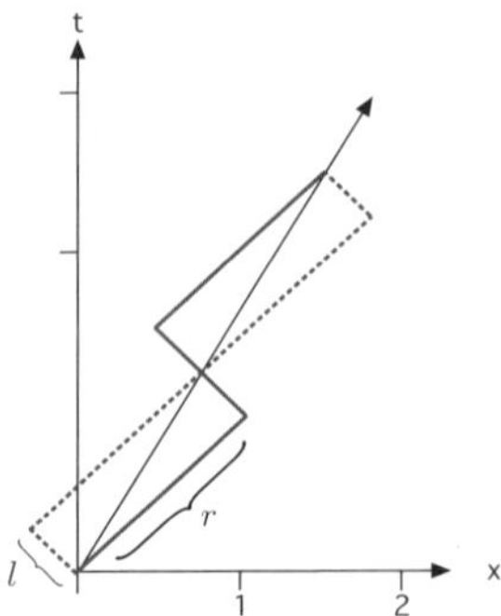

Fig. 7.   A clock cycle in a moving frame. The null boundaries remain null but the left and right projections change length. The signs and magnitudes of the causal areas are invariant.

Here the new transfer matrix T_{m+n} has to function as an $(m+n)$-th root of T_0^2 while weighting the occupation of state 1 more heavily than state 2 $(m > n)$. This means increasing the magnitude of the 1-1 element and decreasing the magnitude of the 2-2 element. To see how this may be done expand $T(t)$ to first order in $t = 1/n$:

$$T(1/n) \simeq \begin{pmatrix} 1 & -\frac{\pi}{2n} \\ \frac{\pi}{2n} & 1 \end{pmatrix}$$

where we assume $n >> 1$. The ones on the diagonal equally weight 1-1 and 2-2 transitions. If we try $T(1/n) \rightarrow T_M(1/n) = T(1/n) + w\frac{\pi}{2n}\sigma_z$ this will effect an increased residence time in the odd states $(w > 0)$ and a decreased residence time in the even states as suggested by Fig. 7. We choose w so that the 'moving clock runs slow' in accordance with the Lorentz transformation and the area preservation of the light clock [8]. To be in accordance with Fig. 7 the first on-worldline event should take place at $t = 2\gamma$ where $\gamma = 1/\sqrt{1 - v^2}$ is the time dilation factor. So we must have:

$$T_v(2\gamma) = \lim_{n\to\infty} T_v(1/n)^{2n\gamma} = -I_2. \tag{2.3}$$

For equation (2.3) to hold, the eigenvalues of T_v raised to the power $2\gamma n$ must be -1 in the limit as $n \rightarrow \infty$. The eigenvalues of T_M to lowest order in $1/n$ are $\lambda = 1 \pm \frac{i\pi\sqrt{1-w^2}}{2n}$ giving

$$\lim_{n\to\infty} \lambda^{2\gamma n} = e^{\mp i\pi\sqrt{1-w^2}\gamma}. \tag{2.4}$$

For this to equal -1 we have $w = \pm v$. Choosing the positive value of v for w we can calculate the transfer matrix for a clock moving at speed v with respect to the lab frame whose transfer matrix was given by (1.7). The matrix is

$$T_v(t) = \cos\left(\frac{\pi t}{2\gamma}\right) I_2 + \gamma(v\sigma_z - i\sigma_y)\sin\left(\frac{\pi t}{2\gamma}\right). \tag{2.5}$$

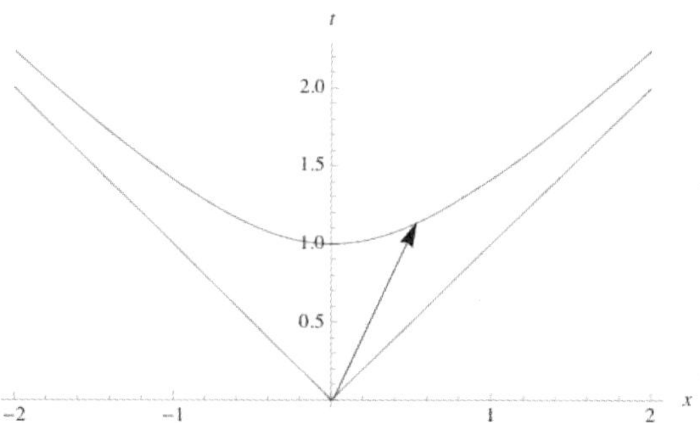

Fig. 8.   The odd part of the transfer matrix is a vector. The unit vector starts at the origin and ends on the first on-worldline event.

Equation (2.5) gives the transfer matrix for a relativistic clock that moves with speed v with respect to the original clock.

We have arrived at equation (2.5) by noticing from Fig. 7 that if the transfer matrix is to accommodate the clock moving at a velocity v with respect to the lab frame, it must re-weight the diagonal elements of the matrix. An alternative method is to recognize that the transfer matrix is a function, or image, defined on a domain that contains a line through the origin parallel to a unit vector. Rewriting equation (1.7) in terms of the Pauli matrices σ_k we have:

$$T(t) = \cos\left(\frac{\pi t}{2}\right) I_2 - i\,\sigma_y \sin\left(\frac{\pi t}{2}\right). \tag{2.6}$$

The matrix $-i\,\sigma_y$ can be thought of as a unit vector of negative norm corresponding to the t-axis. The transfer matrix is then a (multi-vector valued) function defined on that domain, so $T(t)$ is actually an image, defined on the t-axis and the 'point', I_2. The Lorentz transformation then maps the original domain represented by the unit vector $-i\sigma_y$, onto the new unit vector $\gamma(v\sigma_z - i\sigma_y)$, so $L : (-i\sigma_y) \to \gamma(v\sigma_z - i\sigma_y)$ is the domain mapping. The image being mapped onto that domain is the trigonometric function $\sin(\frac{\pi t}{2})$. A literal interpretation of Einstein's first postulate that 'the laws of physics are the same in all inertial frames' would be to say that the image on the proper time-axis is the same in all inertial frames. If t' is the time coordinate on the new axis, $f(t)$ the original image and $g(t')$ is the image on the new axis then, the relativity postulate would imply that

$$g(t') = f(L^{-1}(t')) \tag{2.7}$$

where t is the pre-image of t'. This is exactly what equation(2.5) is saying. The input function is $f(t) = \sin(\frac{\pi t}{2})$. The output function is the input function spread over its new domain. $g(t') = f(L^{-1}(t')) = \sin(\frac{\pi L^{-1}(t')}{2}) = \sin(\frac{\pi t'}{2\gamma})$. The image in the moving frame is identical, in terms of its own coordinate system, to the image in the original coordinate system. In Fig. 9 the function defined on the transformed axis of the transfer matrix, $\sin\left(\frac{\pi t}{2\gamma}\right)$, is plotted as a greyscale map over the future

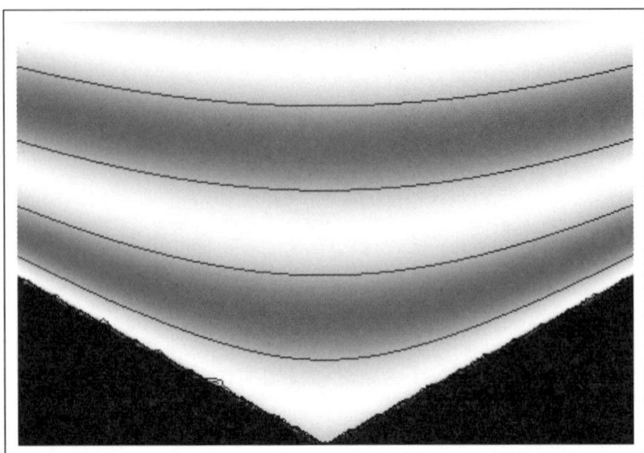

Fig. 9.   The coefficient of the odd term in equation (2.5) plotted as a greyscale map. This illustrates the connection of phase with path that is inherent in any relativistic clock. The black contours correspond to the on-worldline events of the discrete clock. We can think of the image as a raster scan based on the one-dimensional lines through the origin into the future cone of the clock.

cone from the origin. It represents an ensemble of clocks with all possible values of $-1 < v < 1$.

It is worth noting that Fig. 7 and equation (2.5) both indicate why we need a four-state process and interleaved bit-strings to describe the relativistic clock. In the rest frame, the projections of the clock path onto the two light cones mark the two cones at equal intervals. However with non-zero v the intervals on the left and right cones are different. *The Lorentz transformation requires the two projections to be distinguished, so we need the even and odd states to be partitioned.* We cannot have just a single two-state process to keep track of time evolution in a moving frame. We need at least two such processes, and the choice of the encoding scheme illustrated in Fig. 4 makes sense in light of the necessity of allowing for the Lorentz transformation. The left and right densities in that figure are equivalent in the rest frame but different otherwise. While Newton's clock appeared to have redundant information in the form of two complex eigenvalues, these prove necessary for Minkowski's clock. The inter-leaved bit-strings are required to allow the description of boosts.

Equation (2.5) shows that one of the effects of the Lorentz transformation is to replace $-i\sigma_y$ with $\gamma(v\sigma_z - i\sigma_y)$. If we think of the Pauli matrices as unit vectors, then this change looks like a rotation which is accomplished by a similarity transformation. That is, we might expect that

$$\gamma(v\sigma_z - i\sigma_y) = B(-i\sigma_y)B^{-1} \tag{2.8}$$

where B is some invertible matrix.

It may be easily verified that the requisite matrix is

$$B = \exp(-i\alpha\sigma_y\sigma_z/2) = \begin{pmatrix} \cosh\left(\frac{\alpha}{2}\right) & \sinh\left(\frac{\alpha}{2}\right) \\ \sinh\left(\frac{\alpha}{2}\right) & \cosh\left(\frac{\alpha}{2}\right) \end{pmatrix} \tag{2.9}$$

with $\alpha = \mathrm{arctanh}(v)$. This confirms the interpretation that we can indeed regard

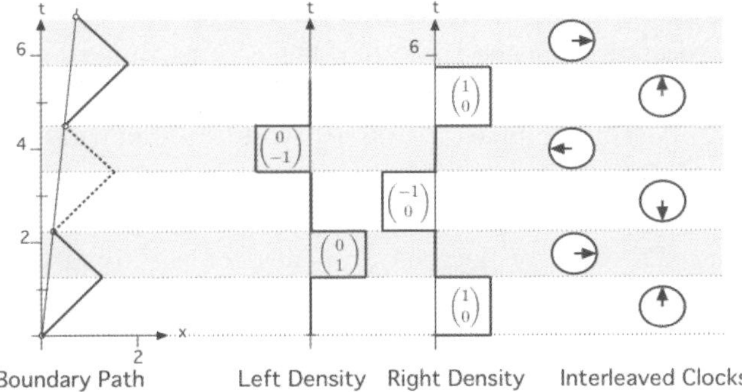

Fig. 10. To accommodate boosts, the two-photon clock uses *both* complex eigenvalues of the transfer matrix. The extra information encodes the two different residence times in the even and odd states. This is illustrated in the figure where the tilted time axis gives rise to different projection intervals on the left and right cones.

the odd part of the transfer matrix equation(2.5) as a function times a unit vector. If we take $-i\sigma_y$ as lying on the t-axis and σ_z lying on the x-axis then the velocity at the point (x,t) is just x/t and the subspace of vectors along $\gamma(v\sigma_z - i\sigma_y)$ lie along the ray $x/t = v$. Thus the two dimensional vector space with unit vectors $(\sigma_z, -i\sigma_y)$ serves a dual purpose.

On one hand, the space constitutes part of the domain of the set of transfer matrices that evolve the light clock. Its presence there is dictated by the need for the clock to preserve *spacetime* area under inertial transformations. It is through the role of the transfer matrix as an *operator on states* that we find the link to the vector space with basis vectors $(\sigma_z, -i\sigma_y)$ and with special subspaces along the rays specified by $\gamma(v\sigma_z - i\sigma_y)$.

On the other hand, as a representation of *spacetime*, it encodes the notion of proper time in the algebra of matrix multiplication. This happens because the odd signature, $\sigma_z^2 = 1$, $(-i\sigma_y)^2 = -1$ automatically encodes the invariant area $t^2 - x^2$. Recalling Minkowski's famous statement [10]

> *Henceforth space by itself, and time by itself, are doomed to fade away into mere shadows, and only a kind of union of the two will preserve an independent reality.*

we can see a justification for this from the clock model. The transfer matrix (1.7) contains two types of information.

$$T_v(t) = \cos\left(\frac{\pi t}{2\gamma}\right) I_2 + \underbrace{\gamma(v\sigma_z - i\sigma_y)}_{\text{kinematics}} \overbrace{\sin\left(\frac{\pi t}{2\gamma}\right)}^{\text{dynamics}} \tag{2.10}$$

The 'dynamic' information includes the clock frequency and phase while the 'kinematic' information involves the *spacetime* metric. To extract time as a real number,

Newton's clock *neglects* both the dynamic information and the dependence of the kinematic information on v. Space-time then has a time dimension that is independent of space. However, in Minkowski's clock the kinematic information is retained. The continuum limit still produces a smooth worldline but time duration is path dependent and *spacetime* has an odd signature.

It is worthwhile noting that by extracting space-time or spacetime from Newton's and Minkowski's clock respectively, we neglect the dynamic information in their transitions to a continuum. This might seem harmless in that the dynamic information pertains to a single clock and one might expect frequency and phase to be irrelevant for an ensemble of clocks. This may indeed be the case, however by ignoring the dynamic information we are ignoring the fact that the 'parameter' t is the inverse function of the clock signal produced by $T_v(t)$. The latter is periodic so t and the signal itself are constrained by an intrinsic uncertainty principle. If this uncertainty principle is well below the scale of any physical restrictions on the clock, we may expect that Minkowski *spacetime* is all we need. If however the natural uncertainty principle involving t and the signal is important, we shall have to take into account the dynamic information as well as the kinematic information.

3. The Dirac Clock

We see from equation (2.10) that periodicity is manifest in two ways. Primarily, the periodicity manifests itself in the kinematics of special relativity by implicating *spacetime* algebra. In the previous section we accepted the argument that all periodic relativistic clocks will contain this kinematic information and we ignored the more detailed 'dynamic' information of the clock that explicitly displays period and phase. This would make sense for an ensemble of clocks in which the dynamic information was inaccessible.

But what about a single clock in which the transfer matrix actually represents the clock's signal? The clock must operate with a representation of the information necessary to tick periodically in its rest frame. The clock is locally guided by the null geodesics, as these are invariant under boosts. However, the clock does not *choose* the reference frame that we decide to observe it in. The only special reference frame for the clock *is* its rest frame. In this frame the continuum limit was motivated by replacing periodic major ticks by uniformly distributed minor ticks:

$$T_0 \rightarrow \underbrace{T_0\, I_2\, I_2 \cdots I_2}_{n-\text{terms}} \rightarrow \underbrace{T_n\, T_n \cdots T_n}_{n-\text{terms}}. \qquad (2.1)$$

This was an averaging procedure that we might expect represents an ensemble average. However, the information *contained* in the ticking clock must be sufficient to provide for *any inertial frame from which the clock is read*. The period and phase information contained in Fig. 9 that we have ascribed to an ensemble of clocks, must be available to be read *from a single clock!* For this to be the case, the uniform replacement scheme specified above is too specific. It restricts the clock to a specific worldline by specifying detailed instructions on all scales *in a specific frame.* When we specified $T_v(t)$, we gave the transfer matrix in a specific boosted frame when in fact the clock itself cannot 'know' the velocity of the lab frame in which we are to view the clock.

In the event that clocks contain a blueprint for their paths through *spacetime*, and not just a prescription for a specific frame, a more sensible interpolation scheme for the clock would be *to bring the averaging procedure outside the fundamental scale of the clock*. This could be implemented by replacing the transfer matrix with a *configuration generating function*. So for example the above interpolation scheme could be replaced by a generating function

$$\underbrace{T_0\,I_2\,I_2\cdots I_2}_{n-\text{terms}}\underbrace{T_0\,I_2\,I_2\cdots I_2}_{n-\text{terms}}\cdots \to \underbrace{(T_0+I_2+I_2+\cdots I_2)}_{n-\text{terms}}\cdots\underbrace{(T_0+I_2+I_2+\cdots I_2)'}_{n-\text{terms}}$$

$$(3.1)$$

where we replace a sequence by a product of terms representing the number of cycles of the clock, with the prime on the product refering to a choice of inertial frame. Only those product configurations that contain events at the origin and at the *spacetime* point in the observer frame are *selected* from the generating function. The point here is that the clock itself has to have instructions encoded in its own rest frame that, none-the-less, may be *readable* in any inertial frame. Instead of assuming that spacetime informs the clock, we assume the clock contains sufficient information that it may be read from any frame. The form of (3.1) is also suggestive. Each term in the product is a sum of just two distinct terms, one a 'continue' term (I_2) and one a 'switch' term (T_0). There is a simple binary choice for each term in the product.

One way to implement the transfer matrix as a generating function is to use complex exponentials to count ticks of the clock [11, 12] and let the observer read this information in whatever frame he happens to be. In terms of the transfer matrix, it becomes a generating function if we just replace I_2 by $\exp[-ip\sigma_z\epsilon]$ and then take the continuum limit. To find the state of the clock at a specific *spacetime* point we then invert what has become a Fourier transform. To illustrate how this works let us look at, say, the fourth power of the transfer matrix, so modified to count configurations. We start with the short time transfer matrix eqn. (2.3) modified to count configurations

$$T = \begin{pmatrix} e^{-ip\epsilon} & -m\epsilon \\ m\epsilon & e^{ip\epsilon} \end{pmatrix}$$

where for clarity we have replaced $\pi/2$ by m and $1/n$ by ϵ. The $(1,1)$ component of the fourth power of this is

$$T_{11}(p) = m^4\epsilon^4 - m^2\epsilon^2 e^{2ip\epsilon} - 3m^2\epsilon^2 e^{-2ip\epsilon} - 2m^2\epsilon^2 + e^{-4ip\epsilon} \qquad (3.2)$$

Here we notice that powers of $m\epsilon$ count state changes and hence 'age' of the clock and the power of the exponent counts displacement in space Fig. 11.

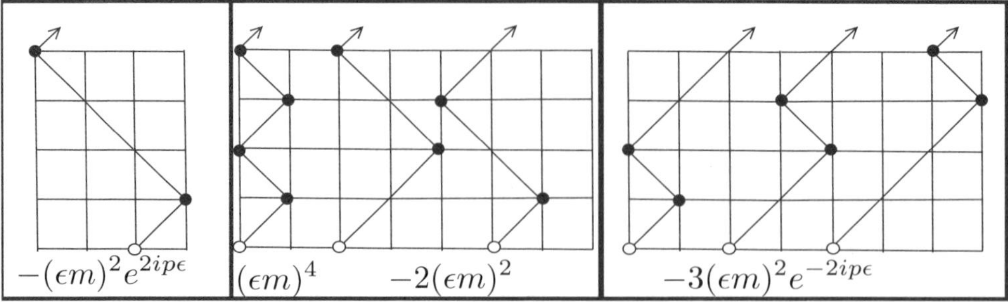

Fig. 11. Four steps with the transfer matrix give rise to the eight configurations of equation (3.2), seven of them sketched here. The open circle is the origin and closed circles are events. The argument of the exponent gives the displacement of the boundary in terms of $p\epsilon$. The power of $m\epsilon$ counts the number of events. The first pane is the single configuration that arrives two units to the left. The second pane has three configurations that arrive at the origin and the third pane has three configurations that arrive two steps to the right. The configuration not sketched has no events and arrives four steps to the right. One can see the beginning of a relativistic ageing process. The missing configuration stays on the lightcone and does not age. The configurations moving two units to the right have three events and have aged somewhat. The configurations at the origin have a maximum number of four events (maximal ageing).

The continuum limit of the generating function equation (3.2) takes the form

$$T(p, t) = \lim_{n \to \infty} \begin{pmatrix} \exp[ip\epsilon] & -m\epsilon \\ m\epsilon & \exp[-ip\epsilon] \end{pmatrix}^{t/\epsilon}$$

$$= \cos(Et)I_2 + \frac{-1}{E}(ip\sigma_z + im\sigma_y)\sin(Et) \tag{3.3}$$

where $E = \sqrt{p^2 + m^2}$.

Notice here the generating function form of the transfer matrix still has a multivector structure that is a combination of a scalar and a unit vector since

$$(\frac{-1}{E}(ip\sigma_z + im\sigma_y))^2 = \frac{1}{E^2}(-p^2 - m^2)I_2 = -I_2. \tag{3.4}$$

The generating function has still singled out one-dimensional subspaces of a 2-dimensional domain via a unit vector. To see how we return to 'configuration' space notice that we can extract the coefficients of powers of $exp(-ip\epsilon)$ in equation (3.2) by using a Kronicker delta of the form

$$\delta_{kk'} = \frac{1}{2\pi} \int_{-\pi}^{\pi} e^{i(k-k')\phi} \, d\phi. \tag{3.5}$$

In the context of equation (3.2) this is

$$T_{11}(k) = \frac{1}{2\pi} \int_{-\pi}^{\pi} e^{ik(p\epsilon)} \, (m^4\epsilon^4 - m^2\epsilon^2 e^{2ip\epsilon} - 3m^2\epsilon^2 e^{-2ip\epsilon} - 2m^2\epsilon^2 + e^{-4ip\epsilon})d(p\epsilon) \tag{3.6}$$

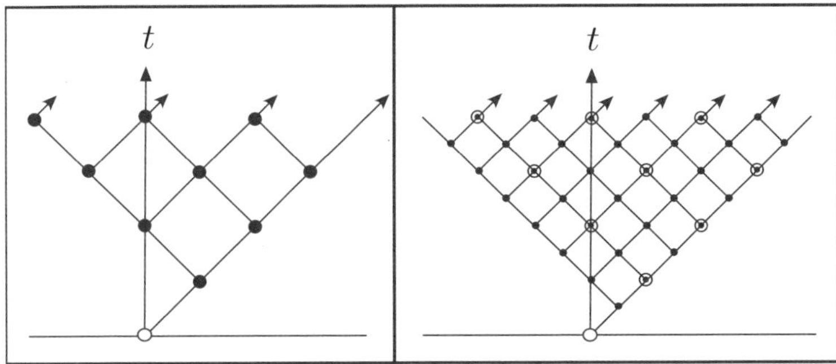

Fig. 12.   In Dirac's clock, when we refine the lattice, the configuration generating function allows for more configurations throughout the entire future light cone. It has to do this since it has to be 'readable' from any inertial frame in the continuum limit. In the first pane, the discrete clock contains configurations corresponding to the precision of the original digital clock and illustrates the configurations counted in equation (3.2) and Fig. 11. In the second pane the frequency is doubled. Note the lack of dimensional collapse in comparison to Minkowski's or Newton's clock Fig. 5.

which will extract the contributions at the integers $k' \in \{-2, 0, 2, 4\}$. Proceeding formally with equation (3.2) we see that the configuration space version of the transfer matrix is

$$
\begin{aligned}
T(x,t) &= \lim_{n \to \infty} \frac{1}{2\pi\epsilon} \int_{-\pi/\epsilon}^{\pi/\epsilon} e^{ix(p\epsilon)} \begin{pmatrix} \exp[ip\epsilon] & -m\epsilon \\ m\epsilon & \exp[-ip\epsilon] \end{pmatrix}^{t/\epsilon} d(p\epsilon) \\
&= \frac{1}{2\pi} \int_{-\infty}^{\infty} e^{-ipx} \left(\cos(Et) I_2 + \frac{1}{E}(ip\sigma_z + im\sigma_y)\sin(Et) \right) dp \qquad |x| < t
\end{aligned}
$$

$$(3.7)$$

the Fourier transform of $T(p,t)$. Here the division of the integral by ϵ in the continuum limit gives $T(x,t)$ as a density and we see that our infinitesimal configuration generating function equation (3.2) leads to a Fourier transform. The transform may be calculated explicitly leading to a replacement of the trigonometric functions by Bessel functions [6].

It is worthwhile noting at this point that the above limit is conceptually very different from the continuum limit employed in Minkowski's clock. That clock contained a dimensional collapse as illustrated in Fig. 5. To obtain high frequencies, the clock placed events ever closer together on the t-axis, shrinking the inter-arrival areas to zero. This produced a smooth worldline in a specific lab frame.

In contrast, the transfer matrix as a generating function (3.3) still singles out one dimensional subspaces, but does so as a Fourier transform. As illustrated in Fig. 12 this does not employ a dimensional collapse in configuration space provided x and t are different from 0. In fact, *it decouples the continuum limit from an infinite event density limit*. It preserves the finite number of domains in the bit-string picture of chessboard paths.

To see why we call this Dirac's clock, notice from equation (3.3) that

$$\frac{\partial T(p,t)}{\partial t} = -E\sin(Et)I_2 - (ip\sigma_z + im\sigma_y)\cos(Et) \tag{3.8}$$

so

$$ip\sigma_z T(p,t) = ip\sigma_z \cos(Et) + \frac{p}{E}\sin(Et)\,(I_2 p + im\sigma_x)$$

and

$$im\sigma_y T(p,t) = im\sigma_y \cos(Et) - \frac{m}{E}\sin(Et)\,(-I_2 m + ip\sigma_x)$$

thus

$$(\frac{\partial}{\partial t} + ip\sigma_z + im\sigma_y)T(p,t) = 0.$$

Collecting terms this gives

$$i\frac{\partial T(p,t)}{\partial t} = H\,T(p,t) \tag{3.9}$$

with

$$H = p\sigma_z + m\sigma_y \tag{3.10}$$

This is a form of the Dirac equation in a two dimensional *spacetime* in units with $c = \hbar = 1$.

4. Discussion

Even in the classical world of Newton, access to arbitrarily precise clocks is a physical impossibility. Real numbers have infinite decimal expansions and to mark an event in time to arbitrary precision would require a device with infinite storage capacity. Time as a real number has always been an idealization. In reality, the measurement of time is digital, reflecting the limited precision of all measuring devices, if not a digital aspect to time itself. Since real numbers *are* an idealization, their value in physical models has to be attuned to the limiting process that gives rise to them.

Newton's clock sketches how one might interpolate between periodic events to extract time as a real number. Newton's space-time followed on the assumption that clock frequency was independent of frame velocity, and this independence encouraged the view of space-time as a container. Absolute time as an independent dimension made sense since fundamental clock frequencies could vary, but their eventual link to a single continuous parameter was common to all clocks, and the worldline was a natural extension of the sequence of periodic ticks of a digital clock.

The invariance of the velocity of light fundamentally changed this picture. The Lorentz transformation between moving frames forces space and time to be coupled in such a way that tick rates are not independent of relative velocities. Instead of the invariant tick rates of Newton's clock, Minkowski's clock preserves inter-tick areas through boosts. This manifests itself in the transfer matrix by extracting Clifford (geometric) algebra, a matrix encoding of linear subspaces, as kinematic information. The subspaces extracted by Minkowski's clock corresponded to rays through the origin, the worldlines of free clocks. The continuum limit for Minkowski's clock is still an idealization but, like Newton's clock, shows commonality for all digital-analog transitions. All such clocks have transfer matrices that display a *spacetime* of odd signature. The odd signature is a remnant of the periodicity of the clock, but one that appears differently in the kinematic and dynamic information in the transfer matrix. For this reason it makes sense to extract the kinematic *spacetime* as a description of all clocks, dropping the period and phase information of individual clocks and inheriting the worldline and spacetime-as-a-container features of Newton's clock.

This latter feature is not as strongly motivated as for Newton's clock since tick rate is not constant for moving clocks and it is not obvious that phase information of individual clocks is irrelevant. Both Newton's and Minkowski's clock have 'clock-faces' that appear a little like 'wavefunctions'. In both the classical theories, we ultimately ignore these column vectors as being irrelevant to the program of extracting time parameterized by real numbers. Although neither clock has a superposition principle, the transfer matrix for Minkowski's clock shows that individual clocks do have a path dependent phase, a feature that occurs in path-integral formulations of quantum mechanics. This phase is ultimately ignored in the transition to *spacetime*, as is the natural uncertainty principle associated with the signal generated by the transfer matrix.

Dirac's clock arises when we consider the averaging procedure for the continuum limit to be associated with a reading of a *single* clock rather than an ensemble. It presupposes that if a single clock is to tick periodically, whatever the actual machinery of the clock, it must contain enough information that it may be 'read' from any frame that an observer might choose. That is, the clock must encode time in its own proper frame, but must also encode enough information that it can be read from *any* inertial frame.

Clock	Mass	Spacetime	Event Frequency	Worldline
Dirac	Clock Frequency	Intrinsic	Finite	No
Minkowski	Background	Ensemble Property	Infinite	Yes
Newton	Background	Independent Dimensions	Infinite	Yes

Table 1.   The three clocks compared as increasingly coarse approximations. Mass is an intrinsic property of the Dirac clock but is lost in the Minkowski and Newton clocks.   This is because both of these clocks mimic a smooth worldline with an infinite event density. For these two clocks, mass is reinstated as a background feature.   *Spacetime*, and the Clifford algebra that describes it, appear intrinsically in Dirac's clock and as an ensemble feature for Minkowski's clock.   Newton's clock assumes Gallilean transformations so time becomes a separate dimension from space.

Instead of a transfer matrix that encodes the rest frame of the clock in a specific coordinate system, must act more like a generating function that encodes all the information necessary to allow a reading from any frame. The generating function employed by Dirac's clock is, in the continuum limit, a Fourier transform.

As we saw in the last section, the transfer-matrix-as-generating-function satisfied the Dirac equation and was a form of the Dirac propagator. As such, the dynamic information encoded includes the period and phase information. The extra information shows that the frequency information ignored in Minkowski's clock characterizes inertial mass in Dirac's clock. Furthermore, the intrinsic uncertainty principle relating time and frequency, ignored by Newton and Minkowski, appears in Dirac's clock. The three clocks are compared in Table 1.

The propagator as it appears in Dirac's clock is not, by itself, quantum mechanics. Indeed, its derivation here is *completely* in the domain of classical statistical mechanics. There has been no formal analytic continuation or quantization procedure. We always know what is being counted, and why. It is consequently interesting to see the Dirac equation appear in a context that addresses the single issue of how a periodic clock can keep time in a relativistic world. It is also interesting to see that the distinction between classical *spacetime* and Dirac propagation involves the scale at which the continuum limit is taken. In the Newton and Minkowski cases, the limit is taken at scales below the clock period. The limit involves a dimensional collapse to place events densely on the worldline. The Dirac clock avoids the dimensional collapse by allowing the clock to encode all possible configurations so that the time can be 'read' from any frame. This essentially postpones the continuum limit to the reading process, thereby bringing the wave propagation, lurking under the continuum limit in Minkowski's clock, to scales above the continuum limit at which Dirac's clock is read.

5. Conclusions

Technical details aside, our exploration probes how Nature might ultimately handle the continuum limit that is traditionally a *given* in physics. Since Newton's time, the convenience of the differential calculus has encouraged us to think in terms of smooth manifolds and real numbers. Even the manifestly discrete nature of atoms has not persuaded us to switch from the continuum to completely discrete models, the 'mathematical overhead' of such a switch being too high. However, from an informational perspective the existence of true continua in Nature is questionable. In principle, a single real number could code for the positions of a countable number of objects with finite precision and so is an immensely powerful object by itself. Densely packed in a continuum, the real numbers make the configurational information of the entire visible universe at fixed precision seem small by comparison. Does Nature really require the infinite depths of continua to evolve, or is she more parsimonious. Does she draw or simply count? DPDE suggests the latter.

The evolution of Pierre's thought in relation to the Combinatorial Hierarchy [13] demonstrates courage, persistence and foresight. The CH and bit-string physics tackle difficult issues that are conventionally ignored in the rush to calculate eigenvalues by ever more complicated routes. Bit-string physics looks for simplicity and finds it underneath the rumpled cloak of Dirac propagation. The CH offers a glimpse of even deeper digital structure that, once fully disclosed, may change how we think about the physical universe.

References

[1] L. H. Kauffman and H. P. Noyes. Discrete physics and the Dirac equation. *Phys. Lett. A*, 218:139, 1996.

[2] H. Pierre Noyes, Edited by J. C. van den Berg. *Bit String Physics*. World Scientific Pub. Co., 2001.

[3] G. N. Ord. Three clocks, spacetime, and the Dirac equation. *Unpublished*, 2013.

[4] M. Kac. A stochastic model related to the telegrapher's equation. *Rocky Mountain Journal of Mathematics*, 4(3):497–509, 1974.

[5] B. Gaveau, T. Jacobson, M. Kac, and L. S. Schulman. Relativistic extension of the analogy between quantum mechanics and brownian motion. *Physical Review Letters*, 53(5):419–422, 1984.

[6] T. Jacobson and L. S. Schulman. Quantum stochastics: the passage from a relativistic to a non-relativistic path integral. *J. Phys. A*, 17:375–383, 1984.

[7] G. N. Ord and R. B. Mann. Entwined paths, difference equations and the Dirac equation. *Phys. Rev. A*, 67, 2003.

[8] G. N. Ord and R.B. Mann. How does an electron tell the time? *International Journal of Theoretical Physics*, 51(2):652–666, September 2011.

[9] G. N. Ord. Quantum propagation from the twin paradox. In G. Grössing, editor, *Emergent Quantum Mechanics*, volume 361 of *Conference Series*. Journal of Physics, 2012.

[10] Hermann Minkowski. *The Principle of Relativity: A Collection of Original Memoirs on the Special and General Theory of Relativity*, chapter "Space and Time", pages pp. 75–91. Dover: New York, 1952.

[11] Colin J. Thompson. *Mathematical Statistical Mechanics*. Princeton University Press, 1979.

[12] G. N. Ord. A reformulation of the feynman chessboard model. *J. Stat. Phys.*, 66:647–659, 1992.

[13] H. Pierre Noyes. A short introduction to bit-string physics. *arXiv:hep-th/9707020*, 1997.

Via Aristotle, Leibniz, Berkeley & Mach to Necessarily Fractal Large-scale Structure in the Universe

D. F. Roscoe

School of Mathematics & Statistics
Sheffield University, Sheffield S10 2TN
E-mail: D.Roscoe@shef.ac.uk

The claim that the large scale structure of the Universe is hierarchical has a very long history going back at least to Charlier's papers of the early 20th century. In recent years, the debate has largely focussed on the works of Sylos Labini, Joyce, Pietronero and others, who have made the quantitative claim that the large scale structure of the Universe is quasi-fractal with fractal dimension $D \approx 2$. There is now a concensus that this is the case on medium scales, with the main debate revolving around what happens on the scales of the largest available modern surveys.

Apart from the (essentially sociological) problem that their thesis is in absolute conflict with any concept of a Universe with an age of ≈ 14 billion years or, indeed, of any finite age, the major generic difficulty faced by the proponents of the hierarchical hypothesis is that, beyond hypothesizing the case (*c.g.* : Nottale's *Scale Gravity*), there is no obvious mechanism which would lead to large scale structure being non-trivially fractal. This paper, which is a realization of a worldview that has its origins in the ideas of Aristotle, Leibniz, Berkeley and Mach, provides a surprising resolution to this problem: in its essence, the paper begins with a statement of the primitive self-evident relationship which states that, in the universe of our experience, the amount of material, m, in a sphere of redshift radius R_z is a monotonic increasing function of R_z. However, because the precise relationship between any Earth-bound calibration of radial distance and R_z is unknowable then fundamental theories cannot be constructed in terms of R_z, but only in terms of a radial measure, R say, calibrated against known physics. The only certainty is that, for any realistic calibration, there will exist a monotonic increasing relationship between R_z and R so that we have $m = M(R)$ for m and R inceasing monotonically together. But the monotonicity implies $R = G(m)$ which, *in the absence of any prior calibration of R*, can be interpreted as the *definition* of the radius of an astrophysical sphere in terms of the amount of mass it contains – which is the point of contact with the ideas of Aristotle, Leibniz, Berkeley and Mach. The development of this idea leads necessarily to the final result, which is that a definitive signature of a Leibnizian universe is the *perception* that large scale structure in the Universe of our experience is quasi-fractal of dimension $D = 2$.

1.   Introduction

We can summarize the major results of this paper as follows:

- *The conflict between the idea that matter in the universe is, on large enough
  scales, distributed homogeneously and the idea that it is distributed in a quasi-
  fractal fashion with $D \approx 2$ is, at source, the conflict between two opposing views
  of the nature of space, the one having its origins in the ideas of Democritus
  and traceable through Newton to Einstein and the other having its origins in
  the ideas of Aristotle and traceable through Leibniz & Berkeley to Mach. The
  perception that matter in the universe of our experience is distributed as $D \approx 2$
  on large scales can be interpreted directly as a signature that this universe is
  structured according to the Leibniz-Berkeley ideal.*
- *The foregoing considerations lead naturally into a discussion concerning the nature of physical time, and equally naturally to its quantitative definition which
  can be qualitatively described as "ordered process". When this quantitative definition is considered in the idealized case of an exactly $D = 2$ universe, we
  find that the Lorentz transformations emerge naturally as those transformations
  which keep invariant this quantitative definition of physical time.*

The arguments of this paper serve to emphasize that we will never be able to say
how the universe *is* – being forever constrained merely to say how it *looks*.

We begin with a brief review of the phenomenological arguments which led to the
basic hypothesis that large-scale structure in the Universe is non-trivially quasi-
fractal, and comment briefly on the ramifications to canonical cosmology *if* this
hypothesis proves to be correct. We then give a brief account comparing and con-
trasting the two major world-views which have dominated thinking about the nature
of the cosmos for over two thousand years before, finally, providing the quantitative
development which leads to the main results of this paper.

1.1.   *Brief technical note on "fractals"*

Throughout this analysis, we constantly refer to large scale structure as being fractal
$D \approx 2$. Strictly speaking, this is loose language, since the term "fractal" refers
to scale-invariant behaviour on all scales whereas, in astrophysical parlance, only
astrophysical scales are implied. Additionally, in this paper we use the term "fractal"
to describe *classical distributions* which behave as $M \sim R^2$ when, strictly speaking,
the power-law behaviour of fractal distributions refers to an average behaviour when
averaged over all points of the distribution. For this reason, and from time to time,
we use the phrase "quasi-fractal" as a reminder of that fact.

1.2.   *The hierarchical Universe*

Prior to the 1920s, the Universe was generally conceived to be of infinite extent –
in time, space and mass content – and to be regulated by Newton's Laws. This,

combined with the apparently reasonable cosmological assumption of a uniform distribution of matter gave rise to two problems - Olber's paradox (1825) and its gravitational version, Seeliger's paradox (1895), according to which every massive body experiences an indefinite gravitational acceleration. Charlier [1–3] showed that if the assumption of a uniform matter distribution was replaced by that of an hierarchical distribution of matter (with, in modern terms, a fractal dimension of $D \leq 2$), then Olber's paradox became resolved. It is also the case – as Charlier similarly argued – that such an hierarchical distribution of matter also resolves Seeliger's gravitional paradox.

However, Charlier's work was overtaken by Eddington's recognition that Le Maitre's dust-cosmology solution of Einstein's equations provided a natural explanation for the redshift discoveries of Slipher & Hubble. Le Maitre's rudimentary cosmology has evolved, over the years, into the *Standard Model,* a basic assumption of which is that, on some scale, the universe is homogeneous; however, in early responses to suspicions that the accruing data was more consistent with Charlier's conceptions of an hierarchical universe than with the requirements of the Standard Model, de Vaucouleurs [4] showed that, within wide limits, the available data satisfied a mass distribution law $M \approx r^{1.3}$, whilst Peebles [5] found $M \approx r^{1.23}$. The situation, from the point of view of the Standard Model, continued to deteriorate with the growth of the data-base to the point that, Baryshev *et al.* [6] state

> *...the scale of the largest inhomogeneities (discovered to date) is comparable with the extent of the surveys, so that the largest known structures are limited by the boundaries of the survey in which they are detected.*

For example, several redshift surveys of the late 20th century, such as those performed by Huchra *et al.* [7], Giovanelli and Haynes [8], De Lapparent *et al.* [9], Broadhurst *et al.* [10], Da Costa *et al.* [11], and Vettolani *et al.* [12], *etc.* discovered massive structures such as sheets, filaments, superclusters and voids, and showed that large structures are common features of the observable universe; the most significant conclusion drawn from all of these surveys was that the scale of the largest inhomogeneities observed in the samples was comparable with the spatial extent of those surveys themselves.

In the closing years of the century, several quantitative analyses of both pencil-beam and wide-angle surveys of galaxy distributions were performed: three examples are given by Joyce, Montuori & Labini [13] who analysed the CfA2-South catalogue to find fractal behaviour with $D = 1.9 \pm 0.1$; Labini & Montuori [14] analysed the APM-Stromlo survey to find fractal behaviour with $D = 2.1 \pm 0.1$, whilst Labini, Montuori & Pietronero [15] analysed the Perseus-Pisces survey to find fractal behaviour with $D = 2.0 \pm 0.1$. There are many other papers of this nature, and of the same period, in the literature all supporting the view that, out to $30 - 40h^{-1}Mpc$ at least, galaxy distributions appeared to be fractal with $D \approx 2$.

This latter view became widely accepted (for example, see Wu, Lahav & Rees [16]), and the open question became whether or not there was transition to homogeneity on some sufficiently large scale. For example, Scaramella *et al.* [17] analyse the ESO Slice Project redshift survey, whilst Martinez *et al.* [18] analyse the Perseus-Pisces, the APM-Stromlo and the 1.2-Jy IRAS redshift surveys, with both groups claiming to find evidence for a cross-over to homogeneity at large scales.

At around about this time, the argument reduced to a question of statistics (Sylos Labini & Gabrielli [19], Gabrielli & Sylos Labini [20], Pietronero & Sylos Labini [21]): basically, the proponents of the fractal view began to argue that the statistical tools (*e.g.* correlation function methods) widely used to analyse galaxy distributions by the proponents of the opposite view are deeply rooted in classical ideas of statistics and implicitly assume that the distributions from which samples are drawn are homogeneous in the first place. Recently, Hogg *et al.* [22], having accepted these arguments, applied the techniques argued for by the pro-fractal community (which use the *conditional density* as an appropriate statistic) to a sample drawn from Release Four of the Sloan Digital Sky Survey. They claim that the application of these methods does show a turnover to homogeneity at the largest scales thereby closing, as they see it, the argument. In response, Labini *et al.* [23] have criticized their paper on the basis that the strength of the conclusions drawn is unwarranted given the deficiencies of the sample – in effect, that it is not big enough.

To summarize, the proponents of non-trivially fractal large-scale structure have won the argument out to medium distances and the controversy now revolves around the largest scales encompassed by the SDSS.

1.3. *Theoretical implications*

The notion of non-trivially quasi-fractal large-scale structure in the Universe is problematic for proponents of any form of big-bang cosmology for the following reason: if there is non-trivially fractal structure in the Universe then, ideally, the mechanism for the formation of such structure should be open to a theoretical understanding. But the equations of General Relativity are hyperbolic so that global structure at any epoch is always going to be determined primarily by initial conditions – in other words, large-scale fractal structure (should it exist) in the Universe cannot be explained within the confines of General Relativity, but only in terms of *initial* conditions which are external to it. Whilst this *might* actually be the case, the primary function of science must always be to attempt the understanding of that which is observed. Thus, any attempt to explain away vast and spectacular structures observed today in terms of "conditions before the universe came into being" would be to evade the most significant of all questions – in other words, big-bang cosmology would fail to meet the most basic requirement of a scientific theory.

It follows that the hypothesis that large-scale structure in the Universe is non-trivially fractal and the hypothesis that General Relativity is the fundamentally correct theory by which the Universe and its properties can be understood, are in direct opposition to each other. This implies that, if we are to take the fractal hypothesis seriously (as the phenomenology suggests we should), then only a radical review of our ideas of space and time can hope to provide any understanding how such structure can occur.

2. A brief history of ideas of space and time

The conception of space as the container of material objects is generally considered to have originated with Democritus and, for him, it provided the stage upon which material things play out their existence – *emptiness* exists and is that which is devoid of the attribute of *extendedness* (although, interestingly, this latter conception seems to contain elements of the opposite view upon which we shall comment later). For Newton [24], an extension of the Democritian conception was basic to his mechanics and, for him:

> ... *absolute space, by its own nature and irrespective of anything external, always remains immovable and similar to itself.*

Thus, the absolute space of Newton was, like that of Democritus, the stage upon which material things play out their existence – it had an *objective existence* for Newton and was primary to the order of things. In a similar way, time – *universal time,* an absolute time which is the same everywhere – was also considered to possess an objective existence, independently of space and independently of all the things contained within space. The fusion of these two conceptions provided Newton with the reference system – *spatial coordinates* defined at a *particular time* – by means of which, as Newton saw it, all motions could be quantified in a way which was completely independent of the objects concerned. It is in this latter sense that the Newtonian conception seems to depart fundamentally from that of Democritus – if *emptiness* exists and is devoid of the attribute of *extendedness* then, in modern terms, the *emptiness* of Democritus can have no *metric* associated with it. But it is precisely Newton's belief in *absolute space & time* (with the implied virtual clocks and rods) that makes the Newtonian conception a direct antecedent of Minkowski spacetime – that is, of an empty space and time within which it is possible to have an internally consistent discussion of the notion of *metric.*

The contrary view is generally considered to have originated with Aristotle [25, 26] for whom there was no such thing as a *void* – there was only the *plenum* within which the concept of the *empty place* was meaningless and, in this, Aristotle and Leibniz [27] were at one. It fell to Leibniz, however, to take a crucial step beyond the Aristotelian conception: in the debate of Clarke-Leibniz (1715∼1716) [28] in which Clarke argued for Newton's conception, Leibniz made three arguments of which the second was:

> *Motion and position are real and detectable only in relation to other objects ...*
> *therefore empty space, a void, and so space itself is an unnecessary hypothesis.*

That is, Leibniz introduced a *relational* concept into the Aristotelian world view –
what we call *space* is a projection of *perceived relationships* between material bodies
into an inferred world whilst what we call *time* is the projection of ordered *change*
into an inferred world. Of the three arguments, this latter was the only one to which
Clarke had a good objection – essentially that *accelerated motion,* unlike uniform
motion, can be perceived *without* reference to external bodies and is therefore, he
argued, necessarily perceived with respect to the *absolute space* of Newton. It is
of interest to note, however, that in rebutting this particular argument of Leibniz,
Clarke, in the last letter of the correspondence, put his finger directly upon one of
the crucial consequences of a relational theory which Leibniz had apparently not
realized (but which Mach much later would) stating as absurd that:

> *... the parts of a circulating body (suppose the sun) would lose the* vis cen-
> trifuga *arising from their circular motion if all the extrinsic matter around*
> *them were annihilated.*

This letter was sent on October 29th 1716 and Leibniz died on November 14th 1716
so that we were never to know what Leibniz's response might have been.

Notwithstanding Leibniz's arguments against the Newtonian conception, nor Berkeley's contemporary criticisms [29], which were very similar to those of Leibniz and
are the direct antecedents of Mach's, the practical success of the Newtonian prescription subdued any serious interest in the matter for the next 150 years or so
until Mach himself picked up the torch. In effect, he answered Clarke's response
to Leibniz's second argument by suggesting that the *inertia* of bodies is somehow
induced within them by the large-scale distribution of material in the universe:

> *... I have remained to the present day the only one who insists upon referring*
> *the law of inertia to the earth and, in the case of motions of great spatial and*
> *temporal extent, to the fixed stars ...* [30]

thereby generalizing Leibniz's conception of a relational universe. Mach was equally
clear in expressing his views about the nature of time: in effect, he viewed *time*
(specifically Newton's *absolute time*) as a meaningless abstraction. All that we can
ever do, he argued in [30], is to measure *change* within one system against *change*
in a second system which has been defined as the standard (*e.g.* it takes half of one
complete rotation of the earth about its own axis to walk thirty miles).

Whilst Mach was clear about the origins of inertia (in the fixed stars), he did not
hypothesize any mechanism by which this conviction might be realized and it fell
to others to make the attempt – a typical (although incomplete) list might include
the names of Einstein [31], Sciama [32], Hoyle & Narlikar and Sachs [33, 34] for
approaches based on canonical ideas of spacetime, and the names of Ghosh [35] and

Assis [36] for approaches based on quasi-Newtonian ideas.

It is perhaps one of the great ironies of 20thC science that Einstein, having coined the name *Mach's Principle* for Mach's original suggestion and setting out to find a theory which satisfied the newly named Principle, should end up with a theory which, whilst albeit enormously successful, is more an heir to the ideas of Democritus and Newton than to the ideas of Aristotle and Leibniz. One only has to consider the special case solution of Minkowski spacetime, which is empty but metrical, to appreciate this fact.

3. From Leibniz to inertia as a relational property

In this paper we take the general position of Leibniz about the relational nature of space to be self-evident and considered the question of *spatial metric* within this general conceptualization – that is, how is the notion of invariant *spatial* distance to be defined in the Leibnizian particle universe? To answer this question, we begin by considering the universe of our actual experience and show how it is possible to define an invariant measure for the *radius* of a statistically defined astrophysical sphere purely in terms of the amount of matter it contains (to within a calibration exercise); we then show how the arguments deployed can be extended to define an invariant measure for an *arbitrary* spatial displacement within the statistically defined astrophysical sphere. In this way, we arrive at a theory within which a metrical three-space is projected as a secondary construct out of the primary distribution of universal material. This has the subtle consequence that it becomes misleading to talk about "the distribution of matter in space" – rather, the talk is about "the distribution of space around matter". In practice, as we shall see, this gives rise to a worldview in which the most simple model universe is Euclidean and in which it is *axiomatic* that material appears to be distributed quasi-fractally, $D = 2$.

The question of how *time* arises within this worldview is particularly interesting: the simple requirement that *time* should be defined in such a way that Newton's Third Law is automatically satisfied has the direct consequence that *time* becomes an explicit invariant measure of change within the system, very much as anticipated by Mach. In the most simple model, the overall result is a quasi-classical (that is, one-clock) theory of relational space & time:

- which represents a Euclidean universe within which it is axiomatic that material appears to be distributed quasi-fractally, $D = 2$ and within which there is a condition of global dynamical equilibrium;
- and which is such that point-source perturbations of this equilbrium Euclidean universe recover the usual Newtonian prescriptions for "gravitational" effects.

The first of these two points refers to the universe that Leibniz was effectively considering in his debate with Clarke of 1715~1716 - one within which inertial effects play no part. The second refers to the universe that Clarke used to refute Leibniz's second argument, and the one that Mach had in mind – the universe of rotations and accelerations. Thus, given our Leibnizian worldview, we see that inertial effects themselves have their fundamental source in changed material relationships – they, too, are *relational* in nature.

3.1. *The general argument*

Following in the tradition of Aristotle, Leibniz, Berkeley and Mach we argue that no consistent cosmology should admit the possibility of an internally consistent discussion of *empty* metrical space & time – unlike, for example, General Relativity which has the empty spacetime of Minkowski as a particular solution. Recognizing that the most simple space & time to visualize is one which is everywhere inertial, then our worldview is distilled into:

> **The elemental question:** *is it possible to conceive a globally inertial space & time which is irreducibly associated with a non-trivial global mass content and, if so, what is the inferred distribution of this matter content for any observer?*

In pursuit of an answer to this question, we shall assume an idealized universe:

- *which consists of an infinity of identical, but labelled, discrete "galaxies" which possess an ordering property which allows us to say that galaxy $\mathcal{G}_0$ is nearer/further than galaxy $\mathcal{G}_1$. The redshift properties of galaxies in the real universe are an example of such an ordering property;*
- *within which " time" is to be understood, in a qualitative way, as a measure of process or ordered change in the model universe;*
- *within which there is at least one origin about which the distribution of material particles is statistically isotropic – meaning that the results of sampling along arbitrary lines of sight over sufficiently long characteristic 'times' are independent of the directions of lines of sight.*

3.2. *Astrophysical spheres and a galaxy-count calibrated metric for radial displacements*

It is useful to discuss, briefly, the notion of spherical volumes defined on large astrophysical scales in the universe of our experience: whilst we can certainly give various precise operational definitions of spherical volumes on small scales, the process of giving such definitions on large scales is decidedly ambiguous. In effect, we have to suppose that redshift measurements are (statistically) isotropic when taken from an arbitrary point within the universe and that they vary monotonically with distance on the large scales we are concerned with. With these assumptions, spherical volumes can be defined (statistically) in terms of redshift measurements – however, their radial calibration in terms of ordinary units (such as metres) becomes increasingly uncertain (and even unknown) on very large redshift scales.

With these ideas in mind, the primary step taken in answer to the elemental question above is the recognition that, on large enough scales in the universe of our experience (say $> 30\, Mpc$), we can identify a redshift-defined astrophysical sphere and, having identified it, we can estimate the amount of matter, m say, contained within this sphere by, for example, simply *counting* the galaxies it contains. Once this is done, we can, if we wish, *define* a radial scale for the sphere at any given time (whatever we might mean by *time*) purely in terms of the mass it contains; that is, we can say

$$R = G(m, t) \rightarrow \delta R = G(m + \delta m, t) - G(m, t) \tag{1}$$

where R is the mass-calibrated radius, m is the mass concerned and G is an arbitrarily chosen monotonic increasing function of m.

Thus, for any given G, we have immediately defined an invariant radial measurement such that it becomes undefined in the absence of matter – in effect, we have, in principle, a metric which follows Leibniz in the required sense for any displacement which is purely radial.

3.3. *The time-independent versus the time-dependent development*

If we continue the development from (1) then the result is a two-clock theory of the *frame-time/particle proper time* type. By contrast, a much-simplified one-clock quasi-classical theory results if we make the (cosmological) assumption that all epochs are identical so that t in (1) can be dropped. It transpires that, from the point of view of statements about large-scale structure, the differences between the simple one-clock theory and the more complicated two-clock theory are differences of detail only: for example, suppose that S is a statement concerning large scale-structure in the one-clock theory then, in the two-clock theory, this statement would be qualified as follows: *when observed over time scales that are small compared to the characteristic time scales of the volumes concerned, then S.*

Thus, for the sake of simplicity – so that the basic ideas are most easily revealed – we make the assumption that all epochs *are identical* so that the t-dependence in (1) can be dropped with the consequence that it becomes

$$R = G(m). \tag{2}$$

It is important to emphasize that this latter equation is not an *hypothesis* about reality; it is a non-trivially true statement about the real universe which recognizes that we can, if we so choose, *define* a measure of physical length on a cosmological scale purely in terms of a quantity of matter. It is this which is the non-trivial point of contact with the ideas of Leibniz, Berkeley *et al.* and which allows the development (below) of a quantitative – if rudimentary – cosmology in the mode of Leibniz & Berkeley.

3.4. *A mass-calibrated metric for arbitrary spatial displacements*

The foregoing provides a way of giving an invariant measure in the real universe, defined in terms of mass, for displacements which are purely *radial* from the chosen origin. However, if, for example, a displacement is *transverse* to a radial vector, then the methodology fails. Thus, we must look for ways of generalizing the above ideas so that we can assign a mass-calibrated metric to *arbitrary* spatial displacements within the real universe – that is, we must look for a way of assigning a *metric* to this universe. To this end, we invert (2) to give a mass model

$$Mass \equiv m = M(R) \equiv M(x^1, x^2, x^3),$$

for our rudimentary universe. Note that we make no assumptions about the relation of the spatial coordinates, (x^1, x^2, x^3), to the redshift-defined radial displacement, R.

Now consider the normal gradient vector $n_a = \nabla_a M$ (which does not require any metric stucture for its definition): the change in this arising from a displacement dx^k can be *formally* expressed as

$$dn_a = \nabla_i (\nabla_a M) \, dx^i, \tag{3}$$

where we shall later argue (cf §3.7) that the affinity required to give this latter expression an unambiguous meaning is the usual metric affinity – *except of course, the metric tensor g_{ab} of our curvilinear three-space required to define this latter object is not, itself, yet defined.*

Now, since g_{ab} is not yet defined, then the covariant counterpart of dx^a, given by $dx_a = g_{ai} dx^i$, is also not yet defined. However, we note that, assuming $\nabla_a \nabla_b M$ to be nonsingular, then (3) provides a 1:1 mapping between the contravariant vector dx^a and the covariant vector dn_a so that, in the absence of any other definition, we can *define dn_a to be the covariant form of dx^a.* In this latter case the metric tensor of our curvilinear three-space automatically becomes

$$g_{ab} = \nabla_a \nabla_b M \tag{4}$$

which, through the implied metric affinity, is a highly non-linear partial differential equation defining g_{ab} to within the specification of M. The scalar product

$$dS^2 \equiv dn_i dx^i \equiv g_{ij} dx^i dx^j \tag{5}$$

then provides an invariant measure for the magnitude of the infinitesimal three-space displacement, dx^a, given purely in terms of the mass model, M.

To briefly summarize: equation (4) encapsulates the reality that, *in the universe of our experience,* it is possible to give a real quantitative meaning to the ideas of Leibnitz and Berkeley; in other words, that within this universe, the notion of metrical physical space can be considered as a projection from the relationships that exist between material 'objects' – whatever these might be.

3.5. *This definition of g_{ab} as a model of daily human experience*

In order to emphasize how the foregoing is more than merely an "interesting hypothesis", it is useful to show how it relates directly to the primitive human being's intuitive assessments of "distance traversed" in everyday life without recourse to formal instruments.

In effect, as we travel through a physical environment, we use our changing perspective of the observed scene in a given elapsed time to provide a qualitative assessment of "distance traversed" in that elapsed time. So, briefly, when walking across a tree-dotted landscape, the changing *angular* relationships between ourselves and the trees provide the information required to assess "distance traversed", measured in units of human-to-tree displacements, within that landscape. If we remove the perspective information – by, for example, obliterating the scene with dense fog – then all sense of "distance traversed" is destroyed.

In the above definition of g_{ab}, the part of the tree-dotted landscape is played by the mass-function, M, whilst the instantaneous perspective on this "landscape" is quantified by the normal vector n_a and the change in perspective arising from a coordinate displacement, dx^a, is quantified by the change in this normal vector, dn_a. The invariant measure defined at (5) can then be considered to be based on a comparison between dx^a (for which an invariant magnitude is required) and dn_a.

In other words, g_{ab} as defined acts as a descriptive model for our intuitive sense of physical metric space – the same intuitive sense of space that probably led Aristotle, Leibniz and Berkeley et al to formulate their quantitative ideas on the nature of physical space in the first place.

3.6. *Units*

The units of dS^2 at (5) are easily seen to be those of *mass* only and so, in order to make them those of *length^2* – as dimensional consistency requires – we define the working invariant as $ds^2 \equiv (2r_0^2/m_0)dS^2$, where r_0 and m_0 are scaling constants for the distance and mass scales respectively and the numerical factor has been introduced for later convenience.

Finally, if we want

$$ds^2 \equiv \left(\frac{r_0^2}{2m_0}\right) dn_i dx^i \equiv \left(\frac{r_0^2}{2m_0}\right) g_{ij} dx^i dx^j \tag{6}$$

to behave sensibly in the sense that $ds^2 > 0$ whenever $|d\mathbf{r}| > 0$ and $ds^2 = 0$ only when $|d\mathbf{r}| = 0$, then we must replace the condition of non-singularity of g_{ab} by the condition that it is strictly positive definite; in the physical context of the present problem, this will be considered to be a self-evident requirement.

3.7. *The affine connection*

We have assumed that the affinity required to give unambiguous meaning to (3) can be defined in some sensible way. To do this, we simply note that, in order to define conservation laws (ie to do physics) in a Riemannian space, it is necessary to have a generalized form of the Gauss divergence theorem in that space. This is certainly possible when the affinity is defined to be the metric affinity, but it is by no means clear that it is ever possible otherwise. Consequently, the affinity is defined to be metrical so that g_{ab}, given at (4), can be written explicitly as

$$g_{ab} \equiv \nabla_a \nabla_b M \equiv \frac{\partial^2 M}{\partial x^a \partial x^b} - \Gamma^k_{ab} \frac{\partial M}{\partial x^k}, \tag{7}$$

where Γ^k_{ab} arc the Christoffel symbols, and given by

$$\Gamma^k_{ab} = \frac{1}{2} g^{kj} \left(\frac{\partial g_{bj}}{\partial x^a} + \frac{\partial g_{ja}}{\partial x^b} - \frac{\partial g_{ab}}{\partial x^j} \right).$$

4. The metric tensor given in terms of the mass model

We have so far made no assumptions about the nature of the coordinate system $\left(x^1, x^2, x^3\right)$. However, we now recall our elemental question, originally formulated in §3.1, and note that within the context of the quasi-classical one-clock model now being considered, it can be more explicitly stated as

> ***The elemental question:*** *is it possible to conceive a globally Euclidean space with a universal time which is irreducibly associated with a non-trivial global mass distribution which is in a state of dynamical equilibrium and, if so, what are the properties of this distribution as inferred from within the Euclidean space?*

This question is most directly answered by *assuming* the existence of the globally Euclidean space in the first place for the analysis of (7) and seeing what conclusions about the structure of M follow from this assumption. Thus, we define $\left(x^1, x^2, x^3\right)$ as Cartesian coordinates in a globally defined rectangular frame and suppose the usual Pythagorean relationship, $R^2 = \left(x^1\right)^2 + \left(x^2\right)^2 + \left(x^3\right)^2$.

With this understanding, it is shown, in appendix A, how (7) can be exactly resolved to give an explicit form for g_{ab} in terms of $m \equiv M(R)$: defining the notation

$$\mathbf{R} \equiv (x^1, x^2, x^3), \quad \Phi \equiv \frac{1}{2} (\mathbf{R} \cdot \mathbf{R}) = \frac{1}{2} R^2 \text{ and } M' \equiv \frac{dM}{d\Phi},$$

this explicit form of g_{ab} is given as

$$g_{ab} = A\delta_{ab} + Bx^i x^j \delta_{ia}\delta_{jb}, \tag{1}$$

where

$$A \equiv \frac{d_0 M + m_1}{\Phi}, \quad B \equiv -\frac{A}{2\Phi} + \frac{d_0 M' M'}{2A\Phi},$$

for arbitrary constants d_0 and m_1 where, as inspection of the structure of these expressions for A and B shows, d_0 is a dimensionless scaling factor and m_1 has dimensions of mass. Noting now that M always occurs in the form $d_0 M + m_1$, it is convenient to write $\mathcal{M} \equiv d_0 M + m_1$, and to write A and B as

$$A \equiv \frac{\mathcal{M}}{\Phi}, \quad B \equiv -\left(\frac{\mathcal{M}}{2\Phi^2} - \frac{\mathcal{M}'\mathcal{M}'}{2d_0\mathcal{M}} \right). \tag{2}$$

4.1. *A unique calibration for the radial scale*

We began by assuming only the Leibnizian definition of an astrophysically defined sphere's radius in terms of the sphere's mass content, $R = G(m)$ (cf equation (2)), where G is an unknown monotonic increasing function. Note that, since galaxy-counts can be used as a proxy for m, then $R = G(m)$ is necessarily an *invariant*. We subsequently generalized the discussion using, as a model, a human being's in-tuitive notion of "distance traversed" and requiring, as a crucial detail, that we ended up with a formal structure within which we can define conservation laws – cf §3.7. In the following, we show that, as a direct consequence of this modelling approach, the functional form, G, and hence the Leibnizian definition of R in terms of m, become uniquely determined.

Using (1) and (2) in (6), and using the Pythagorean identity $x^i dx^j \delta_{ij} \equiv R dR$ then, with the notation $\Phi \equiv R^2/2$, we find, for an arbitrary displacement $d\mathbf{x}$, the invariant measure:

$$ds^2 \equiv g_{ij} dx^i dx^j = \left(\frac{R_0^2}{2m_0} \right) \left\{ \frac{\mathcal{M}}{\Phi} dx^i dx^j \delta_{ij} - \Phi \left(\frac{\mathcal{M}}{\Phi^2} - \frac{\mathcal{M}'\mathcal{M}'}{d_0\mathcal{M}} \right) dR^2 \right\},$$

which is valid for the unknown calibration $R = G(m) \longleftrightarrow m = M(R)$. If the displacement $d\mathbf{x}$ is now constrained to be purely *radial*, then we find

$$ds^2 = \left(\frac{R_0^2}{2m_0} \right) \left\{ \Phi \left(\frac{\mathcal{M}'\mathcal{M}'}{d_0\mathcal{M}} \right) dR^2 \right\}.$$

Use of $\mathcal{M}' \equiv d\mathcal{M}/d\Phi$ and $\Phi \equiv R^2/2$ reduces this latter relationship to

$$ds^2 = \frac{R_0^2}{d_0 m_0} \left(d\sqrt{\mathcal{M}} \right)^2 \quad \rightarrow \quad ds = \frac{R_0}{\sqrt{d_0 m_0}} d\sqrt{\mathcal{M}} \quad \rightarrow$$

$$s = \frac{R_0}{\sqrt{d_0 m_0}} \left(\sqrt{\mathcal{M}} - \sqrt{\mathcal{M}_0} \right), \quad \text{where} \quad \mathcal{M}_0 \equiv \mathcal{M}(s = 0)$$

which defines the invariant magnitude of a purely *radial* displacement from the origin solely in terms of the mass-model representation $\mathcal{M} \equiv d_0 M + m_1$. But the invariant magnitude of a purely radial displacement has, until this point, been represented as $R = G(m)$ – remember that this is invariant simply because m it determined via a *counting* process. It follows that $s \equiv R = G(m)$ so that, finally, we have

$$s \equiv R = G(m) \equiv \frac{R_0}{\sqrt{d_0 m_0}} \left(\sqrt{\mathcal{M}} - \sqrt{\mathcal{M}_0} \right), \tag{3}$$

where $\mathcal{M}_0$ is the value of $\mathcal{M}$ at $R = 0$.

Since we are engaged in a discussion about astrophysical distance scales, it is pertinent to ask, at this point, what relation the radial measure R here has to the ordinary distance measures of astrophysics? This question is largely answered at the beginning of §4, where R is defined via the Pythagorean relationship, and completely answered in §4.2 where we show how Euclidean space is an exact special case of the forgoing considerations so that R can be identified definitively as the first rung of the standard cosmic distance ladder.

4.2. *The special case of Euclidean space*

To summarize general points, m is the amount of mass contained within an astrophysical sphere of redshift radius R_z, and $R = G(m)$ is the Leibnizian statement which defines the sphere's radius in ordinary units of length in terms of its mass content.

Remembering $\mathcal{M} \equiv d_0 m + m_1$ (see §4) and noting that $M(R = 0) = 0$ necessarily, then $\mathcal{M}_0 = m_1$ and so (3) can be expressed as

$$R = \frac{R_0}{\sqrt{d_0 m_0}} \left(\sqrt{d_0 m + m_1} - \sqrt{m_1} \right),$$

for the Leibnizian *definition* of the radius R in ordinary units of length of the astrophysical sphere which has mass content m. For the idealized case $m_1 = 0$ this definition becomes

$$R = R_0 \sqrt{\frac{m}{m_0}}. \tag{4}$$

Reference to (1) shows that, with this latter definition and $d_0 = 1$, then $A = 2m_0/R_0^2$ and $B = 0$ so that $g_{ab} \sim \delta_{ab}$ with the consequence that the associated three-space becomes ordinary Euclidean space.

Furthermore, since all points are equivalent in a Euclidean universe, then the choice of origin for R in (4) is arbitrary. In other words, (4) can be understood to describe – in the same statistical sense that we assume the universe to be isotropic from arbitrarily chosen origins – the relationship between the Leibnizian definition of R in terms of m as observed from an arbitrarily chosen origin.

4.3. *Fractality $D \approx 2$ as a signature of the Leibniz-Berkeley ideal*

The relationship (4) can be equivalently stated as

$$4\pi R^2 = 4\pi R_0^2 \frac{m}{m_0}$$

which can be directly interpreted to mean that

> *the area, measured in ordinary Euclidean units, of the astrophysically defined sphere of redshift radius R_z is equal to its mass content to within a scale factor.*

This result has the *axiomatic* consequence that, if an observer in this Euclidean universe chooses to investigate the "spatial distribution of matter" using a Euclidean yardstick then he will necessarily find a quasi-fractal $D = 2$ distribution of matter. Thus, from this point of view, the observation of $D \approx 2$ in the universe of our experience is actually a *signature* that the space of our real experience really is projected as a secondary construct out of its matter content, in the manner of the Leibnitz-Berkeley ideal.

5. The temporal dimension

So far, the concept of "time" has only entered the discussion in a qualitative way in §2 – it has not entered in any quantitative way and, until it does, there can be no discussion of dynamical processes.

Since, in its most general definition, time is a parameter which orders change within a system, then a necessary pre-requisite for its quantitative definition is a notion of change within the universe. The most simple notion of change which can be defined in the universe is that of changing relative spatial displacements of the objects within it. Since our model universe is populated solely by primitive galaxies which possess only the property of enumerability (and hence quantification in terms of the *amount* of material present) then, in effect, all change is gravitational change. Thus, the question of how we define "time" is tightly bound up with the question of how we quantify gravitation in the present schema.

At this point we need to be careful: if we were dealing with the two-clock model (where one of the clocks – ordinary frame-time – is built into the theory from the beginning) then we would be working within a Riemannian *spacetime* and could then incorporate the fact that "all change is gravitational change" by invoking the Weak Equivalence Principle and requiring that all trajectories are geodesic in the spacetime manifold. However, we are actually working with the one-clock model (where this clock has yet to be defined at all) and the manifold is a purely spatial one; consequently gravitation here *cannot* be modelled simply by requiring that all trajectories are geodesic. Notwithstanding this point, we note that within a defined Newtonian gravitating system (say, the solar system) the overall *shape* in ordinary Euclidean space of a body's trajectory is determined purely by the body's instantaneous velocity at any particular defined point on the trajectory – specifically, it is independent of the body's mass. This suggests that the *geometric shapes* of gravitational trajectories *can* be identified as geodesics within our spatial manifold. This leads directly to a perfectly conventional Lagrangian description of particle trajectories in terms of an arbitrarily defined temporal ordering parameter in which the Lagrange density is degree zero in that temporal-ordering parameter. From this, it follows, as a standard theorem, that the corresponding Euler-Lagrange equations form an *incomplete* set.

The origin of this incompleteness property traces back to the fact that, because the Lagrangian density is degree zero in the temporal ordering parameter, it is then invariant with respect to any transformation of this parameter which preserves the ordering. This implies that, in general, such temporal ordering parameters cannot be identified directly with physical time – they merely share one essential characteristic.

So, there exists a symmetry – the invariance of the equations of motion under transformations of the temporal ordering parameter – which must be broken if any notion of physical time is to enter the discussion. Remembering that we are working within a quasi-classical regime, it is then interesting (and gratifying) to find that the Newtonian condition "the action between two particles is along the line connecting them" is exactly what is required to break the symmetry, thereby introducing a notion of "physical time" into the present schema. The net result, in effect, is that orbital periods within gravitating systems become the basis by which systems of physical time are defined within our model universe.

5.1. *The equations of 'motion' on the spatial manifold*

The geodesic equations in the space with the metric tensor (1) can be obtained, in the usual way, by defining the Lagrangian density

$$\mathcal{L} \equiv \left(\frac{1}{\sqrt{2g_0}} \right) \sqrt{g_{ij}\dot{x}^i\dot{x}^j} = \left(\frac{1}{\sqrt{2g_0}} \right) \left(A \left(\dot{\mathbf{R}} \cdot \dot{\mathbf{R}} \right) + B\dot{\Phi}^2 \right)^{1/2} , \tag{1}$$

where $g_0 \equiv m_0/R_0^2$, and $\dot{x}^i \equiv dx^i/dt$, etc., and forming the associated variational principle

$$\mathcal{I} = \left(\frac{1}{\sqrt{2g_0}} \right) \int \sqrt{g_{ij}\dot{x}^i\dot{x}^j}\,dt , \tag{2}$$

where t is the temporal ordering parameter.

The first thing to note here is that this variational principle is of order zero in t so that it is invariant with respect to arbitrary transformations of this parameter. This has the consequences referred to above:

- firstly that the temporal ordering parameter cannot be identified with physical time;
- secondly that the resulting Euler-Lagrange equations for a single particle, given by

$$2A\ddot{\mathbf{R}} + \left(2A'\dot{\Phi} - 2\frac{\dot{\mathcal{L}}}{\mathcal{L}}A\right)\dot{\mathbf{R}} + \left(B'\dot{\Phi}^2 + 2B\ddot{\Phi} - A'\left(\dot{\mathbf{R}}\cdot\dot{\mathbf{R}}\right) - 2\frac{\dot{\mathcal{L}}}{\mathcal{L}}B\dot{\Phi}\right)\mathbf{R} = 0,$$

$$(3)$$

where $\dot{\mathbf{R}} \equiv d\mathbf{R}/dt$, $A' \equiv dA/d\Phi$ and $\Phi \equiv (\mathbf{R}\cdot\mathbf{R})/2 = R^2/2$ etc, cannot form a complete set so that some additional constraint must be applied to close the system.

A similar circumstance arises in General Relativity when the equations of motion are derived from an action integral which is formally identical to (2). In that case, the system is closed by specifying the arbitrary time parameter to be the "particle proper time", so that

$$d\tau = \mathcal{L}(x^j, dx^j) \quad \rightarrow \quad \mathcal{L}(x^j, \frac{dx^j}{d\tau}) = 1,$$

$$(4)$$

which is then considered as the necessary extra condition required to close the system.

5.2. *Newtonian action as the closing constraint*

Given the isotropy conditions imposed on the model universe from the chosen origin, symmetry arguments lead to the conclusion that the net action of the whole universe of particles acting on any given single particle is such that any net acceleration of the particle must always appear to be directed through the coordinate origin. Note that this conclusion is *independent* of any notions of retarded or instantaneous action. This constraint can then be stated as the Newtonian condition that:

> *Any acceleration of any given material particle must necessarily be along the line connecting the particular particle to the coordinate origin.*

This latter condition simply means that the equations of motion must have the general structure

$$\ddot{\mathbf{R}} = G(t, \mathbf{R}, \dot{\mathbf{R}})\mathbf{R},$$

for scalar function $G(t, \mathbf{R}, \dot{\mathbf{R}})$. Equation (3) has this structure if the coefficient of $\dot{\mathbf{R}}$ in that equation is zero; that is, if

$$\left(2A'\dot{\Phi} - 2\frac{\dot{\mathcal{L}}}{\mathcal{L}}A\right) = 0 \quad \rightarrow \quad \frac{A'}{A}\dot{\Phi} = \frac{\dot{\mathcal{L}}}{\mathcal{L}} \quad \rightarrow \quad \mathcal{L} = k_0 A,$$

$$(5)$$

for arbitrary constant k_0 which is necessarily positive since $A > 0$ and $\mathcal{L} > 0$. The condition (5) can be considered as the condition required to close the incomplete set (3) and is directly analogous to (4), the condition which defines 'proper time' in General Relativity.

Thus, from (3) the equations of motion can be finally written as

$$2A\ddot{\mathbf{R}} + \left(B'\dot{\Phi}^2 + 2B\ddot{\Phi} - A'\left(\dot{\mathbf{R}}\cdot\dot{\mathbf{R}}\right) - 2\frac{A'}{A}B\dot{\Phi}^2 \right)\mathbf{R} = 0. \tag{6}$$

5.3. *Physical time defined as process*

Equation (5) can be considered as that equation which removes the pre-existing arbitrariness in the 'time' parameter by *defining* physical time:- from (5) and (1) we have

$$\mathcal{L}^2 = k_0^2 A^2$$
$$\downarrow$$
$$A\left(\dot{\mathbf{R}}\cdot\dot{\mathbf{R}}\right) + B\dot{\Phi}^2 = 2g_0 k_0^2 A^2 \tag{7}$$
$$\downarrow$$
$$g_{ij}\dot{x}^i\dot{x}^j = 2g_0 k_0^2 A^2$$

so that, in explicit terms, physical time is *defined* by the relation

$$dt^2 = \left(\frac{1}{2g_0 k_0^2 A^2}\right) g_{ij}dx^i dx^j, \quad \text{where} \quad A \equiv \frac{\mathcal{M}}{\Phi}. \tag{8}$$

In short, the elapsing of time is given a direct physical interpretation in terms of the process of *displacement* in the model universe. It is gratifying to note that this coincides exactly with (for example) Mach's view of time, expressed in [30], according to which all we ever do when measuring elapsed time is to measure *change* within one system against *change* in a second system which has been defined as the standard (*e.g.* it takes half of one complete rotation of the earth about its own axis to walk thirty miles).

Finally, noting that, by (8), the dimensions of k_0^2 are those of $L^6/[T^2 \times M^2]$, then the fact that $g_0 \equiv m_0/R_0^2$ (cf §4.1) suggests the change of notation $k_0^2 \propto v_0^2/g_0^2$, where v_0 is a constant having the dimensions (but not the interpretation) of 'velocity'. So, as a means of making the dimensions which appear in the development more transparent, it is found convenient to use the particular replacement $k_0^2 \equiv v_0^2/(4d_0^2 g_0^2)$. With this replacement, the *definition* of physical time, given at (8), becomes

$$dt^2 = \left(\frac{4d_0^2 g_0}{v_0^2 A^2}\right) g_{ij}dx^i dx^j.$$

5.4. *Physical time and the Lorentz transformations*

We show that, in the idealized limiting case of $(d_0, m_1) = (1, 0)$ which gives rise to *exactly* Euclidean space, the foregoing definition of "physical time" leads to a natural emergence of what is canonically termed "light cone geometry" together with Bondi's interpretation of light velocity, c.

For this case, the definition of R at (4) together with the definitions of A and B in §4 give

$$A = \frac{2m_0}{R_0^2}, \quad B = 0$$

so that, by (7) (remembering that $g_0 \equiv m_0/R_0^2$ and $k_0^2 \equiv v_0^2/(4d_0^2 g_0^2)$), and writing $\mathbf{R} \equiv (x, y, z)$ for convenience, we have

$$\left(\frac{dx}{dt}\right)^2 + \left(\frac{dy}{dt}\right)^2 + \left(\frac{dz}{dt}\right)^2 = v_0^2 \tag{9}$$

for all displacements in the model universe. It is (almost) natural to assume that the constant v_0^2 in (9) simply refers to the constant velocity of any given particle, and likewise to assume that this can differ between particles. However, each of these assumptions would be wrong since – as we now show – v_0^2 is, firstly, more properly interpreted as a conversion factor from spatial to temporal units and, secondly, is a *global* constant which applies equally to all particles.

To understand these points, we begin by noting that (9) is a special case of (7) and so, by (8), is more accurately written as

$$dt^2 = \frac{1}{v_0^2}\left(dx^2 + dy^2 + dz^2\right) \tag{10}$$

which, by the considerations of §5.3, we recognize as the *definition* of the elapsed time experienced by any particle undergoing a spatial displacement (dx, dy, dz) in the model inertial universe. Since this universe is isotropic about all points, then there is nothing which can distinguish between two separated particles (other than their separateness) undergoing displacements of equal magnitudes; consequently, each must be considered to have experienced equal elapsed times. It follows from this that v_0 is not to be considered as a locally defined particle speed, but is a *globally* defined constant which has the effect of converting between spatial and temporal units of measurement; that is, it is more accurately thought of as being a conversion factor from length scales to time scales – which is identical to Bondi's interpretation of the parameter c, which is most commonly interpreted as the speed of light.

Since, in the Universe of our experience, the only global constant with dimensions of velocity is c, then it is natural to make the identification $v_0 \equiv c$ in (10). The Lorentz transformations, which now follow automatically, then have the natural interpretation as those transformations which keep invariant the definition of physical time.

6. The axiomatically $D = 2$ inertial universe

By the considerations of §4.2, the limiting case of the axiomatically $D = 2$ Euclidean universe is recovered by the parameter choice $(d_0, m_1) = (1, 0)$ and, as is also shown in §4.2, this implies $A = const > 0$ and $B = 0$ in (6). In other words, in this limiting case

$$\ddot{\mathbf{R}} = 0.$$

Thus, this universe is necessarily also a globally inertial equilibrium universe, and the questions originally posed in §3.1 are finally answered.

6.1. *Seeliger's gravitational paradox*

This paradox boils down to the statement that, within a Euclidean universe with a uniform distribution of matter and within which Newton's Universal Law is assumed, it is possible to assign an *arbitrary* gravitational acceleration to *any* particle depending on how one evaluates the (non-uniformly convergent) integrals which arise. Charlier showed that, within his proposed hierarchical universe, the paradox disappears.

However, Charlier's argument is not relevant here since Newton's Universal Law is not assumed. The relevant point here is simply that, as shown above, the axiomatically $D = 2$ material universe is necessarily a globally inertial equilibrium universe so that Seeliger's gravitational paradox does not arise within the Leibnizian cosmology developed here.

6.2. *Olber's paradox*

It has already been pointed out in §1.2 that Charlier argued in 1908 how Olber's Paradox disappeared if matter in the universe was assumed distributed hierarchically (fractally in modern parlence) such that $M \sim R^a$ for $a \leq 2$. Thus, by Charlier's arguments, Olber's Paradox – like Seeliger's paradox – cannot exist within the Leibnizian cosmology developed here.

6.3. *Implications for theories of gravitation*

Given that gravitational phenomena are usually considered to arise as mass-driven perturbations of flat inertial backgrounds, then the foregoing result – to the effect that the inertial background is axiomatically associated with the *appearance* of a $D = 2$ "distribution of matter in space" – must necessarily give rise to completely new perspectives about the nature and properties of gravitational phenomena. However, as we showed in §5.4, the structure of the idealized "inertial background" suggests that it can be properly interpreted as a "photon sea" – reminiscent of the microwave background - out of which (presumably) ordinary material can be considered to condense in some fashion.

7. Summary

7.1. *The significance of "Large Scale Structure"*

The main result of this paper amounts to the statement that the conflict between the idea that material in the Universe "should be" distributed homogeneously and the claim that it appears (on large scales) to be fractal $D \approx 2$ is, at source, the conflict between two opposing notions of "space":

- one as an objective reality which functions as a container of material - a notion which can be traced from Democritus, through Newton to Einstein (in practice);
- the other as a secondary construct projected out of relationships between material objects – a notion which can be traced from Aristotle, Leibniz & Berkeley to, most recently, Mach.

The problem for those subscribing to the Leibniz-Berkeley ideal has always been how to provide it with quantitative expression – the problem which this paper set out to address. We began by noting the empirical fact that in the universe of our experience, and on large enough scales, we can use the distance-ordering property of galactic redshifts to define astrophysical spheres and can subsequently estimate the amount of matter, m say, contained within any given such sphere simply by *counting* the galaxies it contains. We then noted that we could *define* a calibration for the radius of such a sphere purely in terms of the estimated contained mass, so that $R = G(m)$ where G is an arbitrarily defined monotonic increasing function if m. It is this latter step which is the non-trivial point of quantitative contact with the Leibniz-Berkeley ideal. Such a definition of R implies a simple definition of a radial displacement, dR, in terms of the matter contained between two shells, radius R and $R + dR$.

There then followed a technical discussion about how this idea of *an invariant radial displacement being defined in terms of an amount of matter* could be generalized for arbitrary non-radial displacements. This discussion took its cue primarily from the human being's sense of how relative displacement is perceived (such "sense" being at the root of the Leibniz-Berkeley ideal), and led to a worldview within which quantitative questions can be formulated and answered. Thus, given that the most simple conception of space that we have is one which is everywhere Euclidean, the fundamental quantitative question formulated, in effect, was:

> *What is the quantitative form of the definition $R = G(m)$ in a Euclidean universe?*

The worldview replied $R \sim \sqrt{m}$. It is then axiomatic that if an observer asks the Newtonian question "what is the distribution of matter in space?" and sets out to answer this question using a predetermined Euclidean ruler, he will then find that it appears to be fractal, $D \approx 2$. In fact, according to the arguments presented here, this result should be interpreted as a signature that the universe is "constructed" according to the ideals of Leibniz & Berkeley.

7.2. *The nature of "time"*

The discussion which led to the foregoing conclusion also unavoidably entailed a corresponding discussion concerning the nature of "time" - as Mach himself pointed out, so far as "time" is concerned, the most we can ever do is to define the "time required for process A to occur" in terms of the "time required for process B to occur". For example, "I can walk 100 miles between one sunrise and the next." From this viewpoint, it is arguable, for example, that the individual proton – never observed to decay, and therefore a stranger to change - exists *out of time*. But the internal distribution of an assembly of (labelled) protons, on the other hand, does change and a sequence of snapshots of such a changing assembly could be considered as an evolutionary sequence *defining of itself the passage of time*. In effect, "time" is defined as a metaphor for "ordered change within a physical system" and it is this definition of "time" which arises automatically from the considerations which lead to the $R \sim \sqrt{m}$ Euclidean Universe above – there is a set of particles which possess only the property of *enumerability* from which the concept of "time" arises as a metaphor for "ordered change".

It was then extremely interesting to find that, in the limiting case of an exactly Euclidean universe, our considerations about the nature of time led automatically to the Lorentz transformations as those transformations which keep invariant the definition of temporal process within the Euclidean universe.

7.3. *Implications for gravitation theory*

The foregoing development amounts to that of providing the essential framework for a theory of gravitation in a universe conforming to the Leibniz-Berkeley ideal and so obvious questions to be answered are:

- can the classical theory of Newton for a point source (like the Sun) be recovered?
- does the relativistic form of the theory (the two-clock model) for a point source pass the standard tests, up to and including those associated with the binary pulsar?

For the Newtonian case, it is sufficient to consider point-source perturbations of the $(d_0, m_1) = (1, 0)$ limiting case of the $D = 2$ Euclidean universe. All Newtonian structure can then be recovered in a straightforward way. The relativistic case is currently the subject of investigation.

7.4. *Implications for the observations*

Probably the major difficulty for both sides of the structure debate is that the measuring rods used on large scales cannot be reliably nor independently calibrated

– a problem that becomes increasingly severe with increasing scale. For that reason alone then, as things stand, there is probably no reasonable prospect that the debate can ever be unambiguously concluded. However, *even if* the debate could be concluded to the satisfaction of the pro-homogeniety proponents, an open question would remain: *how are the quasi-fractal structures, which are agreed to exist on medium scales, to be explained?* Since recourse to "initial conditions" cannot constitute an illuminating explanation, then there is no answer even to this limited question within standard big-bang cosmology.

A universe "constructed" according to the Leibniz-Berkeley ideal provides a rational escape route: according to this ideal, it is axiomatic that when reliably calibrated Euclidean measuring rods are used to determine the "distribution of material in space", then a fractal $D = 2$ matter distribution will be inferred. In practice, the measuring rods we use are reliably defined to be Euclidean on small scales and become progressively less reliably calibrated as the scales increase. The actually inferred $D \approx 2$ structure out to medium scales is readily understood on these terms. Accepting this as circumstantial evidence favouring the Leibniz-Berkeley ideal, it then becomes natural to progress by calibrating the measuring rods in such a way that the inferred "structure" is $D = 2$ on all scales, by definition. The gain is a new way to calibrate distance scales and therefore, for example, a new way to investigate the redshift phenomenon itself.

Appendix A.   *A Resolution of the Metric Tensor*

The general system is given by

$$g_{ab} = \frac{\partial^2 M}{\partial x^a \partial x^b} - \Gamma^k_{ab} \frac{\partial M}{\partial x^k},$$

$$\Gamma^k_{ab} \equiv \frac{1}{2} g^{kj} \left(\frac{\partial g_{bj}}{\partial x^a} + \frac{\partial g_{ja}}{\partial x^b} - \frac{\partial g_{ab}}{\partial x^j} \right),$$

and the first major problem is to express g_{ab} in terms of the mass function, M. The key to this is to note the relationship

$$\frac{\partial^2 M}{\partial x^a \partial x^b} = M' \delta_{ab} + M'' x^i x^j \delta_{ia} \delta_{jb},$$

where $M' \equiv dM/d\Phi$, $M'' \equiv d^2 M/d\Phi^2$, $\Phi \equiv R^2/2$ and $R^2 \equiv \left(x^1 \right)^2 + \left(x^2 \right)^2 + \left(x^3 \right)^2$, since this immediately suggests the general structure

$$g_{ab} = A \delta_{ab} + B x^i x^j \delta_{ia} \delta_{jb}, \tag{1}$$

for unknown functions, A and B. It is easily found that

$$g^{ab} = \frac{1}{A} \left[\delta^{ab} - \left(\frac{B}{A + 2B\Phi} \right) x^a x^b \right]$$

so that, with some effort,

$$\Gamma^k_{ab} = \frac{1}{2A} H^k_{ab} - \left(\frac{B}{2A(A + 2B\Phi)} \right) G^k_{ab}$$

where

$$H^k_{ab} = A'(x^i \delta_{ia} \delta^k_b + x^j \delta_{jb} \delta^k_a - x^k \delta_{ab})$$
$$+ B' x^i x^j x^k \delta_{ia} \delta_{jb} + 2B x^k \delta_{ab}$$

and

$$G^k_{ab} = A'(2x^i x^j x^k \delta_{ia} \delta_{jb} - 2\Phi x^k \delta_{ab})$$
$$+ 2\Phi B' x^i x^j x^k \delta_{ia} \delta_{jb} + 4\Phi B x^k \delta_{ab}.$$

Consequently,

$$g_{ab} = \frac{\partial^2 M}{\partial x^a \partial x^b} - \Gamma^k_{ab} \frac{\partial M}{\partial x^k} \equiv \delta_{ab} M' \left(\frac{A + A'\Phi}{A + 2B\Phi} \right)$$
$$+ x^i x^j \delta_{ia} \delta_{jb} \left(M'' - M' \left(\frac{A' + B'\Phi}{A + 2B\Phi} \right) \right).$$

Comparison with (1) now leads directly to

$$A = M' \left(\frac{A + A'\Phi}{A + 2B\Phi} \right) = M' \left(\frac{(A\Phi)'}{A + 2B\Phi} \right),$$
$$B = M'' - M' \left(\frac{A' + B'\Phi}{A + 2B\Phi} \right).$$

The first of these can be rearranged as

$$B = \frac{M'}{2\Phi} \left(\frac{(A\Phi)'}{A} \right) - \frac{A}{2\Phi}$$

or as

$$\left(\frac{M'}{A + 2B\Phi} \right) = \frac{A}{(A\Phi)'},$$

and these expressions can be used to eliminate B in the second equation. After some
minor rearrangement, the resulting equation is easily integrated to give, finally,

$$A \equiv \frac{d_0 M + m_1}{\Phi}, \quad B \equiv -\frac{A}{2\Phi} + \frac{d_0 M' M'}{2A\Phi}.$$

. . . .

References

[1]   Charlier, C.V.L., 1908, Astronomi och Fysik 4, 1 (Stockholm)
[2]   Charlier, C.V.L., 1922, Ark. Mat. Astron. Physik 16, 1 (Stockholm)
[3]   Charlier, C.V.L., 1924, Proc. Astron. Soc. Pac. 37, 177
[4]   De Vaucouleurs, G., 1970, Sci 167, 1203

[5]   Peebles, P.J.E., 1980, The Large Scale Structure of the Universe, Princeton University Press, Princeton, NJ.

[6]   Baryshev, Yu V., Sylos Labini, F., Montuori, M., Pietronero, L. 1995 *Vistas in Astronomy* 38, 419

[7]   Huchra, J., Davis, M., Latham, D.,Tonry, J., 1983, ApJS 52, 89

[8]   Giovanelli, R., Haynes, M.P., Chincarini, G.L., 1986, ApJ 300, 77

[9]   De Lapparent, V., Geller,M.J., Huchra, J.P., 1988, ApJ 332, 44

[10] Broadhurst, T.J., Ellis, R.S., Koo, D.C., Szalay, A.S., 1990, Nat 343, 726

[11] Da Costa, L.N., Geller, M.J., Pellegrini, P.S., Latham, D.W., Fairall, A.P., Marzke, R.O., Willmer, C.N.A., Huchra, J.P., Calderon, J.H., Ramella, M., Kurtz, M.J., 1994, ApJ 424, L1

[12] Vettolani, G., *et al.*, in: Proc. of Schloss Rindberg Workshop: Studying the Universe With Clusters of Galaxies

[13] Joyce, M., Montuori, M., Sylos Labini, F., 1999, Astrophys. J. 514, L5

[14] Sylos Labini, F., Montuori, M., 1998, Astron. & Astrophys., 331, 809

[15] Sylos Labini, F., Montuori, M., Pietronero, L., 1998, Phys. Lett., 293, 62

[16] Wu, K.K.S., Lahav, O., Rees, M.J., 1999, Nature 397, 225

[17] Scaramella, R., Guzzo, L., Zamorani, G., Zucca, E., Balkowski, C., Blanchard, A., Cappi, A., Cayatte, V., Chincarini, G., Collins, C., Fiorani, A., Maccagni, D., MacGillivray, H., Maurogordato, S., Merighi, R., Mignoli, M., Proust, D., Ramella, M., Stirpe, G.M., Vettolani, G. 1998 *A&A* 334, 404

[18] Martinez, V.J., PonsBorderia, M.J., Moyeed, R.A., Graham, M.J. 1998 *MNRAS* 298, 1212

[19] Labini, F.S., Gabrielli, A., 2000, *Scaling and fluctuations in galaxy distribution: two tests to probe large scale structures*, astro-ph0008047

[20] Gabrielli, A., Sylos Labini, F., 2001, Europhys. Lett. 54 (3), 286

[21] Pietronero, L., Sylos Labini, F., 2000, Physica A, (280), 125

[22] Hogg, D.W., Eistenstein, D.J., Blanton M.R., Bahcall N.A, Brinkmann, J., Gunn J.E., Schneider D.P. 2005 ApJ, 624, 54

[23] Sylos Labini, F., Vasilyev, N.L., Baryshev, Y.V., Archiv.Astro.ph/0610938

[24] Newton, I., *Principia: Mathematical Principles of Natural Philosophy,* translated by I.B. Cohen & A. Whitman, University of California Press, 1999

[25] Aristotle, *Categories,* in R. McKeon, *The Basic Works of Aristotle,* Random House, New York, 1941.

[26] Aristotle, *Physics,* translated by P.H. Wicksteed & F.M. Cornford, Leob Classical Library, Harvard University Press, Cambridge, 1929

[27] Leibniz, G.W., *Philosophical Essays,* Edited and translated by R. Ariew & D.Garber, Hackett Publishing Co, Indianapolis, 1989

[28] Alexander, H.G., *The Leibniz-Clarke Correspondence,* Manchester University Press, 1984.

[29] Berkeley, G., *Of motion, or the principle and nature of motion and the cause of the communication of motions* (Latin: *De Motu)* in M.R.Ayers (ed) *George Berkeley's Philosophical Works,* Everyman, London, 1992.

[30] Mach, E., *The Science of Mechanics - a Critical and Historical Account of its Development* Open Court, La Salle, 1960

[31] Einstein, A., *The Foundations of the General Theory of Relativity,* in A. Einstein, H.A.Lorentz, H.Weyl & H.Minkowski, *The Principle of Relativity,* Dover, New York, 1952

[32] Sciama, D.W., 1953, *On the Origin of Inertia,* Mon. Not. Roy. Astron. Soc, 113, 34

[33] Sachs, M., *General Relativity and Matter,* Reidal, Dordrecht, 1982

[34] Sachs, M., *Quantum Mechanics from General Relativity*, Reidal, Dordrecht 1986
[35] Ghosh, A., *Origin of Inertia*, Apeiron, Montreal, 2000
[36] Assis, A.K.T., *Relational Mechanics*, Apeiron, Montreal, 1999

A Dual Space as the Basis of Quantum Mechanics and Other Aspects of Physics

Peter Rowlands

Department of Physics, University of Liverpool
Oliver Lodge Laboratory
Oxford St, Liverpool, L69 7ZE, UK
E-mail: P.Rowlands@liverpool.ac.uk

Space is the only quantity which can be directly observed. Its main characteristics, in its 3-dimensional Euclidean form, are describable by a Clifford algebra. There have been attempts to construct physics on the basis of this parameter alone, involving extensions to extra dimensions and non-Euclidean forms of geometry. However, it seems that adding a second Clifford algebra, identical in structure to the first, but commutative to it, and dual to it in that it contains no extra physical information, with the duality determined by a zero totality condition, gives us the mathematical construction that we need for both quantum mechanics and other forms of physics. The results can be explored from many angles, for example, Clifford algebra, group theory, geometry and topology, to produce significant results in fundamental physics.

Keywords: dual vector space, singularities, Dirac algebra, fermions, *zitterbewegung*, nilpotent
PACS: : 03.65.Fd, 03.65.Pm, 03.65.Vf, 03.65.-w, 04.20.Gz, 11.10.Gh, 11.10.Hi, 11.30.Er, 11.30.Pb, 12.40.-y

1. Defining dual spaces

All physical measurements are mediated via the concept of space. We talk about measurements of such things as time and mass, but, ultimately, these measurements can all be reduced to the equivalent of a pointer moving over a spatial scale, or to a process of counting the number of times movement over the same scale is repeated. The special nature of space has always been recognized, and this has led to more than one attempt by physicists to reduce all other physical parameters to a version of space in one form or another. A very early attempt was made by Descartes who defined matter in terms of extension and created a space which was filled with matter manifesting itself as spatial extension. In the early twentieth century, Minkowski proclaimed that Einstein's special theory of relativity meant that time could be incorporated as a fourth dimension of space, and that the separate existences of space and time would henceforth be abolished. Einstein himself then interpreted the gravitational effect of mass-energy as equivalent to a curvature in the Minkowski space-time. Following this, the Kaluza-Klein theory incorporated electromagnetism by creating an additional fifth dimension for the curved space-time, while string theory has set about defining the whole Standard Model, based on four fundamental interactions, as a space-time constructed from ten dimensions. In these theories point-like particles become 'strings', constructed from one dimension of space and one of time, existing in an eight-dimensional substratum. Five classes of string theory are known and it has been hypothesized that they can be related to each other by adding an extra dimension to the strings, so that they become 'membranes' with two dimensions of space and one of time, again within an eight-dimensional substratum, so requiring eleven dimensions in all.

Such theories have received a great deal of support in the last three decades, but they are undoubtedly faced with major difficulties. There are so many possible string or membrane theories available (loose estimates have suggested figures like 10^{500}) that the chances of hitting on the 'correct' one by accident seem impossibly remote. They also face the difficulty of trying to explain why the 'real' (observed) world seems to be based on a space with just three dimensions, and they have been famously unable to come up with anything resembling experimental predictions. The whole development, from special to general relativity, to Kaluza-Klein, and on to string and membrane theory, seems to produce a fit with nature which is more awkward and less natural at each successive stage. Already there is a problem with making time a fourth dimension of space. This does not seem to be entirely compatible with quantum mechanics, where space is an observable, but time is not, and the problem does not appear to be resolved by increasing the number of dimensions. Clearly, the theories can be made to 'work' to some extent up to the Kaulza-Klein level, but the increasing awkwardness and complication seem to suggest that this is not theory at a truly fundamental level, where we would expect increasing simplification.

There is, however, an alternative, which incorporates ideas which have proved to be of fundamental significance to physics over the last few years, in particular the concept of symmetry, and even more particularly that of duality. The ultimate basis of this approach is a set of principles long held by the author in one form or another, of which the most important is that Nature exhibits zero totality in all of its aspects, material and conceptual, and it does this via a fundamental principle of duality, which can be inferred, but not observed directly, from within the system [1]. If space is the concept through which all physical observation is mediated, then we can only perceive its existence because a dual space, which cannot be observed, operates simultaneously in such a way that the sum total of all of 'Nature' is precisely zero, though this cannot be observed from within the system. The difficulties in describing Nature using a single space are overcome once we recognize that it has a dual partner, even though the dual space contains no new information. Equally significant is the fact that we can construct a model for physics without combining observed space with anything other than its dual, without going beyond 3 dimensions, and without assuming distortion or curvature at a fundamental level.

All the difficulties that arise with trying to construct a more and more complicated single space (often with the feel of adding the modern equivalent of epicycles to the system) are overcome by merging it with its dual partner.

2. Clifford algebra

The only full description of a physical vector space that includes such aspects as its ability to generate areas and volumes as well as lengths and directions is that of Clifford or geometrical algebra. The Clifford algebra of 3-dimensional space can be described by $+$ and $-$ versions of 8 base units:

i	**j**	**k**	vector		
i**i**	i**j**	i**k**	bivector	pseudovector	quaternion
i			trivector	pseudoscalar	
1			scalar		

A bivector is the direct product of two orthogonal vector units and a trivector the direct product of three orthogonal vector units. The vector units **i**, **j**, **k** are identical to the complexified quaternion units i**i**, i**j**, i**k**, and are isomorphic to the Pauli matrices $\boldsymbol{\sigma}_x$, $\boldsymbol{\sigma}_y$, $\boldsymbol{\sigma}_z$. They follow the multiplication rules

$$\mathbf{ij} = -\mathbf{ji} = i\mathbf{k}$$
$$\mathbf{jk} = -\mathbf{kj} = i\mathbf{i}$$
$$\mathbf{ki} = -\mathbf{ik} = i\mathbf{j}$$
$$\mathbf{i}^2 = \mathbf{j}^2 = \mathbf{k}^2 = 1$$
$$\mathbf{ijk} = i$$

Also, two vectors **a** and **b,** made up of summations of the three unit values multiplied by arbitrary scalar coefficients, have a full product which is a combination of the ordinary vector and scalar products:

$$\mathbf{ab} = \mathbf{a}.\mathbf{b} + i\mathbf{a} \times \mathbf{b}$$

It is notable that, while 1 is the fourth component of the quaternion system, with $\boldsymbol{ijk} = -1$, the quantity that plays that rôle in the vector system is i. If we multiply the quaternion units $\boldsymbol{i}$, $\boldsymbol{j}$, $\boldsymbol{k}$, 1, by i to create a 4-vector system, we obtain **i**, **j**, **k**, i.

What we are now proposing is that the reason why all of physics is mediated through the concept of 3-dimensional vector space alone is that this space is supplemented by a dual construct, commutative to itself, which contains no new information but which is necessary to obtain the fundamental totality zero condition which we believe applies to the natural world. Let us describe this dual vector space using $+$ and $-$ versions of the following symbols:

I	**J**	**K**	vector		
i**I**	i**J**	i**K**	bivector	pseudovector	quaternion
i			trivector	pseudoscalar	
1			scalar		

If the two spaces are commutative, then the full algebra combining them will be a tensor product, with 64 units which can be represented as $+$ and $-$ versions of the following:

i	**j**	**k**	i**i**	i**j**	i**k**	i	1
I	**J**	**K**	i**I**	i**J**	i**K**		
iI	**iJ**	**iK**	i**iI**	i**iJ**	i**iK**		
jI	**jJ**	**jK**	i**jI**	i**jI**	i**jK**		
kI	**kJ**	**kK**	i**kI**	i**kJ**	i**kK**		

The algebraic structure represented by these units is clearly a group of order 64, which can be shown to be isomorphic to the algebra of the Dirac equation, or γ matrices, just as the algebra of a single vector space is recognisably that of the Pauli or $\boldsymbol{\sigma}$ matrices. It is not difficult to show that all possible versions of the γ matrices can be derived from a commutative combination of two sets of $\boldsymbol{\sigma}$ matrices, say $\boldsymbol{\sigma}_1$, $\boldsymbol{\sigma}_2$, $\boldsymbol{\sigma}_3$ and $\Sigma_1 \Sigma_2$, Σ_3.

3. The nilpotent condition

So far we have established a group structure made from a combination of two vector spaces. We now need to establish that these two spaces are dual, and we can accomplish this by imposing the *nilpotency condition*. First of all, we identify the generators of the group. The minimal number is 5, and so the minimal number of units for describing physics in this way is also 5. It is possible to generate the group from a combination such as $i, i\mathbf{j}, i\mathbf{I}, i\mathbf{J}, i$, but all the combinations which include the eight base units $1, i, \mathbf{i}, \mathbf{j}, \mathbf{k}, \mathbf{I}, \mathbf{J}, \mathbf{K}$, have the same structure, typically represented by

$$\mathbf{K} \qquad i\mathbf{I} \quad i\mathbf{Ij} \quad i\mathbf{Ik} \qquad i\mathbf{J}$$

The variations within this pattern involve introducing symbols with $-$ signs, switching upper- and lower-case letters, switching the individual vector units within each set of three, and multiplying by the pseudoscalar term i. However, none of these changes affects the overall pattern, which significantly distinguishes between the two sets of commuting vector units, in that one group (here, $\mathbf{i}$, $\mathbf{j}$, $\mathbf{k}$) remains rotation-symmetric because each of the units is multiplied by the same algebraic object, while the other group ($\mathbf{I}$, $\mathbf{J}$, $\mathbf{K}$) loses this symmetry because each of the units is multiplied by a quite different kind of algebraic object. As represented above, $\mathbf{K}$ is multiplied by a scalar unit, $\mathbf{I}$ by a bivector or pseudovector, and $\mathbf{J}$ by a pseudoscalar.

The algebra is unchanged by multiplying the units by arbitrary scalar values, positive or negative. Here we will represent them using the respective symbols E, p_x, p_y, p_z, m. So, incorporating these arbitrary scalar values, we can represent the 5 generators of our group formed from two commuting vector spaces by the expressions:

$$(\mathbf{K}E + i\mathbf{Ii}p_x + i\mathbf{Ij}p_y + i\mathbf{Ik}p_z + i\mathbf{J}km)(\mathbf{K}E + i\mathbf{Ii}p_x + i\mathbf{Ij}p_y + i\mathbf{Ik}p_z + i\mathbf{J}km) = 0$$

which works out as

$$E^2 - p_x^2 - p_y^2 - p_z^2 - m^2 = 0.$$

This is the nilpotent condition: the bracketed object squares to zero. The zeroing ensures, as we can show, that the information in the two spaces represented by the respective units $\mathbf{i}, \mathbf{j}, \mathbf{k}$ and $\mathbf{I}, \mathbf{J}, \mathbf{K}$ is identical. It also defines, in principle, the meaning of a *point* in either of the two spaces as the norm 0 crossover between them. Mathematicians discuss points in ordinary 3-dimensional space, but physically they have no meaning, as space is a nonconserved quantity whose units have no definable identity because they have rotation symmetry. In effect, we cannot identify anything in a single space, but identification becomes possible if we have two spaces. But if nonconservation can be thought of as denying identification, then identification can be thought of as suggesting conservation. In effect, the units represented by the second space ($\mathbf{I}, \mathbf{J}, \mathbf{K}$) become identifiable because they are associated with different algebraic objects, and so the 3-dimensionalty of this second space is somehow structurable as a conserved dimensionality because of its rotation asymmetry.

Gradually, we see that our two spaces are beginning to suggest a picture of physics as we know it. We have an emerging idea of two spaces, which, though containing the same information and though presumably symmetric at some very deep level, look very different to the physical observer because the creation of a system of 5 generators necessarily breaks a symmetry between the way we perceive them. Essentially, from our position *within* the system, we have been forced to 'privilege' one space over the other, to maintain the symmetry of one while losing that of the other. This can be seen as similar to the way in which our most primitive form of numbering, binary arithmetic, 'privileges' 1 over -1, making them dual in summing to 0, but appearing very different in the way they are perceived from within a system defined by unit 1.

4. A physical realization of two spaces

So far we have based our conclusions on physics as we perceive it as mediated *via* the concept of space, but nevertheless requiring a dual concept to ensure that the fundamental condition of the universe remains at zero totality. We have not justified, on fundamental grounds, why we use space at all and where the other space might come from. Clifford algebra, significantly, has three subalgebras, which we can describe as scalar, complex and quaternion, or scalar, trivector and bivector. Each of these is an algebra in its own right, and it is difficult to see why only the full Clifford algebra should have a physical meaning. In fact, previous work suggests that all of the subalgebras have physical meanings on the same level as Clifford algebra, and that they represent the respective physical concepts of mass, time and charge. That is, besides the vector algebra of space, we have three independent algebras which have a physical representation on the same level as space. If we combine these three physical concepts as representing everything that is excluded from space as represented by **i**, **j**, **k**, then the total structure is equivalent to a single vector space represented by the units **I**, **J**, **K**, but *without anything which directly corresponds to these units* :

charge	i**I**	i**J**	i**K**	bivector	pseudovector	quaternion
time	i			trivector	pseudoscalar	
mass	1			scalar		

In other words, we are unable to observe the space represented by **I**, **J**, **K**, because it has no single physical representation. It is, instead, a mathematical combination of three physical quantities which are not part of the space represented by **i**, **j**, **k**. Our second space has physical meaning, but cannot be accessed as a physical quantity. It is conveniently called 'vacuum space', as opposed to 'real space'. It can also be described as 'antispace' because it combines with real space to produce zero totality [1]. Also, to emphasize the independence of charge from the 'space' in which it is incorporated, it is convenient to represent it directly using quaternion, rather than bivector, notation, so the components of 'vacuum space' now become:

charge	i	j	k	bivector	pseudovector	quaternion
time	i			trivector	pseudoscalar	
mass	1			scalar		

We can now link the generation of space as a Clifford algebra and those of mass, time and charge as independent subalgebras, with the evolutionary universal rewrite process we previously defined for the whole sequence from zero totality, and also with the D_2 group symmetry which can be derived from the most fundamental properties of mass, time, charge and space, which also represent a conceptual zero totality:

mass	real (norm $+1$)	commutative	conserved
time	imaginary (norm -1)	commutative	nonconserved
charge	imaginary (norm -1)	anticommutative (3D)	conserved
space	real (norm $+1$)	anticommutative (3D)	nonconserved

The real / imaginary and commutative / anticommutative properties are directly derived from the algebraic units associated with the respective parameters. The conserved and nonconserved natures of charge and space are related to the way they are combined in the 5 group generators creating the norm 0 overall structure, while the corresponding natures of mass and time are related to the fact that quantities with their algebraic characteristics are needed to complete the quaternion and vector properties of charge and space. While the rewrite structure shows that the evolutionary process creating mass, time, charge and space can be continued to infinity, the creation of nilpotent structures zeroing all higher terms and the perfect group symmetry allowing a complete cancellation ensure that all the higher order structures can be incorporated in Clifford algebra at the first level.

The creation of a norm 0 state out of our double Clifford algebra can be interpreted as the creation of the physical objects known as point-particles or fermions. No other fundamental structures are known to be needed for physics, bosons being expressions of fermion interactions. In addition, a fermion interpretation of the Wheeler one-electron theory of the universe allows us to conceive a representation of the universe as a single localised fermion interacting with its nonlocal vacuum. Our dual space structure in which the basic unit is a point-singularity incorporating some components that are conserved and others that are nonconserved provides an opportunity for explaining the whole of physics using this model. We begin by generating a nilpotent version of quantum mechanics.

Here, it is convenient to rewrite the nilpotent condition using quaternions for the charges:

$$(i\mathbf{k}E + i\mathbf{i}p_x + i\mathbf{j}p_y + i\mathbf{k}p_z + \mathbf{j}m)(i\mathbf{k}E + i\mathbf{i}p_x + i\mathbf{j}p_y + i\mathbf{k}p_z + \mathbf{j}m) = 0$$

We can also collect together the vector terms, so that

$$(i\mathbf{k}E + i\mathbf{p} + \mathbf{j}m)(i\mathbf{k}E + i\mathbf{p} + \mathbf{j}m) = 0.$$

Finally, we can use the identifications we have made for the various algebraic units of mass, time, charge and space, to identify the composite quantities in this equation as energy, momentum and rest mass, and to recognise that the bracketed object $(i\mathbf{k}E + i\mathbf{p} + \mathbf{j}m)$ represents a conserved quantity, with E, $\mathbf{p}$ and m being derived by combining the conserved charge units with those, respectively of time, space and mass.

Following this, we approach quantum mechanics as the most exact way of describing the nonconservation of space and time in relation to the conservation of $(i\mathbf{k}E + i\mathbf{p} + \mathbf{j}m)$. The most complete possible variation in space and time is defined by a phase factor which associates E with time and p with space, and then using the differentials $\partial/\partial t$ and ∇ to recover $(i\mathbf{k}E + i\mathbf{p} + \mathbf{j}m)$ from the phase factor. For a free particle, the most complete set of variations in space and time is given by $e^{(-i(Et - \mathbf{p}.\mathbf{r}))}$, and the expression which will recover $(i\mathbf{k}E + i\mathbf{p} + \mathbf{j}m)$ using this as a phase factor is $(-\mathbf{k}\,\partial/\partial t - i\mathbf{i}\nabla + \mathbf{j}m)$. So we construct the nilpotent quantum mechanical equation for a free particle in the form

$$(-\mathbf{k}\,\partial/\partial t - i\mathbf{i}\nabla + \mathbf{j}m)(i\mathbf{k}E + i\mathbf{p} + \mathbf{j}m)\,e^{(-i(Et - \mathbf{p}.\mathbf{r}))} = 0.$$

Including all possible sign variations of E and $\mathbf{p}$, we obtain

$$(\mp\mathbf{k}\,\partial/\partial t \mp i\mathbf{i}\nabla + \mathbf{j}m)(\pm i\mathbf{k}E \pm i\mathbf{p} + \mathbf{j}m)\,e^{(-i(Et - \mathbf{p}.\mathbf{r}))} = 0,$$

which is equivalent to a nilpotent Dirac equation of the form

$$(\mp\mathbf{k}\,\partial/\partial t \mp i\mathbf{i}\nabla + \mathbf{j}m)\,\psi = 0.$$

We can also express it in operator form

$$(\pm i\mathbf{k}E \pm i\mathbf{p} + \mathbf{j}m)(\pm i\mathbf{k}E \pm i\mathbf{p} + \mathbf{j}m)\,e^{(-i(Et - \mathbf{p}.\mathbf{r}))} = 0,$$

where the operators E and $\mathbf{p}$ become $i\partial\,/\partial t$ and $-i\nabla$ as in the usual canonical quantization. This abbreviated account of the derivation of the nilpotent Dirac equation can be supplemented by more extensive versions. [1]

5.   Dual spaces and nilpotency

The principal significance of a nilpotent wavefunction, say ψ_1, is that it is automatically Pauli exclusive because it will form a zero combination state with an identical particle $\psi_1\psi_1$. In addition, if the universe is a totality zero state, then we can imagine that the creation of a fermion, specified by ψ_1, from *absolutely nothing* requires the simultaneous creation of the vacuum state, $-\psi_1$, which would cancel it. In that case, the superposition of fermion and vacuum, $\psi_1 - \psi_1$, and their combination state, $-\psi_1\psi_1$, will both be precisely 0. Now, the concept of Pauli exclusion applies to fermions in any state, and we can use this to extend nilpotent quantum mechanics in a new direction, by imagining that the operator $(\pm i\mathbf{k}E\pm i\mathbf{p}+\mathbf{j}m) = (\mp\mathbf{k}\partial\,/\partial t\mp ii\nabla+\mathbf{j}m)$ can be extended to include any number of field terms or covariant derivatives, so that E and $\mathbf{p}$ now become, for example, $i\partial\,/\partial t + e\phi + \ldots$ and $-i\nabla + e\mathbf{A} + \ldots$. The same will also be true if the external field terms are defined by expectation values, as with the Lamb shift, or in terms of quantum fields.

In this form, we don't even need an equation, just an operator of the form $(\pm i\mathbf{k}E \pm i\mathbf{p} + \mathbf{j}m)$ because the operator will uniquely determine the phase factor needed to produce a nilpotent amplitude. Rather than using a conventional form of the Dirac equation, we find the phase factor such that, using the defined operator,

$$(\text{operator acting on phase factor})^2 = \text{amplitude}^2 = 0.$$

If the operator has a more complicated form than that of the free particle, the phase factor will, of course, be no longer a simple exponential but the amplitude will still be a nilpotent.

A nilpotent operator, in effect, always splits the universe into two halves, a local part represented by the fermion, and a dual, nonlocal, vacuum part, which incorporates the rest of the universe. So we create a fermion in a particular state, including all its interacting potentials, and then we have to construct the vacuum or 'rest of the universe' which enables the fermion to exist in that state. The fact that the fermion in any state needs to create the entire universe which makes it possible makes a Wheeler-type 'one fermion' theory of the universe a seriously interesting possibility. The fermion and the entire universe are then a dual pair, and so the structure of the universe can be thought of as equivalent to that of a single particle. Of course, this single fermion is necessarily an interacting one, constructing a 'space' in which the vacuum is not localised on itself.

Another way of looking at this is to say that the fermion always exists in the two spaces from which it is constructed, real space and vacuum space, and the non-classical *zitterbewgung* motion, which Schrödinger found in the solution to the free-particle Dirac equation, [2] represents the switching between these spaces which makes it possible to define the fermion as creating a point singularity through the intersection of two spaces. We can here apply a reverse argument from topology. The creation of a particle singularity using its 'intersection' with a dual space can be seen as the creation of a multiply-connected space from a simply-connected space through the insertion of a topological singularity.

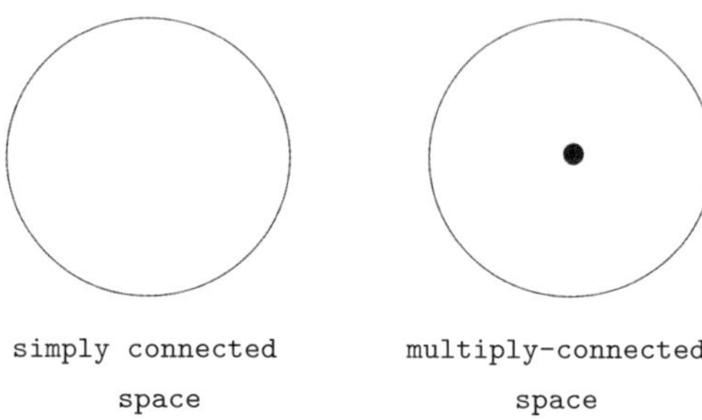

According to a well-known argument, parallel transporting a vector round a complete circuit in a multiply-connected space will produce a phase shift of π or $180°$ in the vector direction, whereas transporting it round a simply-connected space will not, and so, in the first case, the vector will be required to do a double circuit to return to its starting point. [3] This is exactly what happens with a spin 1/2 fermion, which, as a point-singularity, can be regarded as existing in its own multiply-connected space. We can interpret this as meaning that the fermion requires a double circuit because, just as in *zitterbewegung*, it spends only half of its time travelling in the real space of observation.

Spin 1/2 is obtained quite easily from the nilpotent operator $(i\mathbf{k}E + i\mathbf{p} + j m)$ by defining a Hamiltonian specified as $\mathcal{H} = -i\mathbf{k}(i\mathbf{p} + j\mathrm{m}) = -i\mathbf{j}\,\mathbf{p} + i\mathbf{i}\,\mathrm{m}$ and a *mathematical quantity* $\boldsymbol{\sigma} = -\mathbf{1}$, which is a pseudovector of magnitude $-\mathbf{1}$. Then

$$[\boldsymbol{\sigma}, \mathcal{H}] = [\mathbf{1}, -i\mathbf{j}(\mathbf{i}p_1 + \mathbf{j}p_2 + \mathbf{k}p_3) + i\mathbf{i}\,m] = [\mathbf{1}, -i\mathbf{j}(\mathbf{i}p_1 + \mathbf{j}p_2 + \mathbf{k}p_3)]$$
$$= 2\,i\,\mathbf{j}\,(\mathbf{ij}p_2 + \mathbf{ik}p_3 + \mathbf{ji}p_1 + \mathbf{j}p_3 + \mathbf{ki}p_1 + \mathbf{kj}p_2)$$
$$= 2\,\mathbf{j}\,(\mathbf{k}(p_2 - p_1) + \mathbf{j}(p_1 - p_3) + \mathbf{i}(p_3 - p_2))$$
$$= 2\,\mathbf{j}\,\mathbf{1} \times \mathbf{p}\,.$$

If $\mathbf{L}$ is an orbital angular momentum defined by $\mathbf{r} \times \mathbf{p}$, then

$$[\mathbf{L}, \mathcal{H}] = [\mathbf{r} \times \mathbf{p},\ -i\mathbf{j}(\mathbf{i}p_1 + \mathbf{j}p_2 + \mathbf{k}p_3) + i\mathbf{i}\,m] \tag{1}$$
$$= [\mathbf{r} \times \mathbf{p},\ -i\mathbf{j}(\mathbf{i}p_1 + \mathbf{j}p_2 + \mathbf{k}p_3)]$$
$$= i\,[\mathbf{r},\ i\mathbf{j}\,(\mathbf{i}p_1 + \mathbf{j}p_2 + \mathbf{k}p_3)] \times \mathbf{p}$$

But $\qquad [\mathbf{r},\ i\mathbf{j}\,(\mathbf{i}p_1 + \mathbf{j}p_2 + \mathbf{k}p_3)] = i\,\mathbf{1}\ .$

Hence $\qquad [\mathbf{L}, \mathcal{H}] = \mathbf{j}\,\mathbf{1} \times \mathbf{p}\ ,$

and $\mathbf{L} + \boldsymbol{\sigma}/2$ is a constant of the motion, because

$$[\mathbf{L} + \boldsymbol{\sigma}/2, \, \mathcal{H}] = 0 \,.$$

Spin, in this derivation, emerges from the Clifford algebra aspect of the operator $\mathbf{p}$, in effect because of the $i\mathbf{a} \times \mathbf{b}$ aspect of the full vector product $\mathbf{ab} = \mathbf{a.b} + i\mathbf{a} \times \mathbf{b}$. If we use Clifford algebra or *multivariate* vectors, then the term $\mathbf{p}$ will incorporate spin. However, where ordinary vectors are used (with no full product), for example, when we use polar coordinates, an intrinsic spin is no longer structured into the formalism and an *explicit spin* (or total angular momentum) term has to be introduced. Dirac, however, has given a prescription for translating his equation into polar form. [4] Here the momentum operator acquires an additional (imaginary) spin (or total angular momentum) term, and we can easily adapt this to represent a polar transformation of the multivariate vector operator:

$$\nabla \to \left(\frac{\partial}{\partial r} + \frac{1}{r} \right) \pm i\frac{j + 1/2}{r}$$

and use this to define a non-time varying nilpotent operator in polar coordinates:

$$\left(-\mathbf{k}\frac{\partial}{\partial t} - i\mathbf{i}\,\nabla + \mathbf{j}m\right) \to \left(i\,\mathbf{k}E - i\mathbf{i}\,\nabla + \mathbf{j}m\right)$$

$$\to \left(i\,\mathbf{k}E - i\mathbf{i} \left(\frac{\partial}{\partial r} + \frac{1}{r} \pm i\frac{j + 1/2}{r} \right) + \mathbf{j}m \right) .$$

This will become significant when we consider cases involving spherical symmetry.

6. Antisymmetric wavefunctions

The nilpotent structure explains immediately why we have Pauli exclusion between fermions, but the conventional way of explaining this property leads us to a profound insight on the nature of the information available in quantum systems if we structure it in nilpotent form. This is by defining fermion wavefunctions to be antisymmetric, so that:

$$(\psi_1\psi_2 - \psi_2\psi_1) = -(\psi_2\psi_1 - \psi_1\psi_2) \,.$$

In nilpotent terms, we write $(\psi_1\psi_2 - \psi_2\psi_1)$ as

$$(\pm i\mathbf{k}E_1 \pm i\mathbf{p}_1 + \mathbf{j}m_1)(\pm i\mathbf{k}E_2 \pm i\mathbf{p}_2 + \mathbf{j}m_2)$$

$$-(\pm i\mathbf{k}E_2 \pm i\mathbf{p}_2 + \mathbf{j}m_2)(\pm i\mathbf{k}E_1 \pm i\mathbf{p}_1 + \mathbf{j}m_1)$$

$$= 4\mathbf{p}_1\mathbf{p}_2 - 4\mathbf{p}_2\mathbf{p}_1 \;=\; 8\,i\,\mathbf{p}_1 \times \mathbf{p}_2 \;=\; -8\,i\,\mathbf{p}_2 \times \mathbf{p}_1 \,.$$

This result is clearly antisymmetric, but it also has a quite astonishing consequence, for it requires any nilpotent wavefunction to have a $\mathbf{p}$ vector, in real space, the one defined by the axes $\mathbf{i}$, $\mathbf{j}$, $\mathbf{k}$, at a *different orientation* to any other. The wavefunctions of all nilpotent fermions then instantaneously correlate because the planes of their $\mathbf{p}$ vector directions must all intersect. At the same time, the nilpotent condition requires the E, $\mathbf{p}$ and m combinations to be unique, and we can visualize this as constituting a unique direction in vacuum space along a set of axes defined by $\mathbf{k}$, $\mathbf{i}$, $\mathbf{j}$, or $\mathbf{K}$, $\mathbf{I}$, $\mathbf{J}$, with coordinates defined by the values of E, $\mathbf{p}$ and m. The directions of the vectors in each space carry *all the information* available to a fermionic state, and so the information in the two spaces is totally dual, and is equivalent to the instantaneous direction of the spin in the real space. The total information determining the behaviour of a fermion and even of the entire universe is contained in a single spin direction.

7. Symmetry transformations, bosons and baryons

Conventionally, the Dirac wavefunction is a *spinor*, with the four components in $(\pm i\mathbf{k}E \pm i\mathbf{p} + \mathbf{j}m)$ structured as a column vector, incorporating the four combinations of particle and antiparticle, and spin up and spin down. With $\pm E$ and $\pm \mathbf{p}$ (or $\pm\boldsymbol{\sigma}.\mathbf{p}$) representing these possibilities, the respective amplitudes of these four can be identified as, say,

$$
\begin{array}{ll}
(i\mathbf{k}E + i\mathbf{p} + \mathbf{j}m) & \text{fermion spin up} \\
(i\mathbf{k}E - i\mathbf{p} + \mathbf{j}m) & \text{fermion spin down} \\
(-i\mathbf{k}E + i\mathbf{p} + \mathbf{j}m) & \text{antifermion spin down} \\
(-i\mathbf{k}E - i\mathbf{p} + \mathbf{j}m) & \text{antifermion spin up}
\end{array}
$$

each being multiplied by the same phase factor. The helicity or handedness $(\boldsymbol{\sigma}.\mathbf{p})$ is then determined by the ratio of the signs of E and $\mathbf{p}$. So $i\mathbf{p}/i\mathbf{k}E$ has the same helicity as $(-i\mathbf{p})/(i\mathbf{k}E)$, but the opposite helicity to $i\mathbf{p}/(-i\mathbf{k}E)$. The negative energy or antiparticle states in this formalism can also be seen to have the opposite time direction in their differential forms to the positive energy or particle states.

The lead term in the column may be considered as defining the fermion type, and it will often be convenient to represent the entire 4-component structure by just this term. The remaining terms are then automatically derived by sign transformations, becoming equivalent to the lead term, subjected to the respective symmetry transformations, P, T, and C, by pre- and post-multiplication by the quaternion units defining what we have previously described as the vacuum space:

$$
\begin{array}{llccc}
\text{Parity} & P & i(i\mathbf{k}E + i\mathbf{p} + \mathbf{j}m)i & = & (i\mathbf{k}E - i\mathbf{p} + \mathbf{j}m) \\
\text{Time reversal} & T & k(i\mathbf{k}E + i\mathbf{p} + \mathbf{j}m)k & = & (-i\mathbf{k}E + i\mathbf{p} + \mathbf{j}m) \\
\text{Charge conjugation} & C & -j(i\mathbf{k}E + i\mathbf{p} + \mathbf{j}m)j & = & (-i\mathbf{k}E - i\mathbf{p} + \mathbf{j}m)
\end{array}
$$

We can see from these representations that the rules

$$CP \equiv T, \quad PT \equiv C, \quad \text{and } CT \equiv P$$

necessarily apply, as also

$$TCP \equiv CPT \equiv \text{identity}$$

as

$$k(j(i(i\mathbf{k}E + i\mathbf{p} + \mathbf{j}m)i)j)k = kji(i\mathbf{k}E + i\mathbf{p} + \mathbf{j}m)ijk = (i\mathbf{k}E + i\mathbf{p} + \mathbf{j}m)$$

From these rules, it is clear that charge conjugation is effectively defined in terms of parity and time reversal, rather than as an independent operation. This is a consequence of the fact that the variation in space and time is the information that solely determines both the phase factor and the entire nature of the fermion state.

The terms in the nilpotent 4-spinor, other than the lead term are then the states into which it could transform without changing its energy or momentum. They can be seen as vacuum 'reflections' of the real particle state, arising from vacuum operations that can be mathematically defined. A fermion cannot, of course, form a combination state with itself, but it could be imagined as forming a combination state with any of these vacuum 'reflections'. In each of these cases, the combined state will form one of the three classes of bosons or boson-like objects, whose wavefunctions, summed up over 4 terms, yield products which are scalars:

Spin 1 boson:
$$(\pm ik E \pm i\mathbf{p} + jm)(\mp ik E \pm i\mathbf{p} + jm) \qquad T$$

Spin 0 boson:
$$(\pm ik E \pm i\mathbf{p} + jm)(\mp ik E \mp i\mathbf{p} + jm) \qquad C$$

Fermion-fermion combination (Bose-Einstein condensate / Berry phase, *etc.*):
$$(\pm ik E \pm i\mathbf{p} + jm)(\pm ik E \mp i\mathbf{p} + jm) \qquad P$$

A key consequence of this formalization is that a spin 1 boson can be massless, but a spin 0 boson cannot, as then $(\pm ik E \pm i\mathbf{p})(\mp ik E \mp i\mathbf{p})$ immediately reduces to zero: hence Goldstone bosons must become Higgs bosons in the Higgs mechanism. Another consequence is that the fermion and antifermion cannot both be purely left-handed or both purely right-handed — or massless — and act *via* a weak interaction to produce a bosonic state. That is, a left-handed fermion $(\pm ik E \pm i\mathbf{p})$ cannot combine with a left-handed antifermion $(\mp ik E \mp i\mathbf{p})$, *via* a weak interaction, to form a bosonic single state unless a nonzero mass term is introduced. Though the chirality is a direct consequence of the structure of the Dirac equation even in the conventional formalism, it is seen here as an immediate consequence of the nilpotent structure.

Versions of these bosons can also be imagined as being created and annihilated in the switching between a 'real fermion' and its vacuum states, in particular between the positive and negative energy states, or between real and vacuum space (the switching between spin states occurring in real space if particles have nonzero rest mass). Since this is always occurring due to *zitterbewegung*, and the weak interaction, then we can consider weak sources (*i.e.* fermions) as necessarily having a dipole or multipole aspect. Because of the fundamental chirality of the weak interaction, we can also see fermions as characteristic of real space and antifermions of vacuum space. A particle which is a fermion in ordinary space acts for half as its existence as an antifermion in vacuum space, *in exactly the way that the 4-component Dirac spinor would suggest*. Consequently, there is no problem of an antisymmetry between matter and antimatter. There is the same amount of each. Energy, momentum, space, time and charge all cancel overall when we take both spaces into account.

There is, in particular, a backward direction in time and a reverse causality which apply to nonlocal processes. The entire future of the universe could be said to be contained within the vacuum for any fermion, though this does not lead to a deterministic outcome because the nonlocality cannot be defined by an observer any more accurately than the locality associated with the fermion to which it is opposite.

Apart from bosons, the other fundamental composite structures involving fermions are the three-component baryons. Clearly, with nilpotent fermions, a structure such as

$$(ikE \pm i\mathbf{p} + jm)(ikE \pm i\mathbf{p} + jm)(ikE \pm i\mathbf{p} + jm)$$

will automatically zero, but one in which the momentum term is split into components along orthogonal axes will produce a nonzero combination:

$$(ikE \pm i\mathbf{i}p_x + jm)(ikE \pm i\mathbf{j}p_y + jm)(ikE \pm i\mathbf{k}p_z + jm)$$

Spin is defined in a unique direction at any time, so, at any particular instant, the wavefunction will reduce, after normalization, to

$$(ikE \pm i\mathbf{i}p_x + jm)(ikE + jm)(ikE + jm) \rightarrow (ikE \pm i\mathbf{i}p_x + jm)$$
$$(ikE + jm)(ikE \pm i\mathbf{j}p_y + jm)(ikE + jm) \rightarrow (ikE \mp i\mathbf{j}p_y + jm)$$
$$(ikE + jm)(ikE + jm)(ikE \pm i\mathbf{k}p_z + jm) \rightarrow (ikE \pm i\mathbf{k}p_z + jm)$$

with a notable change of sign in the second case.

To maintain the symmetry between the three directions of momentum, and the $+$ and $-$ values of the momentum term, we can define six possible outcomes, resulting in a superposition of six combination states:

$$(ikE + i\mathbf{i}p_x + jm)(ikE + jm)(ikE + jm) \rightarrow (ikE + i\mathbf{i}p_x + jm)$$
$$(ikE - i\mathbf{i}p_x + jm)(ikE + jm)(ikE + jm) \rightarrow (ikE - i\mathbf{i}p_x + jm)$$
$$(ikE + jm)(ikE + i\mathbf{j}p_y + jm)(ikE + jm) \rightarrow (ikE - i\mathbf{j}p_y + jm)$$
$$(ikE + jm)(ikE - i\mathbf{j}p_y + jm)(ikE + jm) \rightarrow (ikE + i\mathbf{j}p_y + jm)$$
$$(ikE + jm)(ikE + jm)(ikE + i\mathbf{k}p_z + jm) \rightarrow (ikE + i\mathbf{k}p_z + jm)$$
$$(ikE + jm)(ikE + jm)(ikE - i\mathbf{k}p_z + jm) \rightarrow (ikE - i\mathbf{k}p_z + jm)$$

The six possible states have exactly the same structure as the six 'colour' combinations in the conventional formalism, to which this structure is exactly isomorphic. Significantly, to maintain gauge invariance between the six possible states, we need a nonlocal exchange of momentum, whose rate does not depend on distance between the components. This is equivalent to a constant force or a potential which increases linearly with distance. A linear potential is an experimentally-observed characteristic of the *strong* interaction between quarks. Here, it would appear to have a fundamental explanation.

Perfect gauge invariance is also only possible if the baryon simultaneously incorporates both left-handed and right-handed components or + and − values of **p**. By the principle which we have previously applied to boson structures, *baryons must therefore have mass*. The principle is, significantly, the same as that which applies in the Higgs mechanism, even though the perfect gauge invariance between the six possible states or switching between the different **p** components is provided by massless gluons. The solution to the so-called mass-gap problem which this incorporates is a significant part of one of the prize challenge problems defined by the Clay Institute.

8. The fundamental interactions

One of the most important aspects of the nilpotent structure is that it offers an immediate separation of the local from the nonlocal. Essentially, everything *inside* a fermion bracket, such as $(\pm i k E \pm i i p_x + j m)$, is local, defined by a Lorentzian structure, while everything *outside it*, such as a combination state or a superposition, is nonlocal. It is possible, then, from the possible nonlocal structures available to fermions, defined by combinations and superpositions, to derive the *local* structures, inside the fermion brackets (essentially the potentials to be added to E or **p**) which would produce the same effect. When we do this we find that the possible local structures lead to just three classes of interaction which give a nilpotent solution, and these interactions correspond to those which, physically, we classify as electric, strong and weak.

The first nonlocal effect is Pauli exclusion, which, in effect, is a prohibition on certain combinations, specified by nilpotency. Here we use Dirac's polar transformation to describe a fermion as a point-particle with spherical symmetry:

$$(i\boldsymbol{k}E - i\boldsymbol{i}\,\nabla + \boldsymbol{j}m) \rightarrow \left(i\,\boldsymbol{k}E - i\boldsymbol{i}\,\left(\frac{\partial}{\partial r} + \frac{1}{r} \pm i\,\frac{j+1/2}{r} \right) + \boldsymbol{j}m \right).$$

To establish Pauli exclusion, we need to define this operator as producing a nilpotent solution, and, in fact, inspection reveals that this is impossible unless the $i\boldsymbol{k}E$ component also includes a term involving $1/r$ to cancel out the terms with this factor multiplied by $i\boldsymbol{i}$.

The minimum operator for a point-particle is therefore of the form

$$\left(\pm i\,\boldsymbol{k}\left(E - \frac{A}{r} \right) \mp i\boldsymbol{i}\,\left(\frac{\partial}{\partial r} + \frac{1}{r} \pm i\,\frac{j+1/2}{r} \right) + \boldsymbol{j}m \right),$$

exactly as we require for the Coulomb interaction. This is easily solved, requiring, in fact, only six lines of calculation. First, we have to find the phase factor ϕ which will make the amplitude nilpotent. As in the standard solution, we assume that it is of the form:

$$\phi = e^{-ar}\, r^{\gamma} \sum_{\nu=0} a_{\nu}\, r^{\nu}.$$

We then apply the operator we have just defined to ϕ and square the result to 0 to obtain:

$$4\left(E - \frac{A}{r}\right) = -2\left(-a + \frac{\gamma}{r} + \frac{\nu}{r} + \ldots + \frac{1}{r} + i\,\frac{j + 1/2}{r}\right)^2$$
$$- 2\left(-a + \frac{\gamma}{r} + \frac{\nu}{r} + \ldots + \frac{1}{r} - i\,\frac{j + 1/2}{r}\right)^2 + 4m^2$$

Equating constant terms produces

$$a = \sqrt{m^2 - E^2}\,.$$

Equating terms in $1/r^2$, following standard procedure, with $\nu = 0$, leads to:

$$\left(\frac{A}{r}\right)^2 = -\left(\frac{\gamma + 1}{r}\right)^2 + \left(\frac{j + 1/2}{r}\right)^2\,.$$

Assuming the power series terminates at n', following another standard procedure, and equating coefficients of $1/r$ for $\nu = n'$, we obtain

$$2EA = -2\sqrt{m^2 - E^2}\,(\gamma + 1 + n')\,,$$

the terms in $(j + 1/2)$ cancelling over the summation of the four multiplications, with two positive and two negative. Algebraic rearrangement of these equations then yields

$$\frac{E}{m} = \frac{1}{\sqrt{1 + \dfrac{A^2}{(\gamma + 1 + n')^2}}} = \frac{1}{\sqrt{1 + \dfrac{A^2}{\left(\sqrt{(j + 1/2)^2 - A^2} + n'\right)^2}}}\,.$$

This is a general formula, but in the particular case where $A = Ze^2$, this becomes the hyperfine or fine structure formula for a one-electron nuclear atom or ion, for example, that of the hydrogen atom, where $Z = 1$.

A second case suggests itself for the strong interaction, which we know requires a linear potential to explain both its experimental characteristics, and also the nilpotent structures of baryons. We have found also that there must be a Coulomb component or inverse linear potential ($\propto 1/r$), just to accommodate spherical symmetry, and this has a known physical manifestation in the strong interaction in the one-gluon exchange. So, we might imagine that the nilpotent operator incorporating Coulomb and linear potentials from a source with spherical symmetry (either the centre of a 3-quark system or one component of a quark-antiquark pairing) could be written in the form:

$$\left(\pm\,i\,\boldsymbol{k}\left(E - \frac{A}{r} + Br\right) \mp i\boldsymbol{i}\left(\frac{\partial}{\partial r} + \frac{1}{r} \pm i\,\frac{j + 1/2}{r}\right) + i\boldsymbol{j}m\right).$$

Again, we need to identify the phase factor, which, by analogy with the pure Coulomb calculation, we might suppose to be of the form:

$$\phi = \exp(-ar - br^2)\,r^\gamma \sum_{\nu = 0} a_\nu\, r^\nu\,.$$

Applying the operator we have defined and the nilpotent condition, we obtain:

$$E^2 + 2AB + \frac{A^2}{r^2} + B^2 r^2 + \frac{2AE}{r} + 2BEr \;=\; m^2$$

$$-\left(a^2 + \frac{(\gamma + \nu + \ldots + 1)^2}{r^2} - \frac{(j + 1/2)^2}{r^2} + 4b^2 r^2 + 4abr \right.$$

$$\left. - 4b(\gamma + \nu + \ldots + 1) - \frac{2a}{r}(\gamma + \nu + \ldots + 1) \right)$$

with the positive and negative $i(j + 1/2)$ terms again cancelling out over the four solutions. Then, assuming a termination in the power series (as with the Coulomb solution), we can equate:

$$
\begin{array}{ll}
\text{coefficients of } r^2 \text{ to give} & B^2 = -4b^2 \\
\text{coefficients of } r \text{ to give} & 2BE = -4ab \\
\text{coefficients of } 1/r \text{ to give} & 2AE = 2a(\gamma + \nu + 1)
\end{array}
$$

These equations lead directly to:

$$b \;=\; \pm\frac{iB}{2}$$

$$a \;=\; \mp iE$$

$$\gamma + \nu + 1 \;=\; \mp iA.$$

The ground state case (where $\nu = 0$) then requires a phase factor of the form:

$$\phi = \exp(\pm iEr \mp iBr^2/2)\, r^{(\mp iqA - 1)}.$$

The imaginary exponential terms in ϕ clearly represent asymptotic freedom, a term like $\exp(\mp iEr)$ being typical for a free fermion. The complex r term can be structured as a component phase, $\chi(r) = \exp(\pm iqA\ln(r))$, varying less rapidly with r than the rest of ϕ. We can therefore write ϕ as

$$\phi \;=\; \frac{\exp(kr + \chi(r))}{r}\,,$$

where
$$k \;=\; \pm iE \mp iBr/2\,.$$

While the first term will dominate at high energies, where r is small, and approximate to a free fermion solution, suggesting asymptotic freedom, the second term, with its confining potential Br, is significant at low energies, when r is large, suggesting infrared slavery. The Coulomb term, required for spherical symmetry, is the one which defines the strong interaction phase, $\chi(r)$, and this can be related to the directional status of $\mathbf{p}$ in the state vector.

One further interaction is built into the structure of the nilpotent operator as a 4-component combination state. This is the weak interaction, and we have already seen that it requires a dipole or multipole term in addition to the standard Coulomb term from spherical symmetry. We can therefore suppose that the nilpotent operator takes a form such as

$$\left(\boldsymbol{k} \left(E - \frac{A}{r} - Cr^n \right) + i\boldsymbol{i} \left(\frac{\partial}{\partial r} + \frac{1}{r} \pm i\frac{j+1/2}{r} \right) + i\boldsymbol{j}m \right),$$

where n is an integer greater than 1 or less than -1, and, as before, look for a phase factor which will make the amplitude nilpotent. Extending our work on the Coulomb solution, we may suppose that the phase factor is of the form:

$$\phi = \exp\left(-ar - br^{n+1} \right) r^\gamma \sum_{\nu=0} a_\nu \, r^\nu \, .$$

Applying the operator and squaring to zero, with a termination in the series, we obtain

$$4 \left(E - \frac{A}{r} - cr^n \right)^2$$

$$= +2 \left(-a + (n{+}1)br^n + \frac{\gamma}{r} + \frac{\nu}{r} + \frac{1}{r} + i\frac{j+1/2}{r} \right)^2$$

$$- 2 \left(-a + (n{+}1)br^n + \frac{\gamma}{r} + \frac{\nu}{r} + \frac{1}{r} + i\frac{j+1/2}{r} \right)^2$$

Equating constant terms, we find

$$a = \sqrt{m^2 - E^2} \, .$$

Equating terms in r^{2n}, with $\nu = 0$:

$$C^2 = -(n+1)^2 b^2 \, ,$$

$$b = \pm\frac{iC}{(n+1)} \, .$$

Equating coefficients of r, where $\nu = 0$:

$$AC = -(n+1)b(1+\gamma) \, ,$$

$$(1+\gamma) = \pm iA \, .$$

Equating coefficients of $1/r^2$ and coefficients of $1/r$, for a power series terminating in $\nu = n'$, we obtain

$$A^2 = -(1+\gamma+n')^2 + (j+1/2)^2 \, ,$$

$$-EA = a(1+\gamma+n') \, .$$

An algebraic combination of these conditions produces:

$$\left(\frac{m^2 - E^2}{E^2} \right) (1+\gamma+n')^2 = -(1+\gamma+n')^2 + (j+1/2)^2 \, ,$$

$$E = -\frac{m}{j+1/2} \left(\pm iA + n' \right) \, .$$

This equation has the form of a harmonic oscillator, with evenly spaced energy levels deriving from integral values of n'. If we make the additional assumption that A, the phase term required for spherical symmetry, derives from the random directionality of the fermion spin, we may assign to it a half-unit value $(\pm 1/2\, i)$, or $(\pm 1/2\, i\hbar c)$, and obtain the complete formula for the fermionic simple harmonic oscillator:

$$e = -\frac{m}{j + 1/2}\,(1/2 + n')\,.$$

Now the potential of the form Cr^n added to the Coulomb term made no assumptions about the value of n except that it was an integer greater than 1 or less than -1. In fact, any potential of this form, or any combination of such potentials, will generate a harmonic oscillator solution for the nilpotent operator. Such potentials naturally emerge from systems where there is complexity, aggregation, or a multiplicity of sources. Because, virtually any potential other than the Coulomb or Coulomb plus linear, must be of the force, we have effectively demonstrated that there are only three possible interaction types that can apply to a nilpotent fermionic operator.

9. Partitioning the vacuum

The nonlocal aspect of the fermionic nilpotent state $(\pm ikE \pm i\mathbf{p} + jm)$ is defined by a continuous vacuum $-(\pm ikE \pm i\mathbf{p} + jm)$. However, we can use the operators $\mathbf{k}$, $\mathbf{i}$, $\mathbf{j}$ effectively to partition this state into discrete components with a dimensional structure, which can then be identified as the weak, strong and electric components responding respectively to the weak strong and electric charges. We can, for example, postmultiply $(\pm ikE \pm i\mathbf{p} + jm)$ by the idempotent $\mathbf{k}(\pm ikE \pm i\mathbf{p} + jm)$ any number of times, without changing its state

$$(\pm ikE \pm i\mathbf{p} + jm)\mathbf{k}(\pm ikE \pm i\mathbf{p} + jm)\mathbf{k}(\pm ikE \pm i\mathbf{p} + jm)\dots$$
$$\rightarrow (\pm ikE \pm i\mathbf{p} + jm)$$

The idempotent acts as a vacuum operator. The same is also true of postmultiplication by $\mathbf{i}(\pm ikE \pm i\mathbf{p} + jm)$ or $\mathbf{k}(\pm ikE \pm i\mathbf{p} + jm)$. Of course, these operations are also equivalent to applying T, P or C transformations to every even bracket. For example,

$$(\pm ikE \pm i\mathbf{p} + jm)(\mp ikE \pm i\mathbf{p} + jm)(\pm ikE \pm i\mathbf{p} + jm)\dots$$
$$\rightarrow (\pm ikE \pm i\mathbf{p} + jm)$$

Here, every alternate state becomes an antifermion, which combines with the original fermion state to become a spin 1 boson $(\pm ikE \pm i\mathbf{p} + jm)(\mp ikE \pm i\mathbf{p} + jm)$ Effectively, repeated post-multiplication of a fermion operator by any of the discrete idempotent vacuum operators creates an alternate series of antifermion and fermion vacuum states, or an alternate series of boson and fermion states without actually changing the character of the real particle state. A fermion becomes *its own* boson by combining with any of its vacuum 'images'. A boson can be similarly shown to be its own fermion or antifermion. Nilpotent operators are thus intrinsically supersymmetric, with supersymmetry operators $Q = (\pm ikE \pm i\mathbf{p} + jm)$, converting boson to fermion, and $Q^\dagger = (\mp ikE \pm i\mathbf{p} + jm)$, converting fermion to boson, and we can represent the infinite post-multiplication sequences by the supersymmetric expressions such as $Q\,Q^\dagger\,Q\,Q^\dagger\,Q\,Q^\dagger\,Q\,Q^\dagger\,Q \ldots$ It is even possible to interpret this as creating the series of boson and fermion loops, with identical energy and momentum values, which an exact supersymmetry would need to cancel the self-energy term in renormalization, and remove the hierarchy problem completely. Such a calculation is now available. [5]

The identification of $i(ikE + i\mathbf{p} + jm)$, $k(ikE + i\mathbf{p} + jm)$ and $j(ikE + i\mathbf{p} + jm)$ as vacuum operators and the corresponding identification of $i(ikE + i\mathbf{p} + jm)i$, $k(ikE + i\mathbf{p} + jm)k$ and $j(ikE + i\mathbf{p} + jm)j$ as their respective vacuum 'reflections' at interfaces provided by P, T, and C transformations provides a new insight into the meaning of the Dirac 4-spinor. The three terms following the lead term which is identified with the particle can be seen as the vacuum 'reflections' that are created with the particle in the three coordinate axes of vacuum space. The four components then become creation and annihilation operators acting on their respective vacua: gravitational (or inertial), strong, weak and electric. We can additionally see the three vacuum coefficients k, i, j as originating in, or being responsible for, the concept of discrete, point-like, charge, which generates the particle state. The operators, k, i, and j act like weak, strong and electric 'charges' or sources, acting to partition the continuous vacuum represented by $-(ikE + i\mathbf{p} + jm)$, into discrete components, with special characteristics determined by the respective pseudoscalar, vector and scalar natures of their associated terms iE, $\mathbf{p}$ and m. The nature of gravity as acting like a summation of the other three forces, long predicted by this theory, is now a fundamental component also of many string theories under the name of 'gravity-gauge theory correspondence'. In addition, this theory also predicted the existence of the 'dark energy' long before its discovery, and fixed it as being equivalent to two-thirds of the energy of the universe, in line with recent results from the Planck Collaboration. [6]

10. The duality of real and vacuum spaces

We began with a dual structure involving two vector spaces, and, though the combination of the two spaces leads to chirality and an asymmetry in the vacuum space in terms of observation, at the deepest level the symmetry is retained. We can see this, if we construct the set of spinors, using the double vector notation, to generate the four components of the nilpotent structure from the basic $(\mathbf{K}E + i\mathbf{I}\mathbf{i}p_x + i\mathbf{I}\mathbf{j}p_y + i\mathbf{I}\mathbf{k}p_z + i\mathbf{J}m)$. As primitive idempotents, they are orthogonal (with zero products) sum to 1 and are given by

$$(1 - \mathbf{I}\mathbf{i} - \mathbf{J}\mathbf{j} - \mathbf{K}\mathbf{k})/4$$
$$(1 - \mathbf{I}\mathbf{i} + \mathbf{J}\mathbf{j} + \mathbf{K}\mathbf{k})/4$$
$$(1 + \mathbf{I}\mathbf{i} - \mathbf{J}\mathbf{j} + \mathbf{K}\mathbf{k})/4$$
$$(1 + \mathbf{I}\mathbf{i} + \mathbf{J}\mathbf{j} - \mathbf{K}\mathbf{k})/4$$

Here, we see that the status of the two spaces at this level is *exactly identical*, but that, as soon as they are applied in the nilpotent structure, the perfect symmetry is broken. The zero product of the spinors

$$(1 - \mathbf{i}\mathbf{i} - \mathbf{j}\mathbf{j} - \mathbf{k}\mathbf{k})(1 - \mathbf{i}\mathbf{i} + \mathbf{j}\mathbf{j} + \mathbf{k}\mathbf{k})(1 + \mathbf{i}\mathbf{i} - \mathbf{j}\mathbf{j} + \mathbf{k}\mathbf{k})(1 + \mathbf{i}\mathbf{i} + \mathbf{j}\mathbf{j} - \mathbf{k}\mathbf{k}) = 0$$

interestingly recalls a structure from a quartic Finsler geometry, the Berwald-Moor metric

$$(x_1 - x_2 - x_3 - x_4)(x_1 - x_2 + x_3 + x_4)(x_1 + x_2 - x_3 + x_4)(x_1 + x_2 + x_3 - x_4)$$

where x_1, x_2, x_3, x_4 can be regarded as base units of the dual vector space formed by 1, $\mathbf{i}\mathbf{I}$, $\mathbf{j}\mathbf{J}$, $\mathbf{k}\mathbf{K}$ (with volume unit -1). [7] Physical singularities (fermions or their products) appear to require a perfect dual vector space which nevertheless produces an asymmetry or chirality in the space of observation because it combines with the unobserved dual vacuum space in an asymmetric nilpotent structure.

The nilpotent structure itself incorporates many forms of duality: operator and wavefunction; fermion and vacuum; fermion and vacuum boson; operator and amplitude; nilpotent and idempotent; broken and unbroken symmetries. Essentially, these all originate in the idea of the fermion state as defining a localized singularity, at the same time as we define what is nonlocal or excluded from the singularity. The fermion has a half-integral spin because it requires simultaneously splitting the universe into two halves which are mirror images of each other at a fundamental level, but which appear asymmetric at the observational level because observation privileges the fermion singularity. *Zitterbewegung* is an obvious manifestation of the duality, but, in observational terms, it privileges the creation of positive rest mass.

Though the duality results in fermion and vacuum occupying separate 3-dimensional 'spaces', which are combined in the double Clifford algebra defining the singularity state, these 'spaces', though seemingly different in observational terms, are truly dual, each containing the same information, and the duality manifests itself directly in many physical forms. For example, we have alternative methods for defining the following phenomena using either real spaces axes ($\mathbf{i}$, $\mathbf{j}$, $\mathbf{k}$) or vacuum space axes ($\mathbf{I}$, $\mathbf{J}$, $\mathbf{K}$):

	$\mathbf{i}$, $\mathbf{j}$, $\mathbf{k}$	$\mathbf{I}$, $\mathbf{J}$, $\mathbf{K}$
Pauli exclusion	antisymmetric wavefunction	nilpotency
spin 1/2	anticommuting $\mathbf{p}$ components	Thomas precession
SR velocity addition	using 2 D of space	using space-time
holographic principle	area = space × space	area = space × time
fermion state	direction of $\mathbf{p}$	E, $\mathbf{p}$, m

The relativistic connection between space and time notably exists in a different vector space to the connection between the three spatial components. It is not strictly 4-dimensional at all, though it appears as such when we take the scalar product of a massless object. Assuming an intrinsic 4-D connection gives us a problem with quantum mechanics, where time is not an observable, and also for Penrose's twistors, which have to assume a massless world, with the intrinsic motion of the particles at the speed of light. Though 3-D vector space incorporates a duality of its own in requiring vectors and pseudovectors, quantum mechanics really requires an additional duality. It uses a dual dual space, which does not require an arbitrary extension to 4-D. Mass is a natural consequence of this extra duality even if we assume that the intrinsic motion of the particles is at the speed of light.

Higher dimensionalities naturally result from this double Clifford algebra. Thus, the nilpotent operator $(\pm i\mathbf{k}E \pm i\mathbf{p} + j m)$ can be regarded as a 10-D object in vacuum space: 5 for iE, $\mathbf{p}$, m and 5 for the unit axes $\mathbf{k}$, $\mathbf{i}$, $\mathbf{j}$ (or $\mathbf{K}$, $\mathbf{I}$, $\mathbf{J}$). Six of the ten (all but iE and $\mathbf{p}$) are compactified. The ten also reduce to 8 or 2×4 in a nilpotent structure when the intrinsic redundancy of m and the scalar 1 are considered. The nilpotent structure creates a self-duality in phase space which determines vacuum selection in exactly the way required for a perfect string theory, and, as we have seen, it automatically generates a gravity-gauge theory correspondence. The significant feature of this process, and in fact of our entire discussion, is that only 3-dimensional objects are needed to generate the entire structure, and that, in fundamental terms, this reduces to a single 3-dimensional object and its dual partner.

. . . .

References

[1] P. Rowlands, *Zero to Infinity: The Foundations of Physics*, World Scientific, Singapore. (2007)

[2] E. Schrödinger, 'Über die kräftefreie Bewegung in der relativistischen Quantenmechanik', *Sitz. Preuss. Akad. Wiss. Phys.-Math. Kl.* 24, 418–28, (1930)

[3] M. V. Berry, *Proc. R. Soc. Lond.* **A 392** 45–57 (1984)

[4] P. A. M. Dirac, *The Principles of Quantum Mechanics*, fourth edition. Clarendon Press, Oxford. (1958)

[5] P. Rowlands, 'Physical Interpretations of Nilpotent Quantum Mechanics', arXiv: 1004.1523.

[6] P. Rowlands, 'A critical value for dark energy', arXiv:1306.1420 (and references therein).

[7] P. Rowlands, 'A null Berwald-Moor metric in nilpotent spinor space', *Symmetry*, 23, no. 2, 179–188, (2012)

Discrete Motion
and the Emergence of Space and Time

Richard Shoup

Boundary Institute
P.O. Box 10336 San Jose, CA 95157 USA
E-mail: rshoup@boundary.org

In this short informal paper we give a simple and primitive definition of discrete motion beginning prior to the usual notions of space and time. We show how velocity, the relativistic addition of velocities, and the Lorentz factor naturally emerge from simply counting steps in a sequence of discrete abstract motions.

Time is nature's way of keeping everything from happening at once. Space is what prevents everything from happening to me. − (attributed to) John Archibald Wheeler, physicist

Introduction

In this informal paper, we explore the simplest possible definition of motion, namely a change in abstract position. There is no assumption of pre-existing space or time, and these are seen to emerge naturally from simple considerations. Natural consequences include a maximum velocity, and velocity addition that is consistent with Special Relativity.

Consider a simple discrete motion in one dimension consisting of a sequence of steps to the left (-) or the right (+) along a line. Steps take place in one direction or the other, and have no size. There is no clock present, and so an event occurs (time "passes") only when there is a step.

A typical sequence of steps might look like this:

$$+ \quad + \quad - \quad + \quad - \quad + \quad + \quad + \qquad\qquad (\mathsf{S_1})$$

that is, 6 steps to the right and 2 steps to the left, for a net progress of $6 - 2 = +4$ to the right in a total of 8 steps.

Velocity

It is natural to define a discrete *velocity* just by *counting steps*, as

$$v = \frac{netsteps}{totalsteps} = \frac{n^+ - n^-}{n^+ + n^-}$$

where n^+ is the number of $+$ steps and n^- is the number of $-$ steps. The velocity in this example is thus

$$v = \frac{6-2}{6+2} = \frac{1}{2}$$

The sequence

$$+ \quad + \quad + \quad + \quad - \quad + \quad + \quad + \qquad\qquad (\mathsf{S_2})$$

is another example, in this case with velocity $v = (7-1)/(7+1) = 3/4$.   Again the total number of steps plays the role of time in our definition of velocity.

Note that velocity is independent of the order of the $+$ and $-$ steps, and may be thought of as a fraction of the ultimate speed $c = 1$, a string of all $+$ steps (or $c = -1$, being a string of all $-$ steps). We need not be concerned with the size of the steps in either spatial or temporal terms – they are fundamental indivisible units. Thus c is thus the natural maximum velocity, being one step in "space" for each and every step in "time".

Addition of velocities

Imagine now a sum of two independent motions defined as above, with one displacement relative to the other. Obviously, there are 4 possibilities for the combined motions: both $+$, both $-$, or opposing motions ($+-$ or $-+$). This can be illustrated by combining (adding) the two particular example sequences given above:

$$
\begin{array}{llcccccccc}
\mathsf{S_1} : & & + & + & - & + & - & + & + & + \\
\mathsf{S_2} : & & + & + & + & + & - & + & + & + \\
\mathsf{Sum} : & & + & + & & + & - & + & + & + & .
\end{array}
$$

Thus the result string contains six $+$ and one $-$ motions. In the one case of opposing components in this example, the result is 0, no motion, no event at all. For this example, the velocity of the sum string is then $(6-1)/8 = 5/8$. However, *a viewer of the sum string will only see 7 total steps*, not 8, since $+-$ and $-+$ motions cancel, see below.

We are only interested in the average velocities, so we will next consider addition of velocities in the general case using a normalized statistical argument.

The general case

In general, these strings represent sequences of independent events that have known distributions (number of $+$ and $-$ overall), but which are not in any particular order nor correlated with each other. We will take the probability of a move to the right or to the left as

$$p_1^+ = \frac{n_1^+}{N} \quad \text{and} \quad p_1^- = \frac{n_1^-}{N}$$

respectively, where for two such strings

$$n_1^+ + n_1^- \;=\; n_2^+ + n_2^- \;=\; N$$

and thus

$$p_1^+ + p_1^- \;=\; p_2^+ + p_2^- \;=\; 1\,.$$

First consider the classical sum of two strings in this form given by

$$v_1 + v_2 \;=\; \frac{p_1^+ - p_1^-}{p_1^+ + p_1^-} + \frac{p_2^+ - p_2^-}{p_2^+ + p_2^-} \;=\; \frac{2(\,p_1^+ p_2^+ - p_1^- p_2^-\,)}{p_1^+ p_2^+ + p_1^+ p_2^- + p_1^- p_2^+ + p_1^- p_2^-}$$

For example, summing the two strings S_1 and S_2 above yields

$$v_1 + v_2 \;=\; \frac{2\times\left(\frac{6}{8}\times\frac{7}{8} - \frac{2}{8}\times\frac{1}{8}\right)}{\frac{6}{8}\times\frac{7}{8} + \frac{6}{8}\times\frac{1}{8} + \frac{2}{8}\times\frac{7}{8} + \frac{2}{8}\times\frac{1}{8}} \;=\; \frac{5}{4}$$

in agreement with the common-sense classical notion of additive velocities, in this case

$$v_1 + v_2 \;=\; = \;\frac{1}{2} + \frac{3}{4} \;=\; \frac{5}{4}$$

but which of course exceeds the expected maximum velocity of 1. The difficulty is resolved by realizing that time does not "pass" without an associated event. In other words, time passage itself is derived from events at the lowest level, not *vice-versa*.

Derived time

Suppose we make a simple change to the above classical sum expression based on the postulate of total steps as our measure of time: *When the summed motion is zero no event occurs and thus no time passes.* That is, when there is no net motion, there is no event at all, and our "clock" (the count of total steps) does not increment either. The total number of time steps in the denominator will not include those cases where the two motions are opposite $p^+ p^-$ and $p^- p^+$. In other words, only actual net $+$ or $-$ motions of the result are counted in the total steps, and all motion is the same size – one step. Using the derived time in this way, the effective velocity now becomes

$$v_1 + v_2 \;=\; \frac{\cancel{2}(\,p_1^+ p_2^+ - p_1^- p_2^-\,)}{p_1^+ p_2^+ + \cancel{p_1^- p_2^-} + \cancel{p_1^- p_2^+} + p_1^- p_2^-} \;=\; \frac{p_1^+ p_2^+ - p_1^- p_2^-}{p_1^+ p_2^+ + p_1^- p_2^-}$$

— just that defined by the cases where motion events are present in the sum.

Substituting velocities for probabilities again, a little algebra yields

$$v_{1+2} \;=\; \frac{v_1 + v_2}{1 + v_1 v_2}$$

agreeing with additive velocities under the Lorentz transform of Special Relativity with maximum speed $c = 1$. In the above example, the relativistic summed velocity is then

$$v_{1+2} \; = \; \frac{\frac{6}{8}\times\frac{7}{8} - \frac{2}{8}\times\frac{1}{8}}{\frac{6}{8}\times\frac{7}{8} + \frac{2}{8}\times\frac{1}{8}} \; = \; \frac{10}{11}$$

as expected, and does not exceed the maximum velocity of $c = 1$.

Interpretation

Here we see the real nature of relative motion and the basis for the space and time dilation of Relativity. The "passage" of time is derived from the motion itself, and thus our clock doesn't tick when there's no motion. The usual notions of space and time emerge naturally from discrete events. Further work will be necessary to extend this basic approach to all of Special and General Relativity. For more on the origins of space as distinctions, and time as generated by loops in space, see Shoup (1994).

Acknowledgements

To the best of my knowledge, a similar basic idea was first discussed by mathematician Louis Kauffman in 1987, by physicist Irving Stein at an ANPA meeting (1996) and again by Thomas Etter (1998). A related idea has also been explored in a web page by Kevin Brown (no date). I am grateful for discussions on this subject with Tom Etter, Irving Stein, Andrew Singer, Ken Wharton, and Joseph Depp.

All of our explorations of discrete physics have been inspired and informed by the continuing and courageous work of H. Pierre Noyes and other principals of the Alternative Natural Philosophy Association including Clive Kilmister and Ted Bastin.

Etter, T. and Noyes, H. P.   'Process, System, Causality, and Quantum Mechanics', *Stanford Linear Accelerator Center Pub* 7890 (1998); revised in *Physics Essays*, 12, 4, Dec. 1999,   also available at
`http://www.boundary.org/articles/PSCQM.pdf`.

Stein, I.   *The Concept of Object as the Foundation of Physics*, Peter Lang (1996).

Brown, K.   'Probabilities and Velocities'
`www.mathpages.com/home/kmath216/kmath216.htm`.

Kauffman, L. H.   'Special Relativity and a Calculus of Distinctions', *Proc. 9th Ann. Intl. Meeting of ANPA, Cambridge, (1987)*, also available at
`http://homepages.math.uic.edu/ kauffman/Relativity.pdf`.

Shoup, R.   'Space, Time, Logic, and Things', *PhysComp 1994 Workshop on Physics and Computation*,   IEEE Press (1995), also available at
`http://www.rgshoup.com/prof/pubs/SpaceTime.pdf`.

Brief Biography

Richard Shoup obtained his BSEE degree in 1965 and Ph.D. in Computer Science in 1970 from Carnegie Mellon University in Pittsburgh, where his Ph.D. thesis was the first to explore programmable logic, a precursor technology to today's reconfigurable hardware. Later in 1970, Shoup became one of the first employees at the Xerox Palo Alto Research Center, where he built one of the first digital frame buffers and developed painting and animation software for applications in video and graphic arts. For this work, Shoup later received an Emmy and an Academy Award. Dr. Shoup left Xerox in 1979 to co-found Aurora Systems, an early manufacturer of digital videographics and animation systems. In 1993, Shoup joined Interval Research Corporation, a unique research lab in Palo Alto founded by computer industry pioneers Paul Allen and David Liddle, where he worked in the areas of reconfigurable computing, mathematics of computation, and quantum theory. In early 2000, Shoup and colleagues founded the Boundary Institute, a private nonprofit research group for the study of physics, mathematics, and computation.

Expanding–Contracting Universes

Irving Stein

Rohnert Park, California, USA

Précis

FIRST EXPLICATIONS:

1. One or the other, and both. One **or** the other are opposites of each other; one **and** the other are identities of each other.
2. There are four dimensions: space, time, mass, energy.
3. Time is a fluctuation in space.

SECOND EXPLICATIONS:

4. Every theory is an opinion.
5. Thus, this work is not a theory, the rest of this work being the explication of these explications.

Explication of explications

6. *Distinguishers* are consecutive <u>integers</u>: *conservators* are consecutive <u>fractions</u>.
7. Fractions are *internal* to integers; integers are *external* to fractions.
8. Zeros separate positive and negative numbers.
9. Integers, both positive and negative are unending.
10. Internal to any two fractions are an infinity of other fractions.
11. Internal to any two zeros are an infinity of other zeros.
12. A line on a plane is unending; there is only one line on a plane.
13. A line on a plane, being a single line has two integers; if the line is designated as c^2, its exteriors are designated as $+\vec{c}$ or $-\vec{c}$.
14. Exterior to the line are two exteriors, one on either side; these are $0 \to \infty$ and $0 \to -\infty$; these intersect the line only at zeros or fractions, integers being exterior to fractions and zeros.
15. Between any two opposites (such as fractions and integers) are conservators and distinguishers.
16. The conservators, as in the above case, <u>conserve</u> the fractions and integers; the distinguishers <u>distinguish</u> between fractions and integers.
17. Speeds less than c^2 are v^2; velocities less than $+\vec{c}$ or $-\vec{c}$ are $+\vec{v}$ or $-\vec{v}$.

18. The *identity* of opposites is a line on a plane.

19. The line on a plane is not measurable since, being infinite in both directions means that any point or location on the line can be taken as the end of the line, or any location on the line.

20. Exterior to the line, on either side, is $(0 \to \infty)$ **and** $(0 \to -\infty)$, since there is only a single line on a plane but no end on the plane.

21. *Opposites* can be written as $(+v-)$; *identities* can be written as $(-n+)$; together, they are $[(+v-) \wedge (-n+)]$.

22. Thus a proton <u>or</u> an electron are $(+v-)$; a neutron is $[(+v-) \wedge (-n+)]$.

23. A neutrino is a smaller mass than a neutron (no electric charge).

24. An anti-neutrino and a neutrino has zero charge and zero mass; zero mass does not mean zero energy.

25. Oscillation is explained by the nature of the third dimension, time.

26. Time is an oscillation in time; that is, present time <u>was</u> past present time, and <u>will be</u> future present time; this oscillation takes only an instant of time since the future and past present times cancel each other leaving only the present time. All other oscillations, including energy into mass and mass into energies take only an instant of time.

27. If there is dark matter, there is also its opposite, light matter, one <u>or</u> the other, one <u>and</u> the other, the identity between them.

28. The identity between them is a wormhole, which transmits mass and energy between them in an instant of time.

29. There is an infinity of universes, each universe being created and dying every instant of time.

30. A singularity on a plane where the axes are (x, y) <u>or</u> (y, x) are defined by the equations $(y = \frac{1}{x})$ <u>and</u> $(x = \frac{1}{y}$; that is, such equations approach the x and y axes asymptotically but never reach them.

31. However, a singularity also oscillates in time. Due to this time oscillation it oscillates to "cap" itself but not in the opposite direction, just as a white hole and dark matter do.

32. Furthermore, an ellipse oscillates at an atomic level so that its directions oscillate; in every solar system in which there are 8 planets there must be two suns, one at each focus. Most, but not all of the energy oscillates between the two foci, so that the little energy lost from within the solar system is gained from outside the solar system.

33. A point and locations are opposite of each other; a point is a vector, which goes from a starting location to an ending location. Thus, one vector and two locations cannot be a single vector, only $\longrightarrow$; it cannot be two lines going in opposite directions $\overleftarrow{\longrightarrow}$ since they would cancel.

34. By a *limit* is meant fractions, $\frac{1}{2}, \frac{1}{4}, \frac{1}{8}, \ldots, \frac{1}{2^n}, \ldots, n$ unending, never reaching an integer (location) at either end; the *limit* of a fraction is an integer; the limit however is <u>not</u> a fraction, but two integers are the limits of a fraction.

35. Again a limit of a fraction is distinguished from the location (the integers) of the fractions, and both fractions and integers are each distinguished from each other.

36. Then what is meant by $\frac{\Delta v}{\Delta t} = \frac{dx}{dy}$ as both dx and dy approach zero simultaneously ? Here $v = \frac{dx}{dy}$ is $\frac{0}{0}$ and is not defined; that is, $v < c$ is defined; but c is defined as the maximum velocity and there is a minimum velocity, v , which is larger than zero and smaller than c, — $f(0)1$.

37. If there is $v > 0$ with $v < c$ then there is no greater velocity and speed than c and c^2; that is, the universes (which are in number infinite) : $f(0)1$, $f(1)2$, $f(2)3$, ... define themselves as no larger; that is, if one is to stand on a beam of light, (or any mass), there will be only locations, which are occupied by integers; that is, any mass on a beam of light would still oscillate but only internally. On the other hand, since a singularity is the opposite of expansions of universes, it also is oscillating, but only externally to itself; thus, the oscillation between both is only internally; that is, that which is conserved is the internality of both; thus, there is no heat death of the universe(s), but both an expansion and a contraction, a birth and a death of each and every universe.

38. Each of the integers, positive, negative, zeros are infinite, that is, there is no limit to any of them; they are endless. The only limit to them is their self-limit. The self-limit to expanding universes is its conservation as an internality; the only conservation to singularities is its externality;

this leads to the conclusions that all the universes
are not only expanding but also contracting.

Brief Biography

Education: San Francisco State University, San Francico, CA: Philosophy; Queens College, City University of New York, Flushing, NY: BS – Physics, Math (1942); Stanford University, Stanford, CA: MS, PhD – Mathematics, Theoretical Physics (1949); Dissertation: Mu-Meson/Deuteron Interaction (incomplete) University of Oregon, Eugene, OR: MS – Mathematics (1951).

Employment: Wayne State University, Detroit, MI: Instructor – Physics; Eaton Manufacturing Co., Detroit, MI: Research Physicist; Inventor of 'Cam Slots' for the movement of automobile front seats. Ampex Corporation, Redwood City, CA: Staff Physicist; Merritt College, Oakland, CA: Instructor – Math, Physics, Philosophy; Stanford Linear Accelerator Center, Menlo Park, CA: Collaborator with H. Pierre Noyes in particle physics.

"Now that I have finished my life's work, I will go back to writing :
Poetry, short stories and essays."

Development of a New Approach to Systems Biology and Therapy Design

Fredric S. Young

Chief Scientist Vicus Therapeutics
& Research Collaborator SLAC
E-mail: youngf@slac.stanford.edu

Over the last four decades, the pharmaceutical industry has been in a continuous decline in efficiency with regard to the cost and frequency of approval of new drugs. The pharmaceutical productivity curve is essentially the reciprocal of Moore's law in computer science describing the increase in the speed and efficiency of computing, and has been named Eroom's law to reflect this fact. The human genome project was conceived as a way to accelerate drug development by describing all of the possible targets of therapy in a single data set. In 2010 Nature published a special news feature on the 10th anniversary of the human genome project, and asked about the benefits of the genome sequence to human health. The answer by the experts who led the project was that health and drug development have not benefitted much. Robert Weinberg one of the leading cancer researchers said that there is little to show for all the money invested in genomic studies of cancer.

The surprising complexity of biology has led to the emergence of the new discipline of systems biology which attempts to organize the myriad details of biology into a coherent framework. In this paper, I provide a personalized description of my several decades long project to develop a robust approach to systems biology using a dynamical systems approach. This was initiated in my Ph.D research where I analyzed the regulatory processes that controlled the growth rate of an intact bacterial cell in different growth media using a reverse engineering approach. While investigating the synthesis of the ribosome, a complex structure that catalyzes the synthesis of all proteins, I discovered a general principle that can serve as a core concept for systems biology. This principle has stood the test of time and many of the predictions have been verified. I will describe this principle and its development and show how it anticipated many important aspects of current systems biology. I will also describe some collaborative work with Pierre Noyes and James Lindesay which resulted in an equation for the core control principle, and a paper by Pierre Noyes showing that this approach to biology is an emergent science as described by Philip Anderson in his paper in *Science* entitled *More is Different*.

Despite the sequencing of the human genome which provides a catalog of all the components of the human organism, the pharmaceutical industry remains in a several decades decline in productivity with increasing costs. In this paper, I provide a personalized account of a discovery I made that I believe provides a way to understand biology at a systems level that could provide a way out of this declining productivity. This discovery has stood the test of time and has been corroborated by the work of others, yet remains unknown to the majority of researchers.

In the computer industry, Moore's law is a measure of the increasing efficiency and speed of computing. Moore's law is based on the observation that the number of transistor's on integrated circuits doubles every 18 months. The law is named after Intel co-founder Gordon Moore who noted in a 1965 paper that the number of components in integrated circuits has doubled every year from 1958 to 1965 and predicted that this trend would continue (1).

In contrast, the pharmaceutical industry has been following a curve reciprocal to Moore's law for the last four decades with steadily decreasing productivity and increasing costs. The curve of pharmaceutical productivity has been named Eroom's law to reflect this reciprocity with Moore's law (2, 3). The number of new drug approvals is halved about every nine years.

Several factors were suggested for this decrease in pharmaceutical productivity and efficiency. Most dealt with aspects of the drug discovery process which are beyond the scope of this paper. However, one factor that is directly relevant is the brute force approach. Instead of the functional considerations that guided drug discovery in the past, massive target based screening has been used in the era of genomic biology (4).

This improved single target screening, but the overall drug discovery process became less effective. The problem was that the strategy of optimizing high efficiency binding to a single target failed because the causal link between a single target and the biology of disease states was weaker than commonly thought. In addition, drugs rarely bind to a single target, but often bind promiscuously to many targets, and this is often a factor in their efficacy (5). Furthermore, single targets function as part of complex networks, and a network combination therapy approach may be a more attractive strategy (6).

The popularity of the brute force approach was also due to the way it matched the zeitgeist of the times which revolved around molecular reductionism. Nearly all approaches to biomedical research since the 1970s centered on the gene. This reached culmination in the human genome project which was seen as a way to map all biological components in a single dataset. This molecular reductionism has a natural synthesis as OMICs technologies (genomics. transcriptomics, metabolomics etc), which were considered an advance on the empirical approach of pre-genomic biology. However, most drug leads from 1999 to 2008 were discovered using functional assays, rather than the more widely use large array target screening (7).

The loss of pharmaceutical productivity continued after the sequencing of the human genome was completed. A special feature in Nature on the Human Genome at 10, queried the public and private leaders of the genome project on the benefits of the genome project to human health (8-10). The conclusion was that human health has not yet benefitted much from the sequencing of the human genome.

Furthermore, the leading cancer researcher Robert Weinberg has said that there is little to show for all of the money invested in genomic studies of cancer (11-12).

As the human genome project was nearing completion, researchers began to focus on the need for tools to assemble the facts of biology into functional models resulting in the development of systems biology. Although it attempted to derive quantitative relationships, this new systems biology was in many ways a continuation of the brute force approach described above, studying correlations among the different types of OMICS data sets including genomics, transcriptomics, and metabolomics (13). The mathematical models developed were direct representations of the data rather than conceptual abstractions.

It was pointed out that biology needed a revolution to make it more like physical science which sought after principles rather than the type of post-hoc correlations being analyzed by the new systems biology (14). The attempt to formulate general principles of biology similar to the strategy used in the physical sciences can be traced back to work that began in the 1950s in Denmark. The initial work carried out by Schaecter, Maaloe and Kjeldgaard studied the properties of cells growing exponentially in steady-states (15), and during transitions between different steady-state growth rates (16).

These experiments summarized in Figure 1 showed that the macromolecular composition of the cells and their size varied as a function of the growth rate. These characteristics were determined by the growth rate rather than the composition of the growth medium, since different growth media that achieved the same growth gave rise to cells with same macromolecular composition and sizes. When the growth medium was changed to one allowing a faster growth rate, the cells immediately increased their rates of macromolecular synthesis eventually resulting in rates characteristic of the new growth medium. When the cultures were shifted to a slower growth rate, the cells stopped growing and reorganized macromolecular synthesis to values characteristic of the new growth medium with the slower growth rate.

As shown in Figure 1, the characteristics of macromolecular synthesis as a function of growth rate exhibited fixed properties that could be described through simple relationships. These relationships included the ratio of RNA/Protein, Protein/DNA and RNA/DNA. These fixed relationships can be analyzed mathematically using methods commonly used in the physical sciences, and began the search for principles described above (14).

A later commentary on this work described it as the beginning of the search for a solution to the cell (17). This was considered a different goal than knowing the complete genome sequence of a cell which was referred to as a sterile vision since it is the interaction and relationship of cell components, not the static genome sequence that would be part of a solution of the cell. This integrative approach was called reconstructionist (18) by Michael Savageau, a mathematical biologist who developed the tools of global systems analysis in biology (19). An essay marking the 50th anniversary of the growth rate experiments was recently published (20).

The first attempt to describe the cell as an intact functioning system of inter-actions and relations was carried out by Maaloe (21). The model is based on the finding that growth rate regulation is independent of the specific composition of the growth medium. It starts with the observation that the total gene expression capacity for transcription (RNA synthesis) and translation (protein synthesis) is fixed at a specific growth rate, and must be distributed among all of the active genes in the cell. Because the steady-state rates of transcription and translation are fixed for a specific growth rate, the rate of expression of an uncontrolled gene will depend on the amount of transcription/translation capacity used for the expression of all of the actively controlled genes. The model assumes that the genes for the protein components of the ribosome, the large structure that synthesizes protein are not controlled. The rate of expression of ribosomal proteins will be a function of the sum of transcription/translation capacity used in the expression of all other genes. In rich growth medium allowing rapid rates of growth there are many metabolic end products that would otherwise have to be synthesized. Since Protein/DNA ratio is constant at different growth rates (Figure 1) there will be excess capacity for protein synthesis in rich medium for expression of the uncontrolled ribosomal protein genes. This model of uncontrolled ribosomal protein genes was called passive control.

This is where I began research on this problem. The overall topic I was considering was to understand metabolic regulation, the type of regulation that was coupled to the cellular growth rate, but was independent of the specific signals that controlled the expression of the various genes and operons. Metabolic regulation involved an integrated aspect of cell growth that could not be reduced to any specific signals. The activity of enzymes was controlled by allosteric mechanisms where the bonding of a small molecule induced a conformational change (22). The expression of genes encoding enzymes and other proteins were controlled by specific signals interacting with operators and repressors as described in the operon model (23).

I did not think that the Maaloe model of passive control was tenable for several reasons. The first reason was that all genes whose regulation was known utilized elaborate control mechanisms. Given this fact, it seemed unlikely that the essential genes for ribosomes, the key structures that synthesize proteins and that have a molecular weight of 1 million would not have elaborate controls. The second reason was that experiments with a lactose operon whose controls were inactivated by mutation decreased following a nutritional shift-up, in contrast to the prediction of the Maaloe model (24). The most important reason that the passive control model was inadequate was the fact that growth medium with two different sole sources of carbon, acetate and glucose gave widely different growth rates while neither should bypass the need for gene expression by the supply of metabolic end products as in rich medium (25), contradicting the prediction of the passive control model.

The difference in growth rate achieved with different sole carbon sources seemed related to the problem of catabolite repression where the presence of a rapidly metabolized carbon source represses the expression of the genes for more slowly metabolized carbon sources, even in the presence of signals that normally lead to expression of the genes for the more slowly metabolized carbon source (26).

The process that I was trying to understand was metabolic regulation, a regulatory process that was dependent on the rate of metabolism by different carbon sources yet was independent of any specific signal. This regulatory process translated the chemical composition of the growth medium into a particular steady-state growth rate chosen from a range of allowable steady-states. The curves for macromolecular composition as a function of growth rate shown in Figure 1, suggested that growth rates were changed by changing the number of functioning ribosomes, rather than by varying the rate of function of the ribosomes. This became known as the constant efficiency hypothesis (27).

The hierarchy of processes involved in cell growth is shown in Figure 2. Cell growth requires sources of carbon, nitrogen, phosphorus and sulfate, as well as a number of trace elements. Metabolism of these compounds provides raw materials and energy which are used to synthesize all of the several thousand components of the cell. Medium permitting the most rapid growth rates also contain a number of metabolic end products such as amino acids and nucleotides that would otherwise have to be synthesized by the cell. In steady-state growth, the number of all of these components is doubled during one generation as defined by the steady-state growth rate.

The crucial observation that led to my model of biology was contained in a paper in Journal of Theoretical Biology on the concept of rate effectors (28). The paper suggested that control theory distinguishes two types of feedback effects. In one case a process is sensitive to the concentration of a metabolite, and in the other case the process is sensitive to the rate at which the end product is consumed. The rate effector model is shown in Figure 3. The paper suggested a mechanism for this type of control where the metabolic end product becomes coupled to a carrier molecule that functions in a cycle. A variety of carriers were suggested including tRNA as carriers of amino acids and adenylate nucleotides as carriers of high energy phosphate groups.

Adenylates as carriers of high energy phosphate groups directly connects the rate effector concept to the cellular energy charge. The cellular energy charge is described by the formula ATP + $\frac{1}{2}$ ADP/ATP +ADP +AMP. High energy phosphate bonds in ATP and ADP are derived from electron transport processes in metabolism and provide energy for all cellular reactions (29). The cellular energy charge is similar to a battery charge and represents the degree to which the adenylate nucleotide carrier pool is filled with high energy phosphate bonds. The cellular energy charge has a value of approximately 0.8 in all living cells (30).

In Figure 4, high energy phosphate group flux through the adenylate carrier pool is shown at two different growth rates. The adenylate pool is larger in faster growing cells and the flux/enzyme is similar at both growth rates. The thickness of the arrows is proportional to the number of units contributing to the flux. The faster rates of flux at the faster growth rate are due to there being more enzyme, carrier and substrates contributing to the flux. The mechanisms which maintain the energy charge at the steady-state value are also shown in Figure 4. The energy charge is a function of the relative rates of loading and unloading of the carrier cycle with phosphate groups. These processes are controlled by reciprocal feedback on the enzymes that produce and consume high energy phosphate groups. As shown in the crossover diagram in Figure 4, the producing and consuming enzymes are oppositely regulated by changes in the value of the energy charge. As shown in the diagram, the crossover point at the steady-state value of the energy charge is the only location on the curve where the rates are balanced. Any change in the state value causes the feedback on producing and consuming reactions to change in opposite directions restoring the energy charge to the steady-state value.

In steady-state growth of bacteria the flux through any carrier cycle is maintained by crossover curves in a way that keeps the relevant carrier charge at the steady-state value. The monomer charge for protein synthesis represents the flux of amino acids through the carrier pool of transfer RNA (tRNA). The extent to which tRNA pool is charged with amino acids as aa-tRNA is expected to have the same value at different steady-states. The monomer charge at two different growth rates is shown in Figure 5. The crossover mechanism in Figure 4, acts to damp perturbations in the steady-state value of the energy charge. Since cells transition to other growth rates when there is a long lasting change in flux due to a change in the growth medium, an additional control loop, is required, which changes the number of components participating in the carrier cycle through gene expression. One set of controls maintains flux balance in a steady-state, and a second set of controls coordinates the flux through the carrier cycle with the genes for the components of the carrier cycle and the input and output processes. Thus, flux through the cycle is coupled on a longer time scale to gene expression processes. The two coupled processes, carrier flux and gene expression rates allow the carrier ratios to be maintained at multiple rates of growth.

In protein synthesis, the aa-tRNA and the nucleotide GTP are bound in a ternary complex to the protein elongation factor Tu (EF-Tu) that is the actual substrate for protein synthesis (31). Since rates of protein synthesis are changed by varying the number of functioning ribosomes, the rate of functioning of the individual ribosome does not change. We define the ternary complex charge as the extent to which the pool of EF-Tu molecules is complexed with GTP and aa-tRNA. This additional carrier cycle at two different rates of growth is shown in Figure 6. Since the ternary complex is the actual substrate for protein synthesis we predict that the EF-Tu carrier cycle is involved in the regulation of ribosome synthesis. The ribosomal proteins are regulated by a feedback mechanism on the translation of their mRNA. If sufficient ribosomal RNA (rRNA) is not available to bind to ribosomal RNA during ribosome assembly, ribosomal proteins bind to their mRNA and block translation. The overall rate of ribosome synthesis is controlled through the regulation of rRNA synthesis.

The model I propose for the control of ribosomal RNA synthesis is shown in Figure 7. It is predicted that EF-Tu is a positive regulator of ribosome synthesis and the nucleotide ppGpp has been shown to be a negative regulator. When a ribosome pauses at a codon because an amino acid is missing, ppGpp is synthesized and shuts down ribosome synthesis. Although it is known that products from rRNA genes are somehow involved in the control of ribosome synthesis (32), negative feedback was proposed, and the mechanism of this control has never been completely determined. My general model of biological control indicates that there must be both positive and negative control. Figure 7 also shows some data that supports my model. A mutant in EF-Tu that grows normally in steady-state shows abnormal behavior during a nutritional shift-up. Mutants in the ppGpp system also grow normally in steady-state, but show abnormal behavior in a nutritional shift-down. I propose that these two systems define the reciprocal regulation of ribosomal RNA synthesis (33).

After developing this general model of biological regulation I was sure that I had discovered something of importance. It was possible to understand the regulation of any biological process just by describing the fluxes and expression of the genes involved in these fluxes. I believed that I had discovered the principle of biological integration, a way to understand system level coordination of different control systems. It seemed to me that flux through a carrier cycle coupled to gene expression should replace the gene and protein as the central concept of systems level biology.

I use a bioinformatics method to model any biological system using an object-process description (34) that can represent any functional groupings (35) I also use a top down form of biological control theory which allows a grouping into reaction blocks of separate reactions that produce and consume the same metabolic intermediate (36-37). For example, a bacterial cell can be coarsely modeled dynamically as all reactions that produce phosphate bond energy and all of the reactions that consume it. By focusing on the appropriate carrier cycle, on could easily model physiological systems and their breakdown in diseases for the purpose of drug development. The current focus on molecular reductionism and the gene was just not appropriate for systems level biology causing the pharmaceutical decline described by Eroom's law.

At the time I made my discovery, the field of biological regulation, was centered on a series of catabolic and biosynthetic operons with appropriate signals opening genes for expression by removing a protein repressor from a regulatory sequence known as an operator that blocked gene expression. A few years after finishing my Ph.D. research, findings corroborating my model began to appear in the literature. After beginning work at Genentech, I attended a West Coast bacterial physiology meeting at Asilomar, and noticed that several discussions of global processes in bacteria were showing coupled cycle diagrams. These were examples of what are now called two-component regulators. The two components are a sensor kinase cycle and a response regulator kinase cycle, which activated a regulator by either adding or removing a phosphate group. Figure 8 shows a generalized two-component regulator. Addition of a phosphate group to a sensor kinase activates it to interact with a response regulator which also becomes activated by phosphorylation and triggers a downstream response. An example of the two-component system which controls the assimilation of nitrogen in E. coli is shown in Figure 9. Positive and negative reciprocal controllers regulate both flux and gene expression as my model predicts.

Two-component regulators control most integrated control processes in bacteria that are triggered by environmental signals including chemotaxis, nutrient assimilation, and processes involved in virulence (38-39). My original description of this regulatory logic (40) preceded the discovery of these two-component regulators in bacteria that maintain various supply demand balances and are considered a rudimentary form of perception (41). These regulators can achieve a form of learning through auto-amplification (42), and the sensing processes that they carry out have been suggested as a rudimentary form of cellular intelligence (43). Reciprocal positive and negative feedback creates a tunable motif for gene regulatory networks (44), and the bacterial two-component regulators have been shown to allow tunable stability (45) demonstrating the amazing flexibility of this regulatory module.

Additional corroboration came from the isolation of the first human cancer genes which are now called called oncogenes (46). These gene are named Ras and Src, and their dynamic control behavior is shown in Figure 10. Both oncogenes operate in rate effector cycles as described in my model of biology. In the case of the ras oncogene, the protein uses a switch between conformations of the protein that binds GDP or GTP. The oncogenic form of the protein contains mutations that lock it in the on state with bound GTP leading to constant unregulated cell proliferation. The GTP/GDP switch is also used by the bacterial elongation factor Tu (EF-Tu) discussed above as a potential regulator of rRNA synthesis. EF-Tu has sequence homology to the Ras oncogene protein (47), and both proteins are examples of a universal biological switch (48-50). The Src oncogene is phosphorylated by a protein kinase (51) and dephosphorylated by a protein phosphatase (52).

Rate effectors operating in carrier cycles can be graded allowing multiple steady-states, or act as bistable switches depending upon input and output connections (53). Additional corroboration for the correctness of this approach came from the observation that many biological signaling mechanisms involved positive feedback (54) and could be understood using concepts from dynamical systems theory (55). It was also realized that interlinked positive and negative feedback, a core concept of my approach was a way of simplifying the description of biological signaling (56). It was also realized that carrier cycles with positive and negative regulators acting on different time scales can produce an optimal bistable switch (57). The extreme sensitivity of the rate effector carrier cycles makes them ultrasensitive (58) and able to achieve absolute concentration robustness (59). The ratios in the rate effector carrier cycle act as an internal model which is needed for adaptation in control systems (60).

The crossover point regulation of a carrier cycle as described above for energy charge regulation (30) and interlinked positive and negative feedback regulation (54) makes the steady-state stability point a special point related to non-equilibrium dynamic phase transition point of Bak's model of self-organized criticality (SOC) as an explanation for the ubiquitous $1/f$ noise and a general model for complexity (61-62). There is an interesting relation between the angle of repose in the sandpile model of SOC and the cellular energy charge in that both are models of steady-states exhibiting the property that for each discrete unit added to the sandpile or adenylate carrier, one unit leaves the system as output. An investigation self-organized criticality as a general model of dissipative transport in open systems showed that a conservation law such as the height of the sandpile gives rise self-similarity in space as fractals and time as $1/f$ noise (63). The conserved value of the energy charge is also an example of such a conserved quantity.

The type of phase transition that is found in controlled biological systems is not a freezing of the lattice as in models of catalysis (64), it is a phase transition in amplification versus damping of a controlled system as shown in Figure 11. John Doyle, a Caltech control theorist showed that the rate effector carrier cycle for chemotaxis shown in Figure 12 is an example of integral control a part of PID control in process control theory that has been used for more than 100 years (65-66). Doyle has pointed out that power laws can result from engineering optimization and has suggested that highly optimized tolerance (HOT) is a more useful model for complexity than SOC (67).

As another example of dissipative transport in open systems, Yamamoto has shown that $1/f$ noise in the healthy heart rate (68) is due to opposing positive and negative regulation of the heart rate by the sympathetic and parasympathetic branches of the autonomic nervous system (69), and involves a phase transition providing further corroboration for this approach to systems biology (70). It was also shown that scaling in heart rate regulation with cascades of regulatory signals is related to the Kolmogorov scaling theory of turbulence (71).

The scaling of the mass of all animals with the heart rate can be plotted on a single curve, and is also an example of dissipative transport in open systems. It has been shown that the scaling is due to the transport requirements of steady-states (72-73). Allometric scaling of a lizard at two different masses is shown in Figure 13. The requirement of steady-state flux with a central source and distributed sinks explains the exponents of the scaling law as D/D+1. The mass must grow faster than the size of the system in order to balance the downstream fluxes with the central source at steady-state.

Savageau has shown that the allometric scaling relation can be derived from a power law biochemical systems description through dimensional arguments (74-75). If one considers the heart pumping blood to all of the cells in the tissues, and the cells sending out signals when the ratios in the carrier cycles are not at their steady-state values, the rate effector carrier cycle is sufficient to give rise to the allometric scaling laws. This analysis is an example of top down biochemical systems analysis Rosen in an investigation of similarity in biological systems considered the scaling transformations of D'Arcy Thompson, and concluded that there must be a scaling of the dimensionless numbers that control the dynamics of the system (76-77). Dimensional analysis of biological systems was a research topic in the 1960s. A paper asking what a dimensionless number in biology might represent suggested the ratio of clearance from the blood by the kidney in a single input output cycle (78). In a hierarchical model of a biological system these input output relations will occur on many scales and can be modeled using the rate effector carrier cycles. Thus, the allometric scaling laws can be understood on the basis of the control of steady-states of function by rate effector carrier cycles.

A study revisiting the correlations between macromolecular synthesis and the cellular growth rate showed that the macromolecular ratios in Figure 1 coupled to the partitioning of gene expression between ribosomal proteins and other proteins as calculated from the ratios, allowed the growth rate controls to be understood as a relation between flux and the cellular energy potential allowed by the growth medium (79). The study suggested that the macromolecular ratios can be viewed as microbial growth laws that are analogous to Ohm's law in electric circuit theory (79). The paper suggests that these microbial growth laws can facilitate our understanding of the operation of complex biological systems before the regulatory circuits are mapped in detail. That was the essential point of my Ph.D. thesis completed in 1977 (40). Unfortunately the emphasis on molecular reductionism which led to the human genome project prevented my approach form being considered by most biologists including my own thesis advisor. However, these microbial growth laws together with the rate effector carrier cycles allow biological systems to be investigated using quantitative theoretical tools and phenomenological laws, and not just by molecular reductionism.

Collaborative research I did with James Lindesay and Pierre Noyes in 2004 led to the development of some of these theoretical tools. James Lindesay derived an equation for the steady-state behavior of the carrier cycle.

The equation examines the dynamics of the carrier as a key component of a specific module. The number (or concentration) of carriers will be denoted C, and the carrier in the activated state will be denoted C^*. The ratio of the activated carrier to de-activated carrier will be defined by $R_C = \dfrac{C^*}{C}$. This dimensionless ratio is directly related to the energy charge through its definition

$$\mu = \frac{C^*}{C + C^*} = \frac{R_C}{1 + R_C}.$$

The dynamics of the module is defined by the rates k_X at which each coupled species X in the module is manufactured (or utilized) as described within the module. This can be modeled using the standard rate equation

$$\frac{dX}{dt} = k_X\, X \, ,$$

where the rate parameters k_X generally depend upon those species that are involved in the manufacture or utilization of X. By direct substitution of its definition, the equation for the evolution of the carrier activation ratio is given by

$$\frac{dR_C}{dt} = \left(k_{C^*} - k_C \right) R_C \, .$$

Thus, the carrier activation ratio reaches homeostasis, defining the universal energy charge, only when the rate parameters for the activated and de-activated carriers are equal, $k_{C^*} = k_C$.

Pierre Noyes also worked on the carrier cycle and wrote a paper on biology as an emergent science (80) where he addressed a point made by Phil Anderson in a paper entitled *More is Different* (81). In this paper Anderson argued that the hierarchical structure of science requires that funding be made available for other topics than particle physics which receives an excessive share of research funding. At each new level of the research hierarchy, new laws and concepts are needed. Pierre wrote that he originally disagreed with Anderson, but has now changes his mind after working on my biology model. Anderson's paper can be seen as a preview of the new research discipline of complexity. A more recent paper by Laughlin and Pines that argued for a hierarchical approach to science also started with Anderson's analysis (82).

Acknowledgements

I gratefully acknowledge the collaboration with Pierre Noyes and James Lindesay, and thank them both for hours of discussions which contributed to this work.

References

(1) Moore G.E. (1965). Cramming more components onto integrated circuits. Electronics 38(8)

(2) Scannell J.W., Blanckley A., Boldon H., and Warrington B. (2012). Diagnosing the decline in pharmaceutical R& D productivity. Nat. Rev. Drug Discov. 11:191

(3) Pammolli F., Magazzini L., and Riccaboni M. (2011). The productivity crisis in pharmaceutical R& D. Nat. Rev. Drug Discov. 10:428

(4) Geysen H.M., Schoenen F., Wagner D., and Wagner R. (2003). Combinatorial compound libraries for drug discovery: an ongoing challenge. Nature Rev. Drug Discov. 2:222

(5) Roth B.L., Sheffer D.L., & Kroeze W.K. (2004). Magic shotguns versus magic bullets: selectively non-selective drugs for mood disorders and schizophrenia. Nature Rev. Drug Discov. 3:353

(6) Keith C.T., Borisy A.A., & Stockwell B.R. (2005) Multicomponent therapeutics for networked systems. Nature Rev. Drug Discov. 4:71

(7) Swinney D.C. & Anthony J. (2011). How were new medicines discovered? Nature Rev. Drug Discov. 10:507

(8) The Human Genome at 10. Special News Feature. (2010) Nature 464:649

(9) Collins F. (2010) Has the Revolution Arrived? Nature 464:674

(10) Venter C. (2010) Multiple personal genomes await. Nature 464:676

(11) Weinberg R. (2010) Point: Hypothesis First. Nature 464:678

(12) Golub T. (2010) Counterpoint: DataFirst. Nature 464:679

(13) Ideker T., Galitsky T., and Hood L. (2001) A new approach to decoding life: systems biology. Ann. Rev. Genomics Hum. Biol. 2:343 v(14) Goldenfeld N., and Woese C. (2007) Biology's next revolution. Nature 445:369

(15) Schaechter M., Maaloe O., and Kjeldgaard N.O. (1958) Dependency on medium and temperature of cell size and chemical composition during balanced growth of *Salmonella typhimurium*. J. Gen. Micro. 19:592

(16) M., Maaloe O., and Kjeldgaard N.O. (1958) The transition between different physiological states during balanced growth of *Salmonella typhimurium*. J. Gen. Micro. 19:607

(17) Cooper S. (1993) The origins and meaning of the Schaecter-Maaloe-Kjeldgaard experiment. J. Gen. Micro. 139:1117

(18) Savageau M. (1991) Reconstructionist molecular biology. New Biol. 3:190

(19) Savageau M.A. (1976) *Biochemical Systems Analysis. A Study of Function and Design in Molecular Biology.* Addison-Wesley Advanced Book Program, Boston

(20) Cooper S. (2008) On the fiftieth anniversary of the Schaechter, Maaloe, Kjeldgaard experiments: implications for cell-cycle and cell growth control. Bioessays 30:1019

(21) Maaloe O. (1969) An analysis of bacterial growth. Dev.Biol.Suppl. 3:33

(22) Monod J., Wyman J., and Changeaux J.P. (1965) On the nature of allosteric transitions: a plausible model. J. Mol. Biol. 12:88

(23) Jacob F., and Monod J. (1961) Genetic regulatory mechanisms in the synthesis of proteins. J. Mol. Biol. 3:318

(24) Dalbow D.C., & Bremer H. (1975) Metabolic regulation of ?-galactosidase synthesis in Escherichia coli a test for constitutive ribosome synthesis. Biochem. J. 150:1

(25) Dennis P.P. & Bremer H. (1974) Macromolecular composition during steady-state growth of Escherichia coli B/r. J. Bac. 119:270

(26) Gorke B. & Skulke J. (2008) Carbon catabolite repression in bacteria: many ways to make the most out of nutrients. Nat. Rev. Micro. 6:613

(27) Maaloe O. & Kjeldgaard N.O. (1966) *Control of macromolecular synthesis.* W. A. Benjamin Inc., Amsterdam, New York

(28) Watson M.R. (1972) Rate effectors and their role in metabolic control. J. Theor.Biol. 36:195

(29) Lipmann F. (1941) Metabolic generation and utilization of phosphate bond energy. Advan. Enzymol. 1:99

(30) Atkinson D.E. (1968) The energy charge of the adenylate pool as a regulatory

parameter. Interaction with feedback modifiers. Biochem. 7:4030

(31) Nissen P. et al. The ternary complex of aminoacylated tRNA and EF-Tu-GTP. Recognition of a bond and a fold. Biochimie 78:921

(32) Jinks-Robertson S., Gourse R.L., & Nomura M. (1983) Expression of rRNA and tRNA genes in Escherichia coli: evidence for feedback regulation by products of rRNA operons. Cell 33:865

(33) Young F.S., (1979) Altered Regulation of Ribosome Biosynthesis in tufB mutants of Escherichia coli. Unpublished

(34) Dori D. (1995) Object-process analysis: maintaining the balance between system structure and behavior. J. Logic. Comput. 5: 227 (1995)

(35) Peleg M., Yeh I, & Altman R.B., (2002) Modeling Biological Processes using Workflow and Petri Net Models. Bioinformatics 18, 825, (2002)

(36) Brand M.D. (1996) Top down metabolic control analysis. J. Theor. Biol. 182:351

(37) Brand M.D. (1997) Regulation Analysis of Energy Metabolism. J. Exp. Biol. J. Exp. Biol. 200:193 (1997)

(38) Stock J.B., Ninfa A.J., & Stock A.M. (1989) Protein phosphorylation and regulation of adaptive responses in bacteria. Microbiol. Rev. 53:450

(39) Stock J.B., Stock A.M., & Mottonen J.M. (1990) Signal transduction in bacteria Nature 344:395

(40) Young F.S., (1977) *Bacterial growth Rate and the Metabolic Regulation of Gene Expression.* Ph.D. Thesis University of Michigan

(41) Mascher T., J.D. Helmann J.D., & Unden G. (2006) Stimulus perception in bacterial signal-transducing histidine kinases. Microbiol. Mol. Biol. Rev. 70:910

(42) Hoffer S.M. et al. (2001) Autoamplification of a Two-component regulatory system results in learning behavior. J. Bac. 183: 4914

(43) Hellingwerf K.J., (2005) Bacterial Observations: a rudimentary form of intelligence. Trends Micro. 13: 152

(44) Tian X.-J., Zhang X.-P., Liu F., & Wang W. (2009) Interlinking positive and negative feedback loops creates a tunable motif in gene regulatory networks. Phys. Rev. E 80:11926

(45) Ghim C.-M., & Almaas E. (2009) Two-component genetic switch as a synthetic module with tunable stability. Phys. Rev. Lett. 103:28101

(46) Cooper G.M. (1982) Cellular transforming genes. Science 218:801

(47) Leberman R., & Egner U., (1984) Homologies in the Primary Structure of GTP-binding proteins: the nucleotide binding site of EF-Tu and p21. EMBO J. 3: 339

(48) Wittinghofer A. & Pai E.F. (1991) The structure of Ras protein: a model for a universal molecular switch. Trends Biochem. Sci. 16:382

(49) Weijland A. et al. (1992) Elongation factor Tu: a molecular switch in protein biosynthesis. Mol. Microbiol. 6:683

(50) Wittinghofer A. (1993) From EF-Tu to p21ras and back again. Curr. Biol. 3:874

(51) Erikson R.L. et al. (1980) Molecular events in cells transformed by Rous Sarcomas Virus. J. Cell Biol. 87:319

(52) Zheng X.M., Wang Y. & Pallen C.J. (1992) Cell transformation and activation of pp60c-src by overexpression of a protein tyrosine phosphatase. Nature 359:336

(53) Ninfa A.J. & Mayo A.E. (2004) Hysteresis vs. graded responses: the connections make all the differences. Science STKE 232:pe20

(54) Ferrell J.E. (2008) Feedback regulation of opposing enzymes generates robust, all or none bistable responses. Curr. Biol. 18:R244

(55) Angeli D., Ferrell J.E.jr., and Sontag E.D. (2004) Detection of Multistability, Bifurcation, and Hysteresis in a Large Class of Biological Positive Feedback Systems. PNAS 101:1822

(56) Ferrell J.E. et al. (2009) Simple, realistic models of complex biological processes: positive feedback and bistability in a cell fate switch and cell cycle oscillator. Febs Lett. 583: 3999

(57) Zhang X.-P., Zhang C., Liu F. & Wang W. (2007) Linking fast and slow positive feedback loops creates an optimal bistable switch in cell signaling. Phys. Rev. E 76:31924

(58) Yang Q. & Ferrell J.E. (2013) The Cdk1-APC/C cell cycle oscillator functions as a time delayed ultrasensitive switch. Nat. Cell Biol. 15:519

(59) Shinar G. & Feinberg M. (2010) Structural sources of robustness in biochemical reaction networks. Science 327:1389

(60) Sontag E, (2003) Adaptation and Regulation with Signal Detection implies Internal Model. Systems and Control Lett. 50:119

(61) Bak P., Tang C., & Wiesenfeld K. (1987) Self-organized Criticality: an Explanation for 1/f Noise. Phys. Rev. Lett. 59:381

(62) Bak P., Tang C., & Wiesenfeld K. (1988) Self-organized criticality. Phys. Rev. A 38:364

(63) Hwa T. & Karder M. (1989) Dissipative transport in open systems: an investigation of self-organized criticality. Phys. Rev. Lett. 62:1813

(64) Grinstein G., Lai Z.-W., and Browne D.A. (1989) Critical Phenomena in Nonequilibrium model of Heterogeneous Catalysis. Phys. Rev. A 40:4820

(65) Yi T.M., Huang Y., Simon M.I., & Doyle J, (2000) Robust perfect adaptation in bacterial chemotaxis through integral feedback control. PNAS 97:4649

(66) Csete M., & Doyle J. (2002) Reverse engineering of biological complexity. Science 295:1664

(67) Carlson J.M., and Doyle J. (2000) Highly Optimized Tolerance: a Mechanism for Power Laws in Designed Systems. 97:4649

(68) Kiyono K. et al. (2004) Critical scale invariance in a healthy human heart rate. Phys. Rev. Lett. 93:178103

(69) Kiyono K. et al. (2005) Phase transition in a healthy human heart rate. Phys. Rev. Lett. 95:58101

(70) Struzik Z.R. et al. (2004) 1/f scaling in heart rate requires antagonistic autonomic control. Phys. Rev. E. 70:50901

(71) Lin D.C. & Hughson R.L. (2001) Modeling heart rate variation in healthy humans a turbulence analogy. Phys. Rev.Lett. 86:1650

(72) Dreyer O. (2001) Allometric scaling and central source systems. Phys. Rev. Lett. 87:38101

(73) Banavar J.R., Damuth J., Maritan A. & Rinaldo A. (2002) Supply-demand balance and metabolic scaling. Proc. Nat. Acad. Sci. USA 99:10506

(74) Savageau M.A., (1979) Growth of complex systems can be related to the properties of their underlying determinants. Proc. Nat. Acad. Sci. USA, 76:5413

(75) Savageau M.A., (1979) Allometric morphogenesis of complex systems: Derivation of the basic equation from first principles. Proc. Nat. Acad. Sci. USA 76: 6023

(76) Rosen R. (1978) Dynamical similarity and the theory of biological transformations. Bull. Math. Biol. 40:549

(77) Rosen R., (1983) Role of similarity principles in data extrapolation. Am. J. Physiol. 244, R599

(78) Stahl W.R. (1962) Similarity and dimensional methods in biology. Science 137:205

(79) Scott M. et al. (2010) Interdependence of cell growth and gene expression: origins and consequences. Science 330:1099

(80) Noyes H. P. On biology as an emergent science. arXiv.org, Physics, 0705.4678

(81) Anderson P.A. (1972) More is different. Science 177:493

(82) Laughlin R.B., & Pines D. (2000) The theory of everything. Proc. Nat. Acad. Sci. USA 97:28

. . . .

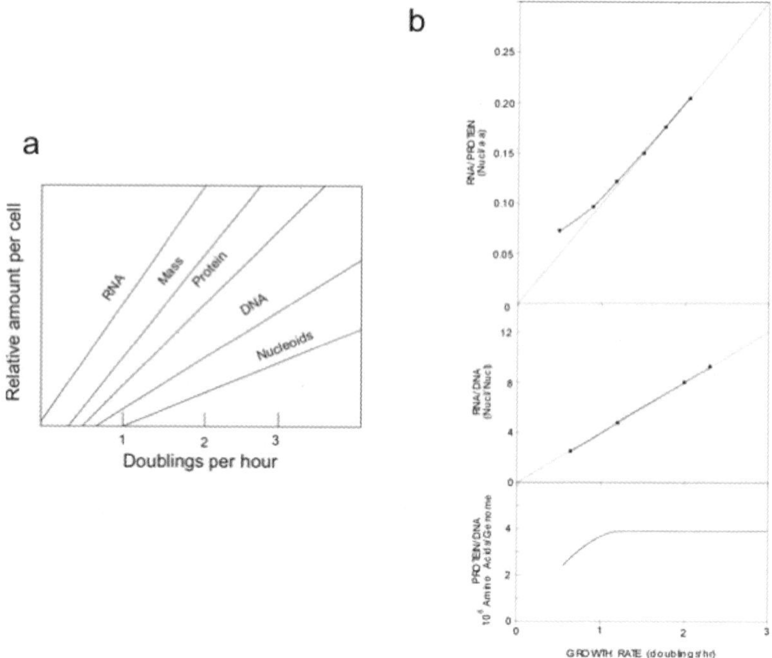

Fig. 1. Macromolecular synthesis as a function of growth in Escherichia coli and Salmonella typhimurium. (a) Quantities of macromolecules DNA, RNA, and protein synthesized as a function of growth. (b) Ratios of macromolecules as a function of growth rate.

Hierarchical Network Construction

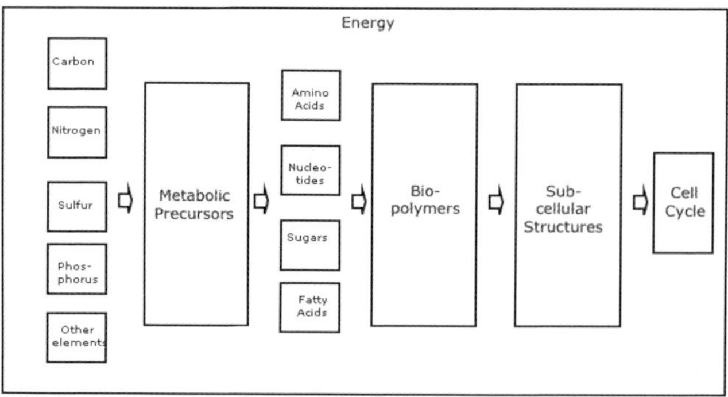

Fig. 2. Hierarchy of processes involved in the growth of the bacteria investigated in Figure 1.

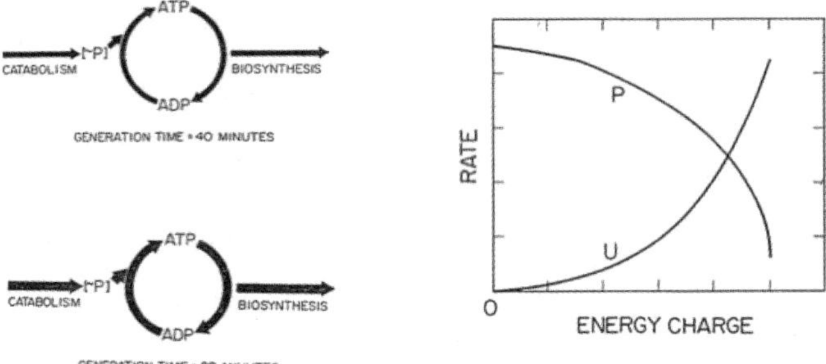

Fig. 3.   The rate effector model. Model showing that control theory requires two types of signals to model dynamic cell cycles; concentration signals and carrier cycle signals including the rate at which concentrations flow into subsequent steps.

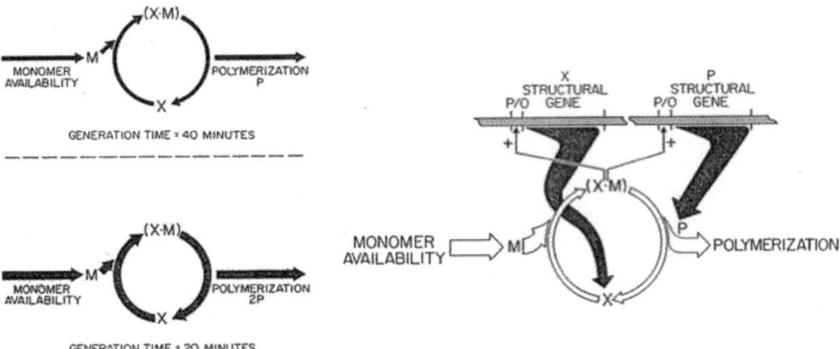

Fig. 4.   Energy charge carrier cycle at two different growth rates and crossover regulation that balances flux in a single steady-state, making balanced energy charge the only stable point.

Fig. 5.   Monomer charge regulation at two different growth rates showing coupling to gene expression allowing multiple steady-states of balanced monomer charge.

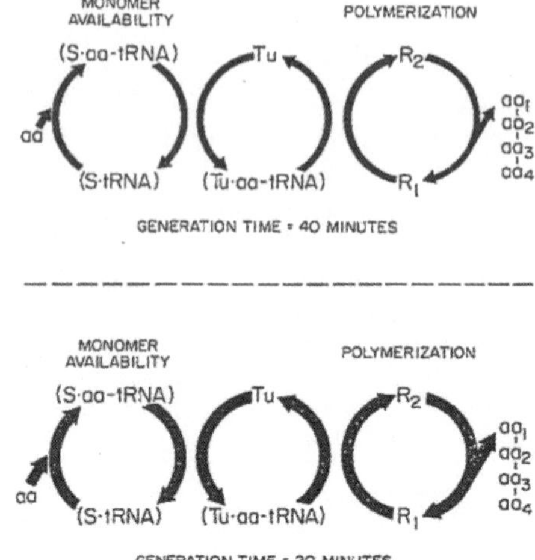

Fig. 6.   Extended monomer charge multiple at two growth rates showing the EF-Tu-GTP-aa-tRNA ternary complex as the substrate for binding to the ribosome.

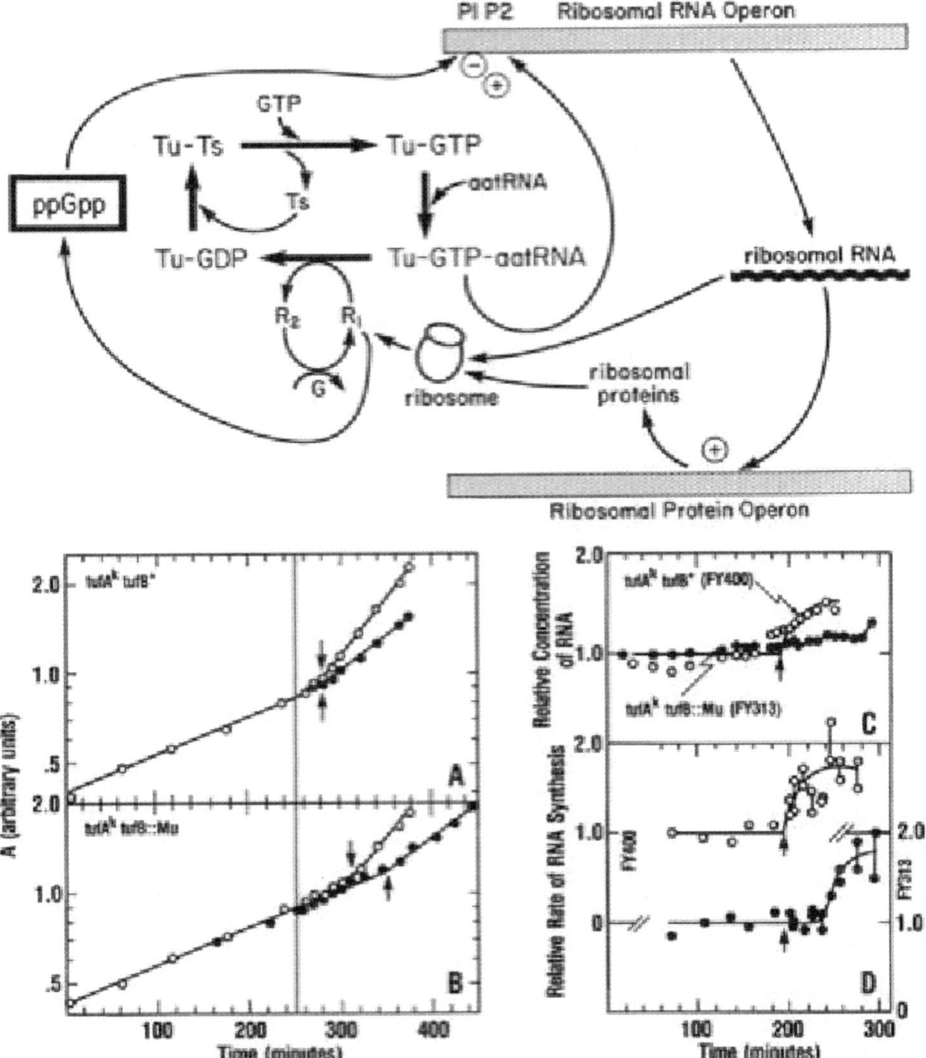

Fig. 7.   Model for the reciprocal regulation of ribosomal RNA synthesis. The negative regulator is the nucleotide ppGpp which is synthesized when the ribosome pauses during translation due to amino acid deprivation. The positive regulator is hypothesized to involve the EF-Tu ternary complex. The data shows that a mutant of EF-Tu exhibits abnormal kinetic during a nutritional shift-up.

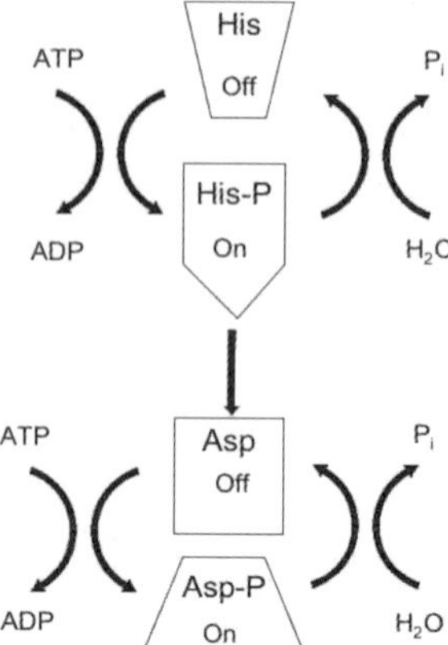

Fig. 8.   General two-component regulator with two carrier cycles, a sensor and a response regulator. These regulators are involved in most global regulatory processes in bacteria.

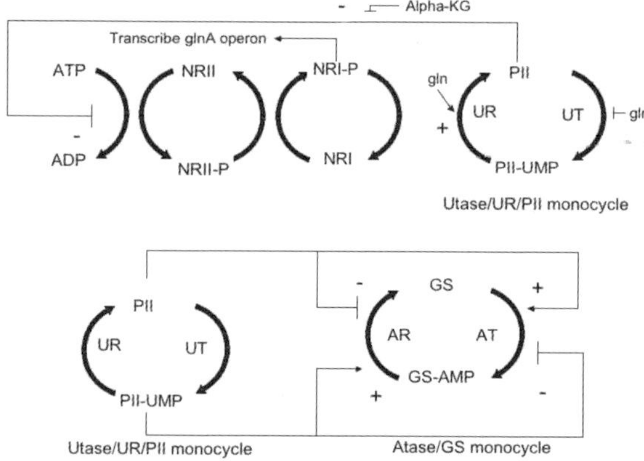

Fig. 9.   Two-component regulation of the nitrogen charge in bacteria. The nitrogen charge involves the ratio of α-ketoglutarate to glutamine and involves reciprocal crossover regulation of flux and gene expression as predicted by the young model.

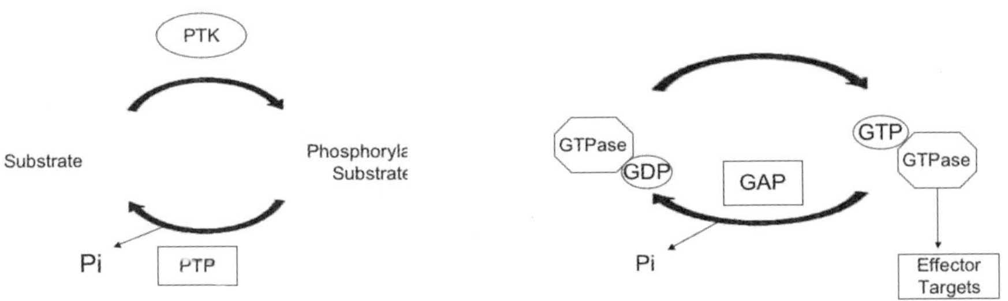

Fig. 10.   Carrier cycles for Src and Ras, the first two identified human oncogenes.
These two carrier cycles define the major carrier cycle switches for all Eucaryotic regulation.
Left : (a) The kinase phosphatase phosphorylation equation. Regulatory proteins are activated by covalent modification.   Right : (b) The GTPase equation defining the GTP/GDP switch. The ras oncogen and Ef-Tu carrier cycles use this switch.

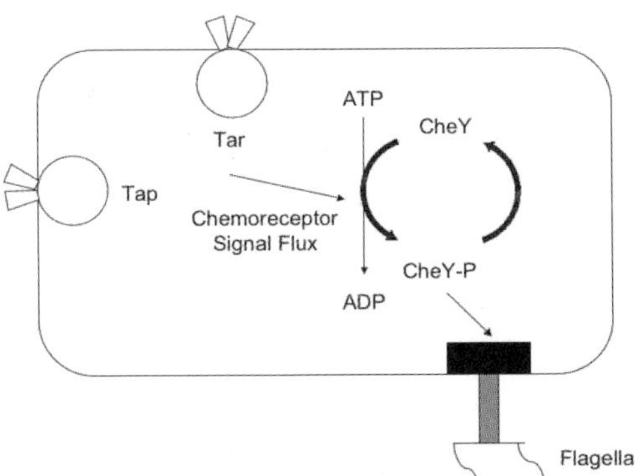

Fig. 11.   Two-component regulation of chemotaxis. Bacteria optimize their swimming trajectories to maximize acquisition of useful nutrients and minimize contact with harmful chemicals. A crossover carrier cycle adjusts straight swimming versus change in direction through control of flagellar rotation direction. This control system has been shown to involve integral process control.

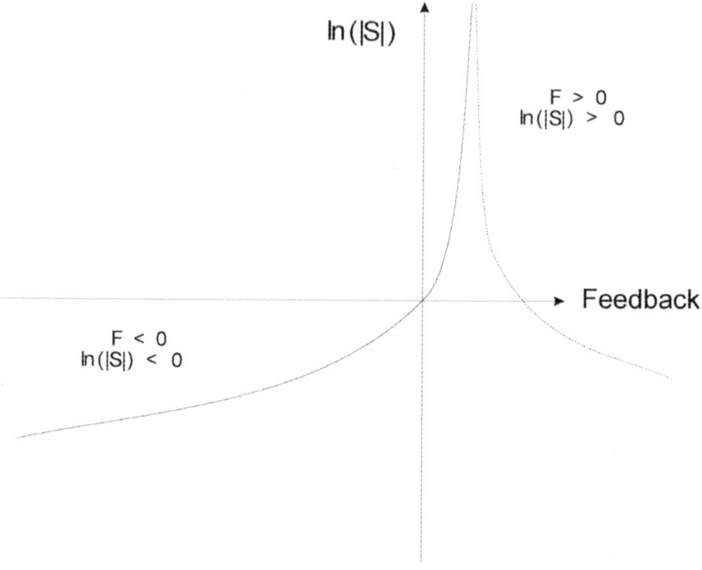

Fig. 12.   Control theory diagram of amplification versus damping of a controlled process illustrating the type of phase transition involved in biological complexity.

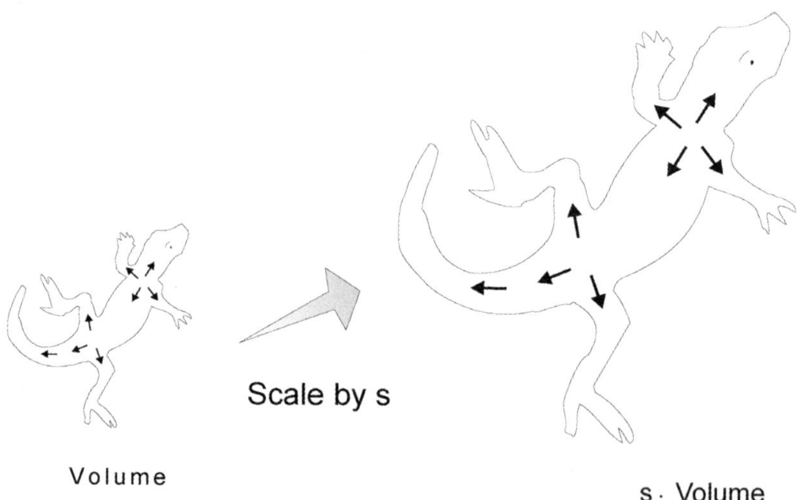

Fig. 13.   Allometric scaling at two different sizes. The allometric scaling laws involve steady-state distribution of fluxes with central sources and distributed sinks.

Brief Biography

Fredric S. Young received a Ph.D in Microbiology from the University of Michigan in 1977. After a four year staff fellowship at the National Institutes of Health in Bethesda Maryland he joined Genentech in 1981 as a scientist in molecular biology. The central focus on the gene culminating in the human genome project indicated that Young's dynamical systems approach was currently outside the main topics of molecular biology. Young left molecular genetics to pursue work as an entrepreneur in fractal geometry and pattern recognition. He started the companies Visual Harmonics, the Chroma group, and Chroma Energy where he developed several patented technologies including a pattern recognition system based on genetics called Image Genetics which was used in oil exploration. In the summer of 1989 Young was visiting scientist in theoretical physics at Brookhaven National Lab working with Per Bak, the originator of Self-organized criticality (SOC) which is related to Young's dynamical systems approach. In 1989 Young began regular discussions with Pierre Noyes, and served as President of the Alternative Natural Philosophy Association (ANPA) from 1990 to 1992, and meeting organizer for ANPA West until it ended in 1998. James Lindesay joined the conversations with Young and Noyes in 2006 leading to the collaboration described in Young's abstract. Young became a research collaborator at SLAC in 2009 which is still in progress. The collaboration with Noyes and Lindesay is also still underway, and James Lindesay is on Sabbatical at Stanford in 2013. After the completion of the human genome project in 1998, biologists began looking for tools to integrate the huge data sets of biology. Young felt the time was right for a dynamical systems approach to biology and founded the pharmaceutical company Vicus Therapeutics where he currently serves as chief scientist. Vicus Therapeutics uses Young's dynamical systems approach as a core technology for drug development.

EPILOGUE

The following essay is a summary of Pierre's most recent thoughts about how,
standing at the gate to his 10th-decade, he foresees what might lie ahead.
The summary was compiled by two of our contributors,
James Lindesay and Fred Young
in the course of conversations during many collaboration sessions
with Pierre this year in his home in Stanford.

The summary is presented as a paper in the usual scientific format.

It stands as an epilogue to this collection of essays honoring Pierre's achievements.

Happy 90-th Birthday, Pierre !

"A New Synthesis of Fundamental Processes" [*]

H. Pierre Noyes

Stanford Linear Accelerator Center,
Stanford University, Stanford, California 94309

"The views the author has expressed in some previous work have changed radically because of ideas developed and results achieved since that work was completed. It has become possible to sketch a new synthesis of fundamental ideas in what follows."

"Connection with Meter, Kilogram, Second, degrees Kelvin units of measurement is made by referring all lengths, masses, times, and temperatures to the four Planck units, and allowing only dimensionless ratios to appear in the theory. Cosmological time can be related to cosmological temperature by using the standard "Big Bang" model for a universe presently made up of about 5% percent ordinary matter, 25% dark matter, and 70% dark energy. Using the present composition of the universe, the currently observed temperature of the cosmic background radiation (about $2.73°$ K) is directly related to the currently accepted time interval since the onset of the Big Bang (about 13.8 Giga-years)[1, 2] Earlier combinatorial hierarchy calculations[3] give the Planck mass/proton mass ratio, the proton/electron mass ratio ($m_p/m_e = 1836.1515$), and the fine structure constant ($1/\alpha = 137.036$), to an accuracy sufficient for our current purposes. Whether the theory can give an acceptable value for the electron/electron-neutrino mass ratio is still open."

"In the opinion of the author the theory will probably provide a specific prediction of three generations of Higgs bosons and three generations of Higgs anti-bosons, but the details remain to be worked out. More speculatively, these might supply the dark matter and dark antimatter needed for our cosmological theory. The biological unification developed from Fred Young's feedback control loop[4, 5] provides a precise specification of the genetic code for the DNA helix (coding strand), the DNA anti-helix (non-coding strand), and the messenger RNA helix, with the immediate consequence that at least some of the basic molecules of "right-handed life" (*i.e.*, with dextrorotary amino acids and levorotary sugars in contrast to the levorotary amino acids and dextrorotary sugars found in all terrestrial life) could be constructed in biochemical laboratories."

"The most direct proof of the matter-antimatter repulsion predicted by the theory would be to show that the anti-hydrogen atoms (*i.e.*, a positron bound to an anti-proton) "fall up" with the same acceleration (g) that neutral hydrogen atoms fall down; such experiments are being vigorously pur-

---

[*]Work supported in part by Department of Energy contract DE-AC03-76SF00515

sued at CERN. The author differs from the approach of Walter Lamb [6] in requiring that both particle-particle pairs and antiparticle-antiparticle pairs attract, while particle-antiparticle pairs repel. This leads to a scale-invariant oscillating universe [7] in which we are now observing the expanding phase. Such an oscillating universe is reminiscent of the astrophysicist Marcello Gleiser's thesis presented in *The Dancing Universe* [8] In this thesis, Gleiser suggests that the fact that all five creation myths can be put in one-to-one correspondence with all five extant scientific cosmologies indicates that we have reached the limits of human imagination in this field of thought. The period of oscillation (yet to be observed) is determined (in the model) only by the total mass-energy in the universe, which the numerical scheme allows to be represented by a finite integer. The author closes by being grateful for his good fortune in having established all of his many collaborations well enough to have been stimulated into making this contribution to what the author hopes will turn out to be a new synthesis. Only the uncertain future will decide whether this effort will make a positive or negative contribution to what happens next."

Acknowledgments

"The author gratefully acknowledges ongoing and fruitful collaborations with E. D. Jones and J. V. Lindesay on cosmology, S. A. Lipinski and H. M. Lipinski on gravitation and elementary particle physics, F. S. Young and J. V. Lindesay on molecular biology and metabolic control, F. S. Young on complexity and catastrophe theory, S. Starson on antigravity and the maximum and minimum limits of applicability of the fundamental units of physics, and W. Lamb on ideas about antigravity and novel modeling of dark energy. The author is indebted to Tom Bakey for his poetic vision of an oscillating universe."

References

[1]   J. V. Lindesay, H. P. Noyes. and E. D. Jones, 'CMB fluctuation amplitude from dark energy partitions', Phys. Lett. B 633, 433 (2006)

[2]   J. Lindesay, 'Consequences of a Cosmological Phase Transition at the TeV Scale', Found. Phys. 37(4-5), 491-531 (2007) (online: March 13, 2007)

[3]   Pierre Noyes, *Bit-String Physics : A Finite and Discrete Approach to Natural Philosophy*, edited by J. C. van den Berg, World Scientific, Singapore, 557 pp. (1991)

[4]   F. S. Young, *Bacterial growth Rate and the Metabolic Regulation of Gene Expression.* Ph.D. Thesis, University of Michigan, (1997)

[5]   F. S. Young, 'A Paradigm for Biology's Next Revolution', *Nature Proceedings*, Available electronically.

[6]   W. Lamb, 'On Dark energy from Antimatter', (completed by Pierre
       Noyes), Physics Essays 21:1, (2008)

[7]   I. Bakey, (private communication)

[8]   M. Gleiser, *Dancing Universe: From Creation Myths to the Big Bang*,
       Norton, New York, (1999)

SERIES ON KNOTS AND EVERYTHING

Editor-in-charge: Louis H. Kauffman *(Univ. of Illinois, Chicago)*

The Series on Knots and Everything: is a book series polarized around the theory of knots. Volume 1 in the series is Louis H Kauffman's Knots and Physics.

One purpose of this series is to continue the exploration of many of the themes indicated in Volume 1. These themes reach out beyond knot theory into physics, mathematics, logic, linguistics, philosophy, biology and practical experience. All of these outreaches have relations with knot theory when knot theory is regarded as a pivot or meeting place for apparently separate ideas. Knots act as such a pivotal place. We do not fully understand why this is so. The series represents stages in the exploration of this nexus.

Details of the titles in this series to date give a picture of the enterprise.

Published:*

Vol. 1:   Knots and Physics (3rd Edition)
          by L. H. Kauffman

Vol. 2:   How Surfaces Intersect in Space — An Introduction to Topology (2nd Edition)
          by J. S. Carter

Vol. 3:   Quantum Topology
          edited by L. H. Kauffman & R. A. Baadhio

Vol. 4:   Gauge Fields, Knots and Gravity
          by J. Baez & J. P. Muniain

Vol. 5:   Gems, Computers and Attractors for 3-Manifolds
          by S. Lins

Vol. 6:   Knots and Applications
          edited by L. H. Kauffman

Vol. 7:   Random Knotting and Linking
          edited by K. C. Millett & D. W. Sumners

Vol. 8:   Symmetric Bends: How to Join Two Lengths of Cord
          by R. E. Miles

Vol. 9:   Combinatorial Physics
          by T. Bastin & C. W. Kilmister

Vol. 10: Nonstandard Logics and Nonstandard Metrics in Physics
          by W. M. Honig

Vol. 11: History and Science of Knots
          edited by J. C. Turner & P. van de Griend

Vol. 12: Relativistic Reality: A Modern View
          edited by J. D. Edmonds, Jr.

**The complete list of the published volumes in the series can also be found at*
http://www.worldscientific.com/series/skae